Advances in Enzyme Biotechnology

Pratyoosh Shukla • Brett I. Pletschke
Editors

Advances in Enzyme Biotechnology

Springer

Editors
Pratyoosh Shukla
Department of Microbiology
Maharshi Dayanand University
Rohtak, Haryana, India

Brett I. Pletschke
Department of Biochemistry,
Microbiology and Biotechnology
Rhodes University
Grahamstown, South Africa

ISBN 978-81-322-1093-1 ISBN 978-81-322-1094-8 (eBook)
DOI 10.1007/978-81-322-1094-8
Springer New Delhi Heidelberg New York Dordrecht London

Library of Congress Control Number: 2013945175

Printed on acid-free paper

Springer is part of Springer Science+Business Media (www.springer.com)

Foreword

This book is a collection of few recent discoveries in enzyme biotechnology by leading researchers in Enzyme Technology and further some selected contributions presented at the 51st Annual Conference of the Association of Microbiologists of India (AMI-2010) which was organized at the beautiful campus of Birla Institute of Technology in Mesra, Ranchi, India, during December 14–17, 2010. The book is edited by Dr. Pratyoosh Shukla, one of the executive members of the Organizing Committee and Prof. Brett I. Pletschke from Rhodes University in Grahamstown, South Africa, who was one of the leading invited speakers of the meeting. The meeting was attended mainly by participants from India but was also made international by a number of invited speakers from abroad. The meeting covered various fields of microbiology, including agricultural and soil microbiology, algal biotechnology, biodiversity, biofuel and bioenergy, bioinformatics and metagenomics, environmental microbiology, enzyme technology, and food and medical microbiology. An important feature of the meeting was participation of industrial researchers which contributed to fruitful interactions between industrial and academic research indispensable for the development of new progressive biotechnologies. The majority of chapters in the book are dedicated to industrially important enzymes modifying plant polysaccharides and lignin. On one hand, the chapters review the current state of the art in the areas of production and application of glycoside hydrolyses, esterases, and lignin-degrading enzymes, while on the other hand, they describe modern trends in the development of enzyme technologies, including the computational enzyme design and enzyme mutations. The fact that most of the chapters originate in India demonstrates rapid emergence of research activity and enormous interest leading to the development of new enzyme technologies in the country. As we know India is a country which is heavily populated, and the sustainability of this country is very strongly dependent on environment friendly biotechnologies. Finally, I would also like to emphasize the general tone of the meeting which was optimistic and enthusiastic about emerging novel applications of enzymes and processes producing usable energy for the future. This book also represents a powerful exposure of important research of the present time to young researchers who filled the meeting/lecture rooms. The hard work of the organizers of the meeting and the editors of this volume is greatly appreciated.

Slovak Academy of Sciences
Institute of Chemistry, Center for Glycomics
Bratislava, Slovakia
September 11, 2012

RNDr. Peter Biely, DrSc

Preface

There has been a rapid expansion of the knowledge base in the field of enzyme biotechnology over the past few years. Much of this expansion has been driven by the bio-discovery of many new enzymes from a wide range of environments, some extreme in nature, followed by subsequent protein (enzyme) engineering. These enzymes have found a wide range of applications, ranging from bioremediation, biomonitoring, biosensor development, bioconversion to biofuels and other biotechnologically important value-added products, etc.

The major goal of this book is to provide the reader with an updated view of the latest developments in the area of enzyme biotechnology. This book presents an exceptional combination of fascinating topics and the reader will be pleased to see that the latest technologies available for an improved understanding of enzymes are included in the book. For example, a thermostable enzyme with sugar metabolic activity is improved by targeted mutagenesis (Chap. 1). The reader will note that there is a significant focus on the role of hydrolases (Chap. 2) and other depolymerising enzymes in this book, as these enzymes form a major component of the annual revenue generated by industrial enzymes. The various other topics ranging from the synthesis of prebiotic galacto-oligosaccharides (Chap. 3), biomass-degrading enzymes, in general, mannanases (Chap. 4), glycoside hydrolases and their synergistic interactions (Chap. 5), manganese peroxidases (Chap. 6) to the modern trends in experimental techniques in enzyme technology (Chap. 7) are also covered in the present book.

Further, the most up-to-date studies related to an overview of the methodologies available for motif finding in biological sequences (Chap. 8), characteristic molecular features and functional aspects of chitin deacetylases (Chap. 9), the role of enzymes in plant–microbe interactions (Chap. 10) and the bioprospecting of industrial enzymes in various grain-processing industries (Chap. 11) have also been included.

Moreover, the readers of the book will be delighted to see that the most up to date technologies available for a better understanding of enzymes are included in this book to enhance the learning skills in key facets of research in enzyme biotechnology.

We hope that the reader will find the information presented here valuable and stimulating. We acknowledge and are indebted to all those who have generously contributed to the completion of this book, and welcome comments from all those who use this book.

Haryana, India Pratyoosh Shukla
Grahamstown, South Africa Brett I. Pletschke

Contents

1 Improvement of Thermostable Enzyme with Sugar Metabolic Activity by Targeted Mutagenesis

Yutaka Kawarabayasi

Abstract

It was well known that improvement of enzymatic activity and stability is very difficult. For most enzymes, introduction of mutation into the amino acid residues located within the reaction center usually disappears their activity. Conversely, it should be useful for application of enzymes if enzymatic activity and stability are artificially enhanced. The enzyme isolated from thermophilic archaea generally possesses absolute stability. The nucleotide-sugar molecule is a powerful material for artificial construction of polymer structure of sugar. The ST0452 protein, an enzyme with sugar-1-phosphate nucleotidylyltransferase activity from *Sulfolobus tokodaii*, was chosen as target for introduction of targeted mutagenesis into the reaction center. All mutant ST0452 enzymes exhibited the same thermostability as shown by the parental ST0452 enzyme. Among 11 mutant ST0452 proteins with substitution of the amino acid residues located at the reaction center by alanine and other amino acids, five mutant ST0452 proteins showed *kcat* values larger than the original value, revealing that in these mutant ST0452 proteins, reactions progress faster than the original enzyme. Even though these mutant ST0452 proteins showed higher *K*m values than that of the original enzyme, these improved mutant ST0452 proteins were capable of exhibiting a higher activity than that of the wild-type ST0452 protein under the presence of high concentration of substrate. These results indicate that thermostable enzymes with higher activity were constructed from *S. tokodaii* ST0452 enzyme by substitution of amino acid residues at the reaction center. These improved enzymes are expected to be useful for application.

Keywords

Thermophilic archaea • Thermostable protein • Sugar-nucleotide • Targeted mutagenesis • Improvement

Y. Kawarabayasi, Ph.D.
Laboratory for Functional Genomics of Extremophiles, Faculty of Agriculture, Kyushu University, Hakozaki 6-10-1, Higashi-ku, Fukuoka 812-8581, Japan

National Institute of Advanced Industrial Science and Technology (AIST), Tsukuba, Japan
e-mail: kyutaka@agr.kyushu-u.ac.jp

P. Shukla and B.I. Pletschke (eds.), *Advances in Enzyme Biotechnology*, DOI 10.1007/978-81-322-1094-8_1, © Springer India 2013

Introduction

Carbohydrate molecules are included within many different types of compounds as a component: outermost structures on a microorganisms' cellular surface which separates the cellular inner and outer environment, source for the energy metabolism which is important for obtaining energy (e.g., TCA cycle), polymer structure for storage of energy (e.g., glycogen), and heredity molecules as a part of DNA and RNA. Also polymer form of carbohydrate molecule modifies the function and stability of protein by binding with the protein molecule (Udenfriend and Kodukula 1995). Many different types of modified sugar molecules are necessary to maintain these biological processes. In most microorganisms, many important modified sugars are synthesized from simple sugar molecules that are incorporated into cells from the surrounding environment.

Among modified sugar molecules, nucleotide-sugar molecules play one of the most important roles for construction of polymer structure of carbohydrate. The nucleotide-sugar, an activated form of sugar molecule, is the sole substrate for construction of polymer structure including a variety of sugar molecules.

Uridine diphosphate *N*-acetyl-D-glucosamine (UDP-GlcNAc) is synthesized by a four-step reaction from fructose-6-phosphate, which is catalyzed by glutamine:fructose-6-phosphate amidotransferase (EC:2.6.1.16), phosphoglucosamine mutase (EC:5.4.2.10), glucosamine-1-phosphate acetyltransferase (EC:2.3.1.157), and *N*-acetyl-D-glucosamine-1-phosphate uridyltransferase (EC:2.7.7.10). UDP-GlcNAc, the final product of this biosynthetic pathway, is an activated form of GlcNAc constructed by combination of GlcNAc-1-phosphate with UTP. This molecule is required for constructing many kinds of polymer structures of carbohydrates. In bacteria, UDP-GlcNAc is required for synthesis of lipopolysaccharides, peptidoglycan, enterobacterial common antigen, and teichoic acid (Frirdich et al. 2004; vanHeijenoort 2001; Harrington and Baddiley 1985). In archaea, the GlcNAc moiety is a major component of the cell surface structure (Niemetz et al. 1997; Kandler and König 1998). In eukarya, the activated molecule is essential for the synthesis of chitin, a major component of the fungal cell wall (Cabib et al. 1982), and the glycosylphosphatidylinositol linker, a molecule anchoring a variety of cell surface proteins to the plasma membrane (Udenfriend and Kodukula 1995). The GlcNAc moiety is found in the polycarbohydrate structure N- or O-linked to the proteins as a posttranslational modification (Guinez et al. 2005; Slawson et al. 2006; Taniguchi et al. 2001; Spiro 2004). As glycosylation is the most important modification for activating peptide drugs, UDP-GlcNAc is thought to be important for future development of effective drugs.

GlcNAc-1-P uridyltransferase activity was identified on the thermostable ST0452 protein from an acidothermophilic archaeon, *Sulfolobus tokodaii* strain 7. The mutation was introduced into this ST0452 enzyme for improvement of the activity.

In this chapter, at first the feature of an acidothermophilic crenarchaeon *S. tokodaii* strain 7, from which the thermostable GlcNAc-1-P uridyltransferase was isolated, will be shown. Then, features of the enzymatic activity of this ST0452 protein and summary on the improvement of the ST0452 protein by targeted mutagenesis will be described.

The Feature of Thermophilic Archaeon *Sulfolobus tokodaii* strain 7

Sulfolobus tokodaii strain 7, an acidothermophilic archaeon, was used for isolation of the enzyme with the sugar-1-phosphate nucleotidylyltransferase activity. This microorganism was isolated from Beppu hot springs located at Kyushu in Japan (Suzuki et al. 2002). As this microorganism was isolated from hot spring, the microorganism is able to grow between 70 °C and 85 °C and between pH 2.5 and 5.0 with the optimal growth condition at 80 °C and 2.5–3.0, respectively. This microorganism was isolated from the geothermal environment; thus, this microorganism can grow under aerobic condition. The phylogenetic analysis showed that this microorganism is included in the kingdom Crenarchaeota of the domain Archaea. The microorganism grows chemoheterotrophically under aerobic respiration condition. Autotrophic growth of this microorganism was not observed in minimal media supplied with elemental sulfur, although several strains isolated as genus *Sulfolobus* are known to be capable of growing autotrophically. Phylogenetic analysis by 16S rDNA sequences indicated that the sequence of this microorganism is most closely related to that of *Sulfolobus yangmingensis* (Suzuki et al. 2002).

The entire genomic sequence of *S. tokodaii* was already determined (Kawarabayasi et al. 2001). The size of the genome of *S. tokodaii* is 2,694,756 bp long, and the G+C content is approximately 32.8 %. Within this genomic sequence, over 2,800 open reading frames (ORFs) were predicted as potential protein-coding regions, and 32.2 % of these are predicted of their functions (annotatable), 32.6 % of these are related to the conserved but unknown ORFs, and 5.1 % of these contain some motif sequences. Among 46 tRNA genes predicted within the genomic sequence, 24 tRNA genes are shown as the interrupted tRNA genes which contain the intron within their genes. The CCA sequence is required for binding with amino acid, and this CCA sequence is not included in most tRNA genes predicted in this genomic sequence of *S. tokodaii*. Also the tRNA nucleotidylytransferase, which is used for addition of CCA sequence posttranscriptionally, was predicted on the genomic sequence of *S. tokodaii*. These features are closely similar to that of eukaryote.

Already entire genomic sequences of two similar species, *Sulfolobus solfataricus* and *Sulfolobus acidocaldarius*, were determined (She et al. 2001; Chen et al. 2005). The genome size and the number of the predicted protein-coding regions of *S. solfataricus* and *S. acidocaldarius* are 2,992,245 bp and 2,977 and 2,225,959 bp and 2,292, respectively. Among these potential protein-coding regions, approximately 1,600 genes are conserved within three *Sulfolobus* species. Approximately from 400 to 900 genes are predicted as that present only in one species (Chen et al. 2005).

The genomic data of *S. tokodaii* was used for identification of the useful enzymes. In the following sections, a brief identification of the enzyme with sugar-1-phosphate nucleotidylyl-transferase activity and improvement of the useful activity are indicated.

Sugar Metabolic Enzyme from an Acidothermophilic Archaeon, *S. tokodaii*

Although many gaps are remaining in the metabolic pathway constructed from the genomic data of *S. tokodaii*, the four genes for TDP-rhamnose biosynthesis pathway from glucose-1-phosphate and TTP were predicted from the genomic data of *S. tokodaii*. Also, genes similar to the first enzyme in this biosynthetic pathway, glucose-1-phosphate thymidylyltransferase, were detected on the genome. Among these genes located at other position than the first enzyme within the TDP-rhamnose biosynthesis pathway, the ST0452 gene was chosen for analysis of its activity and function, because of the presence of the long C-terminal domain which was not present in the other similar genes. Thus, the gene encoding the ST0452 protein was cloned and expressed in *E. coli*. As shown in Fig. 1.1, the forward and reverse direction of Glc-1-phosphate thymidylyl-transferase activity was detected on the purified ST0452 protein. The protein exhibited utilization of multiple metal ions, absolute thermostability with retaining 50 % of maximum activity after 180 min treatment at 80 °C, and relative high activity from pH 5.0 to 8.5 with maximum activity at pH 7.5 (Zhang et al. 2005).

By analysis of substrate specificity, it was indicated that multiple sugar-1-phosphate and NTP plus dNTP substrates were acceptable for the sugar-1-phosphate nucleotidylyltransferase activity of the ST0452 protein as shown in Table 1.1. Among these, GlcNAc-1-phosphate uridyltransferase activity was one of the most important sugar-1-phosphate nucleotidylyltransferase activities, because the GlcNAc moiety is usually found at the most fundamental position of polysaccharide.

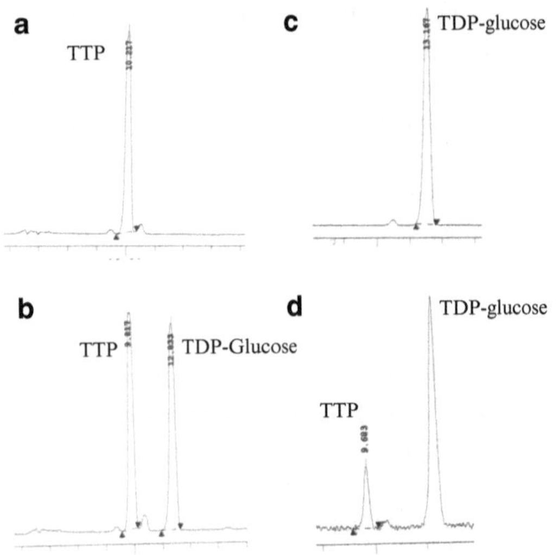

Fig. 1.1 HPLC elution profile of the products by glucose-1-phosphate nucleotidylyltransferase activity of the ST0452 protein. The HPLC elution profiles for the products before (*a and c*) and after (*b and d*) incubation for 20 min at 80 °C with the ST0452 protein. The glucose-1-phosphate was added into the reaction solution as substrate (*a and b*), and TDP-glucose and PPi were added as substrates (*c and d*) for proceeding the reaction

Table 1.1 Substrate specificity of the sugar-1-phosphate nucleotidylyltransferase activity of the ST0452 protein

Substrate A	Substrate B	Relative activity
dTTP	D-Glucose-1-phosphate	100
dATP		35
dCTP		7
dGTP		1
UTP		130
ATP/CTP/GTP		ND[a]
dTTP	*N*-Acetyl-D-glucosamine-I-phosphate	320
	D-Glucosamine-1-phosphate	ND[a]
	D-Galactose-1-phosphate	ND[a]
	D-Mannose-1-phosphate	ND[a]
UTP	*N*-Acetyl-D-glucosamine-I-phosphate	540
	D-Glucosamine-1-phosphate	ND[a]
	D-Galactose-1-phosphate	ND[a]
	D-Mannose-1-phoqsphate	ND[a]

[a]ND: not detected

Therefore, this activity was expected to catalyze the last reaction in the UDP-GlcNAc biosynthesis pathway from fructose-6-phosphate.

The kinetic parameters for the GlcNAc-1-phosphate uridyltransferase activity of the ST0452 protein were obtained. Compared with those of the similar enzyme in *E. coli*, both *K*m and *kcat* values for this activity of the ST0452 protein are lower than those of *E. coli* as shown in Table 1.2. It means that the ST0452 protein is capable of binding with low concentration of substrates, but the turnover rate of reaction is slower than that of the similar *E. coli* enzyme. The low turnover rate is not convenient for production of nucleotide-sugar molecules in application.

Table 1.2 Kinetic properties for the *N*-acetyl-D-glucosamine-1-phosphate uridyltransferase activity of the ST0452 protein

Substrate	*Km*[a]	*kcat*[a]
ST0452 protein		
UTP	1.00	1.00
GlcNAc-1-P	4.65	0.72
E. coli enzyme[b]		
UTP	9.83	6.80
GlcNAc-1-P	10.40	5.95

[a]The relative values are expressed as a proportion of that detected on UTP and the ST0452 protein
[b]The kinetic parameters for *E. coli* enzyme is according to the results described by Gehring et al. (1996)

Conversely, thermostability is beneficial for industrial application; therefore, it was attempted to increase the sugar-1-phosphate nucleotidylyltransferase activity, especially GlcNAc-1-phosphate uridyltransferase activity, of the ST0452 protein.

Improvement of the Archaeal Enzymatic Activity by Targeted Mutagenesis

For increase of the sugar-1-phosphate nucleotidylyltransferase activity of the ST0452 protein, the substitution of amino acid residues without diminishing the thermostability was planned fundamentally according to the expectation that the substitution of the amino acid residues located within the reaction center should not affect the thermostability of the protein, because the reaction center is allocating at the relatively inside of the protein like a pocket. Thus, it was expected that the substitution of the amino acid residues within the reaction center should not affect the overall structure and thermostability of the protein.

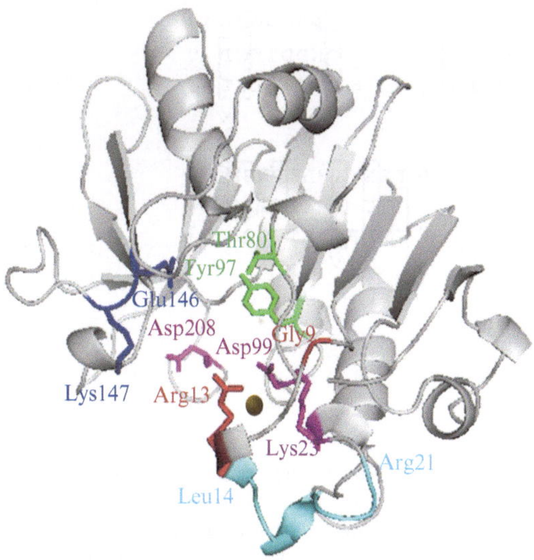

Fig. 1.2 Proposed 3D structure of the sugar-1-phosphate nucleotidylyltransferase reaction center of the ST0452 protein. The amino acid residues participating in binding with nucleoside triphosphate substrates, sugar-1-phosphate substrates, *N*-acetyl portion of GlcNAc-1-phosphate, and metal ions are indicated by *red, blue, green, and magenta*, respectively. The region from Leu14 to Arg21 indicating high conservation with the corresponding sequences of *E. coli* RmlA is indicated by *cyan*. The metal ion is indicated by *brown*

Therefore, the substitution of the amino acid residues located around the reaction center was attempted. As shown in Fig. 1.2, the amino acid residues, shown by color character, surrounding the reaction center of the ST0452 were changed to alanine or other amino acid. Total 11 mutant ST0452 proteins were constructed as shown in Fig. 1.3. Analysis of the thermostability of these mutant ST0452 proteins, SDS-polyacrylamide gel electrophoresis of these proteins after treatment at 80 °C for 30 min, indicated that thermostability of all mutant ST0452 proteins was not affected by substitution of the amino acid residues within reaction center (Fig. 1.4). As all mutant ST0452 proteins exhibited same thermostability as parental wild-type ST0452 protein as expected, all mutant ST0452 proteins were used for detailed analyses of their sugar-1-phosphate nucleotidylyltransferase activity (Zhang et al. 2007).

Relative values of kinetic parameters for GlcNAc-1-phosphate uridyltransferase activity of the mutant ST0452 proteins are shown in Table 1.3. It indicated that five mutant ST0452 proteins exhibited higher *kcat* values than parental wild-type ST0452 protein. However, the *K*m values for the GlcNAc-1-phosphate uridyltransferase activity of these mutant ST0452 proteins also changed to more larger than that of the wild-type ST0452 protein as shown in Table 1.3. These results revealed that these mutant ST0452 proteins required higher concentration of substrate for efficient binding, but reaction proceeds faster than wild-type ST0452 protein. Thus, their activities under presence of the high concentration of substrate were analyzed. The results indicated that when high concentration of GlcNAc-1-phosphate and UTP were supplied into the reaction mixture, five mutant ST0452 proteins exhibited the higher relative activities than that of the parental wild-type ST0452 protein (Fig. 1.5). The result revealed that the substitution of the amino acid residues within reaction center by alanine or other amino acid is effective and useful for improvement of the enzyme with the GlcNAc-1-phosphate uridyltransferase activity from an acidothermophilic archaeon *S. tokodaii*. As similar results were obtained from other enzymes isolated from *S. tokodaii* (data not shown), it can be said that this exclusive feature is common for proteins in this microorganism.

The amino acid residues effective for improvement of *kcat* values of the GlcNAc-1-phosphate uridyltransferase activity of the ST0452 protein by target mutagenesis were shown by enclosure of red lines in Fig. 1.6. These effective residues are located at relatively apart from the reaction center. Thus, it is thought from improvement of the activity of the ST0452 protein, that the amino acid residues located relatively surrounding the area of the reaction center should play an important role for the turnover rate of the activity.

Discussion and Perspective

Some number of improvements of enzymatic activity was already reported (Sun et al. 2011; Qi et al. 2012). However, targets of these experiments are enzymes from mesophilic microorganism. Therefore, the result shown for the ST0452

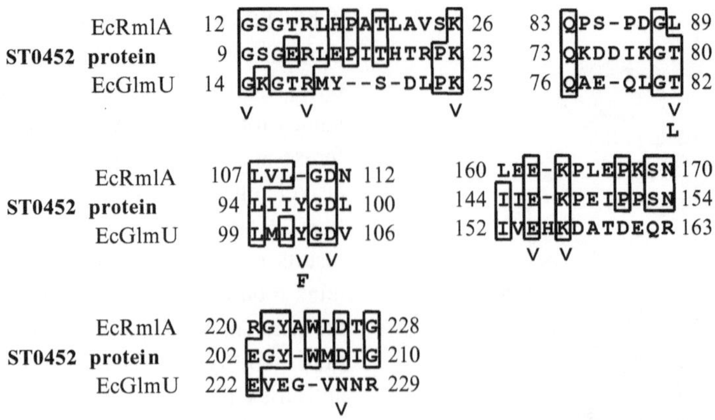

Fig. 1.3 Sequence alignment of five highly conserved domains among the ST0452 protein and *E. coli* glucose-1-phosphate thymidylyltransferase and *N*-acetyl-D-glucosamine-1-phosphate uridyltransferase. EcRmlA and EcGlmU indicate the glucose-1-phosphate thymidylyltransferase from *E. coli* (GenBank accession number P37744) and *N*-acetyl-D-glucosamine-1-phosphate uridyltransferase from *E. coli* (NC_000913). The letters within boxes indicate the residues conserved within three proteins. The amino acid residues chosen for the construction of mutant proteins are indicated by symbol, and amino acid residues introduced into the mutant ST0452 proteins other than alanine are shown below symbols. The numerals indicate the coordinates of the two ends of each domain from the N-terminus of each protein

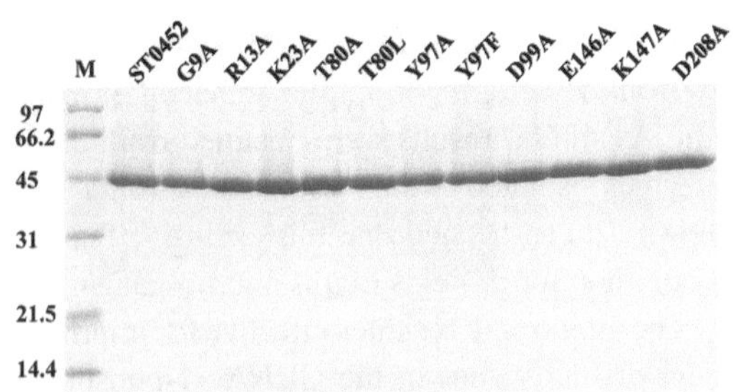

Fig. 1.4 SDS-PAGE analysis of the mutant ST0452 proteins produced in *E. coli*. The wild-type and mutant ST0452 proteins expressed in *E. coli* were subjected to the 12 % of polyacrylamide gel containing 0.1 % of SDS after treatment at 80 °C for 20 min. Lane M: lane for molecular marker

Table 1.3 Kinetic properties for the *N*-acetyl-D-glucosamine-1-phosphate uridyltransferase activity of the wild-type and mutant ST0452 proteins

Proteins	*K*m for UTP	*K*m for GlcNAc-1-P	*kcat*
ST0452	1.00	1.00	1.00
G9A	5.88	271.25	4.07
R13A	1.00	193.75	0.80
K23A	1.53	833.75	0.05
T80A	1.59	43.75	1.51
T80L	8.24	443.75	0.30
Y97A	8.24	8.25	2.11
Y97F	8.82	4.63	4.95
D99A	0.94	1107.5	0.012
E146A	2.94	386.25	0.73
K147A	1.65	312.50	3.23
D208A	3.88	58.75	0.74

The relative values are showed as a proportion of that detected on the wild-type protein

protein was thought to be the first result of improvement of thermostable enzyme.

The results described in this chapter propose the opportunity that activity of the thermostable protein is able to be improved by introduction of the targeted mutagenesis at the amino acid residues allocating around the reaction center. If it is general for the thermostable proteins from archaea, it is convenient for application in industry to provide an enzymatic activity with high turnover rate by introduction of targeted mutagenesis. Therefore, it is planned to attempt to check this possibility for many target proteins from thermophilic archaeal species. If this feature will be detected on many target proteins, introduction of this type of mutation will become a powerful tool for improving the thermostable enzymes isolated from thermophilic archaea. This will be helpful for making a constitutively developing society in this planet for the next generation.

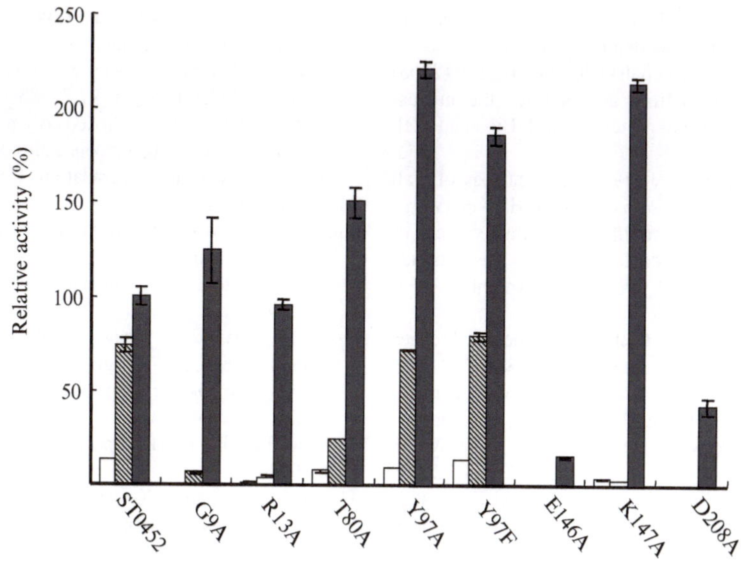

Fig. 1.5 *N*-Acetyl-D-glucosamine-1-phosphate uridyltransferase activity of the mutant ST0452 proteins under three different conditions. *N*-acetyl-D-glucosamine-1-phosphate uridyltransferase activities of each mutant protein indicated were measured in the reaction solution with 5 μM UTP plus 50 μM*N*-acetyl-D-glucosamine-1-phosphate (*open bars*), 100 μM UTP plus 50 μM*N*-acetyl-D-glucosamine 1-phosphate (*hatched bars*), and 100 μM UTP plus 10 mM*N*-acetyl-D-glucosamine-1-phosphate (*closed bars*). The relative activity is expressed as a percentage of the activity detected on the wild-type ST0452 protein under the condition containing 100 μM UTP plus 10 mM*N*-acetyl-D-glucosamine-1-phosphate

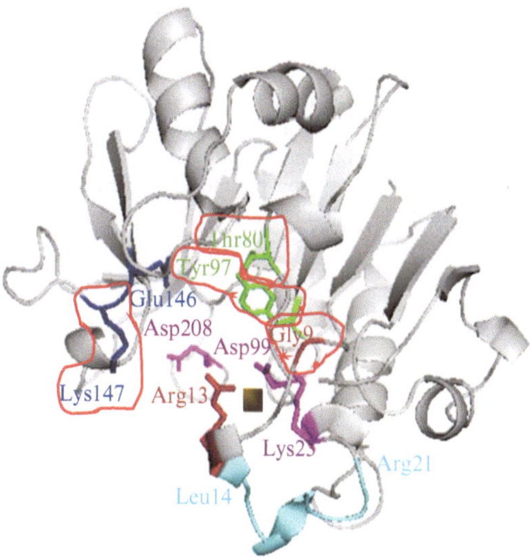

Fig. 1.6 Proposed 3D structure of the sugar-1-phosphate nucleotidylyltransferase reaction center of the ST0452 protein with marking of the amino acid residues working as improving its activity by substitution. The *colored* amino acid residues are shown as legend of Fig. 1.2. The amino acid residues important for improving the sugar-1-phosphate nucleotidylyltransferase activity of the ST0452 are enclosed by *red lines*

Acknowledgements I appreciate five postdoctoral fellows, Tsujimura M., Zhang Z., Akutsu J., Sasaki M., and Md. M. Hossain, working in my laboratory for research in this area. This work was financially supported by special grants for the Protein 3,000 project, a basic knowledge project by New Energy and Industrial Technology Development Organization and a Grant-in-Aid for Research of the Ministry of Education, Culture, Sports, Science and Technology.

References

Cabib E, Roberts R, Bowers B (1982) Synthesis of the yeast cell wall and its regulation. Annu Rev Biochem 51:763–793

Chen L, Brügger K, Skovgaard M, Redder P, She Q, Torarinsson E, Greve B, Awayez M, Zbat A, Klenk HP, Garrett RA (2005) The Genome of *Sulfolobus acidocaldarius*, a model organism of the Crenarchaeota. J Bacteriol 187:4992–4999

Frirdich E, Vinogradov E, Whitfield C (2004) Biosynthesis of a novel 3-deoxy-D-*manno*-oct-2-ulosonic acid-containing outer core oligosaccharide in the lipopolysaccharide of *Klebsiella pneumoniae*. J Biol Chem 279:27928–27940

Gehring AM, Lees WJ, Mindiola DJ, Walsh CT, Brown ED (1996) Acetyltransfer precedes uridylyltransfer in the formation of UDP-*N*-acetylglucosamine in separable active sites of the bifunctional GlmU protein of *Escherichia coli*. Biochemistry 35:579–585

Guinez C, Morelle W, Michalski JC, Lefebvre T (2005) *O*-GlcNAc glycosylation: a signal for the nuclear transport of cytosolic proteins?.Int J Biochem Cell Biol 37: 765–774

Harrington CR, Baddiley J (1985) Biosynthesis of wall teichoic acids in *Staphylococcus aureus* H, *Micrococcus varians* and *Bacillus subtilis* W23. Involvement of lipid intermediates containing the disaccharide N-acetylmannosaminyl *N*-acetylglucosamine. Eur J Biochem 153:639–645

Kandler O, König H (1998) Cell wall polymers in Archaea (Archaebacteria). Cell Mol Life Sci 54:305–308

Kawarabayasi Y, Hino Y, Horikawa H, Jin-no K, Takahashi M, Sekine M, Baba S, Ankai A, Kosugi H, Hosoyama A, Fukui S, Nagai Y, Nishijima K, Otsuka R, Nakazawa H, Takamiya M, Kato Y, Yoshizawa T, Tanaka T, Kudoh Y, Yamazaki J, Kushida N, Oguchi A, Aoki K, Masuda S, Yanagii M, Nishimura M, Yamagishi A, Oshima T, Kikuchi H (2001) Complete genome sequence of an aerobic thermoacidophilic crenarchaeon, *Sulfolobus tokodaii* strain7. DNA Res 8:123–140

Niemetz R, Kärcher U, Kandler O, Tindall BJ, König H (1997) The cell wall polymer of the extremely halophilic archaeon *Natronococcus occultus*. Eur J Biochem 249:905–911

Qi X, Guo Q, Wei Y, Xu H, Huang R (2012) Enhancement of pH stability and activity of glycerol dehydratase from *Klebsiella pneumoniae* by rational design. Biotechnol Lett 34:339–346

She Q, Singh RK, Confalonieri F, Zivanovic Y, Allard G, Awayez MJ, Chan-Weiher CCY, Curtis BA, Moors AD, Erauso G, Fletcher C, Gordon PMK, Heikamp-de Jong I, Jeffries AC, Kozera CJ, Peng X, Thi-Ngoc HP, Redder P, Schenk ME, Theriault C, Theriault C, Tolstrup N, Charlebois RL, Doolittle WF, Douguest M, Gaasterland T, Garrett RA, Ragan MA, Sensen CW, van der Oost J (2001) The complete genome of the crenarchaeon *Sulfolobus solfataricus* P2. Proc Natl Acad Scie USA 98:7835–7840

Slawson C, Housley MP, Hart GW (2006) *O*-GlcNAc cycling: how a single sugar post-translational modification is changing the way we think about signaling networks. J Cell Biochem 97:71–83

Spiro RG (2004) Role of *N*-linked polymannose oligosaccharides in targeting glycoproteins for endoplasmic reticulum-associated degradation. Cell Mol Life Sci 61:1025–1041

Sun Y, Yang H, Wang W (2011) Improvement of the thermostability and enzymatic activity of cholesterol oxidase by site-directed mutagenesis. Biotechnol Lett 33:2049–2055

Suzuki T, Iwasaki T, Uzawa T, Hara K, Nemoto N, Kon T, Ueki T, Yamagishi A, Oshima T (2002) *Sulfolobus tokodaii* sp. nov. (f. *Sulfolobus* sp. strain 7), a new member of the genus *Sulfolobus* isolated from Beppu Hot Springs, Japan. Extremophiles 6:39–44

Taniguchi N, Ekuni A, Ko JH, Miyoshi E, Ikeda Y, Ihara Y, Nishikawa A, Honke K, Takahashi M (2001) A glycomic approach to the identification and characterization of glycoprotein function in cells transfected with glycosyltransferase genes. Proteomics 1:239–247

Udenfriend S, Kodukula K (1995) How glycosylphosphatidylinositol-anchored membrane proteins are made. Annu Rev Biochem 64:563–591

vanHeijenoort J (2001) Formation of the glycan chains in the synthesis of bacterial peptidoglycan. Glycobiology 11:25R–36R

Zhang Z, Tsujimura M, Akutsu J, Sasaki M, Tajima H, Kawarabayasi Y (2005) Identification of an extremely thermostable enzyme with dual sugar-1-phosphate nucleotidyltransferase activities from an acidothermophilic archaeon, *Sulfolobus tokodaii* strain 7. J Biol Chem 280:9698–9705

Zhang Z, Akutsu J, Tsujimura M, Kawarabayasi Y (2007) Increasing in archaeal GlcNAc-1-P uridyltransferase activity by targeted mutagenesis while retaining its extreme thermostability. J Biochem 141:553–562

Glycoside Hydrolases for Extraction and Modification of Polyphenolic Antioxidants

2

Kazi Zubaida Gulshan Ara, Samiullah Khan, Tejas S. Kulkarni, Tania Pozzo, and Eva Nordberg Karlsson

Abstract

Antioxidants are important molecules that are widely used by humans, both as dietary supplements and as additives to different types of products. In this chapter, we review how flavonoids, a class of polyphenolic antioxidants that are often found in glycosylated forms in many natural resources, can be extracted and modified using glycoside hydrolases (GHs). Glycosylation is a fundamental enzymatic process in nature, affecting function of many types of molecules (glycans, proteins, lipids as well as other organic molecules such as the flavonoids). Possibilities to control glycosylation thus mean possibilities to control or modify the function of the molecule. For the flavonoids, glycosylation affect both the antioxidative power and solubility. In this chapter we overview results on in vitro deglycosylation and glycosylation of flavonoids by selected GHs. For optimal enzymatic performance, desired features include a correct specificity for the target, combined with high stability. Poor specificity towards a specific substituent is thus a major drawback for enzymes in particular applications. Efforts to develop the enzymes as conversion tools are reviewed.

Keywords

Glucosidase • Cellulase • Amylase • Glycosynthase • GH • Flavonoid • Quercetin

Introduction

The increased concern about scarcity of fossil resources has lately put the use of renewable resources by biotechnological methods in focus, as these are predicted to have an increased importance in production of food, additives and chemicals. Antioxidants can be foreseen to play a role as

K.Z.G. Ara • S. Khan • T.S. Kulkarni • T. Pozzo • E. Nordberg Karlsson (✉)
Biotechnology, Department of Chemistry, Lund University, P.O. Box 124, SE-22100 Lund, Sweden
e-mail: eva.nordberg_karlsson@biotek.lu.se

P. Shukla and B.I. Pletschke (eds.), *Advances in Enzyme Biotechnology*, DOI 10.1007/978-81-322-1094-8_2, © Springer India 2013

bio-based ingredients in food (as well as other) products, both as preservatives, replacing agents with negative health aspects, and as nutraceuticals. Flavonoids are polyphenolic compounds and a class of secondary metabolites that are widely distributed in nature. The beneficial properties of flavonoids are mainly referred to their ability to counteract oxidative stress caused by oxygen species and metal ions (Lin and Weng 2006; Havsteen 2002), and they are shown to play a protective role against neoplasia, atherosclerosis and neurodegenerative diseases (Lee and Lee 2006; Boudet 2007). Because of these exclusive properties, flavonoids have received great attention and the industrial interest is growing rapidly. Apart from this role, antioxidants can also be added to food and other types of products to prolong their shelf-life. Currently over 6,500 flavonoids have been identified (Corradini et al. 2011), and they are commonly found in plants, fruits, vegetables, ferns, stems, roots, tea, wine and also from bark (Nijveldt et al. 2001). Their role in plants is to protect against UV-radiation diseases, infections and insect invasion (Corradini et al. 2011). The content of flavonoids varies, dependent on the source, but is normally in the mg-range per kg raw material. For example, the content of the flavonoid quercetin is around 300 mg/kg of onion (Griffiths et al. 2002), 100 mg/kg of broccoli, 50 mg/kg of apples, 40 mg/kg of blackcurrants and 30 mg/kg of green tea (Hollman and Arts 2000).

Problems with many flavonoids are, however, low solubility and poor stability (in both polar and nonpolar media) which make their uses difficult in many formulations of food, pharmaceutical and nutraceutical products (Ishihara and Nakajima 2003). Improvement of the hydrophilic nature, biological properties and stability of flavonoids can be achieved by enzymatic structural modification (Haddad et al. 2005). In nature, enzyme function has however evolved according to the conditions in the living cells and may not be perfect in specific biotechnological applications. In this chapter, we review the current use of glycoside hydrolases (GHs) in flavonoid extractions and conversions along with efforts to develop GHs (especially β-glucosidase and endoglucanase) for deglycosylation and glycosylation of these polyphenolic compounds.

Structural Overview of Flavonoids and Different Flavonoid Glycosides

The core structure of a flavonoid is 2-phenyl benzopyranone, also known as 2-phenyl-1, 4-benzopyrone (Fig. 2.1), in which the three-carbon bridge between phenyl groups is cyclised with oxygen (Corradini et al. 2011). Flavonoids are divided into flavones, isoflavones, flavonols, flavanones, flavan-3-ols and anthocyanidins based on their degree of unsaturation and oxidation of the three-carbon segment (Hughes et al. 2001) (Table 2.1). They are generally found as glycosidic conjugates with sugar residues, and sometimes they can also exist as free aglycones (Stobiecki et al. 1999). For example, quercetin exists mostly in the form of glycosides (Fig. 2.1).

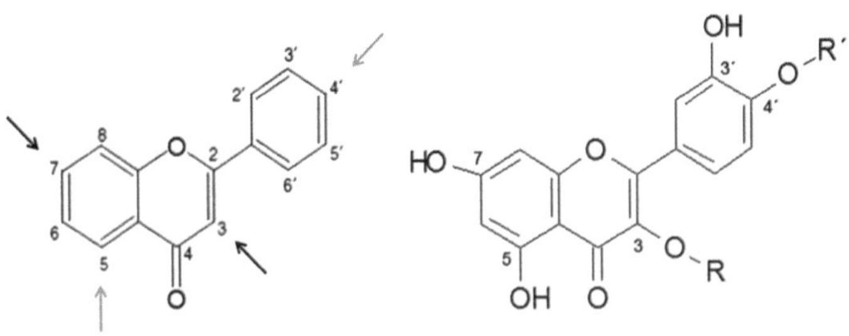

Fig. 2.1 General structure of the flavonoid backbone (*left*), shown with backbone numbering. The most common hydroxyl positions for glycosylation (3 and 7) are indicated with *black arrows*, and the 5 and 4′ hydroxyls that are sometimes glycosylated are indicated with *grey arrows*. A quercetin molecule (*right*) is also shown with the substituents present in this type of flavonoid. R and R′ are hydrogens in the deglycosylated form. In glycosylated forms isolated from onion, R and/or R′ represents glucosyl groups

Table 2.1 Chemical structures of subclasses of flavonoids

Flavonol	*R1*	*R2*
Quercetin	OH	H
Kaempferol	H	H
Myricetin	OH	OH
Isorhamnetin	OMe	H

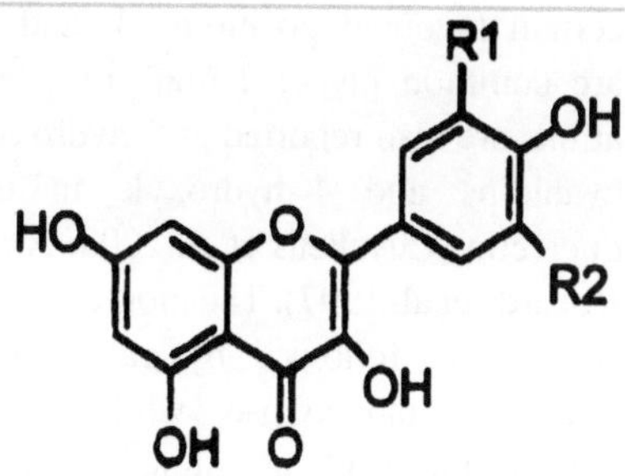

Flavone	*R1*
Apigenin	H
Luteolin	OH

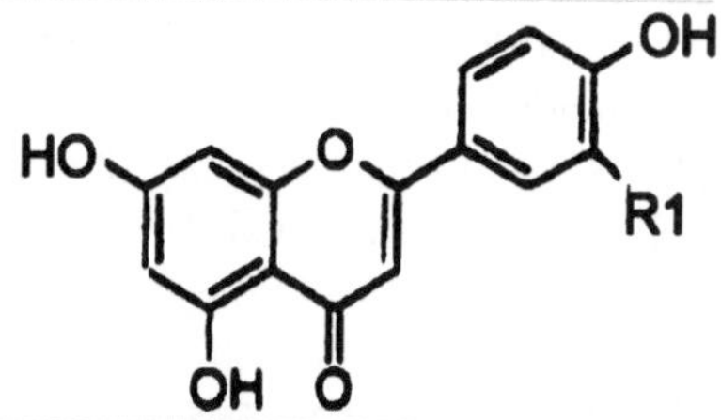

Flavanones	*R1*	*R2*
Eriodictyol	OH	OH
Hesperetin	OH	OMe
Naringenin	H	OH

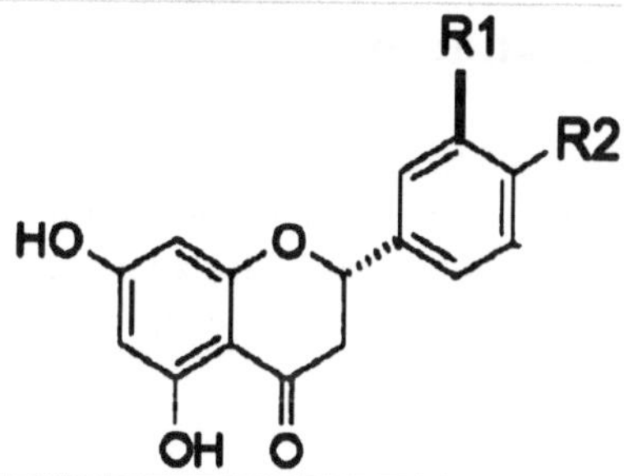

Flavan-3-ols	*R1*
(+) Catechin	H

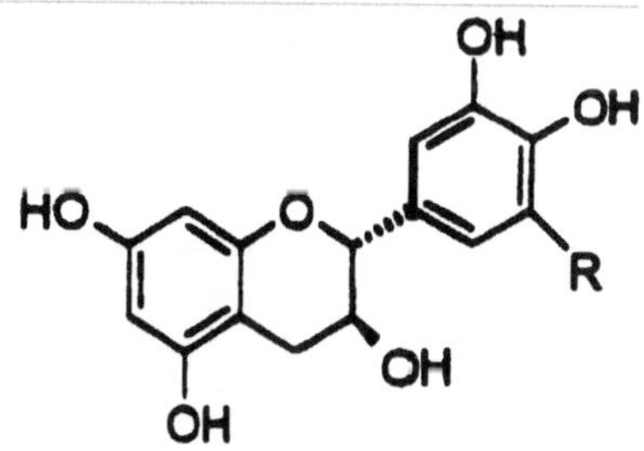

Anthocyanidin	*R1*	*R2*
Cyanidin	OH	H
Delphinidin	OH	OH
Malvidin	OMe	OMe
Petunidin	OMe	OH

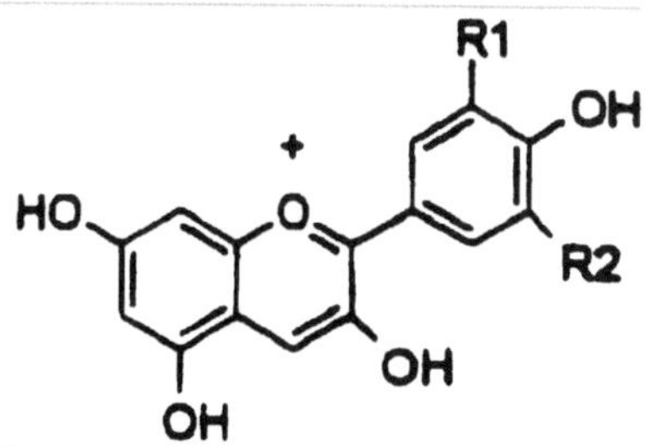

The addition of the glycoside conjugates or glycosylation makes the flavonoid less reactive and more polar, leading to higher water solubility. Hence, this is the most important modification that occurred in plants to protect and store the flavonoids in the cell vacuole (Cuyckens et al. 2003). The development of flavonoid-O-glycosides includes one or more of the aglycone hydroxyl groups bound to a sugar with formation of an O-C acid-labile acetal bond. The glycosylation does not occur in each hydroxyl groups but in certain favoured positions: 3- and 7-hydroxyls are common glycosylation sites, but glycosylations are also reported at 5-hydroxyls in anthocyanidins and 4′-hydroxyls in the flavonol quercetin (Cuyckens et al. 2003; Iwashina 2000; Robards et al. 1997). The most encountered sugar is glucose, followed by galactose, arabinose, rhamnose and xylose, while glucuronic and galacturonic acids are quite rare. Further, some disaccharides are also found in conjugation with flavonoids, like rutinose (6-O-L-rhamnosyl-d-glucose) and neohesperidose (2-O-L-rhamnosyl-d-glucose) (Robards et al. 1997).

Glycoside Hydrolases as Extraction Aids

By-products from agriculture, food and forest industries have the potential to become a major source of flavonoids. Isolation of the polyphenolic compounds from the plant sources is usually done by using different extraction methods (Fig. 2.2). In processing of renewable resources such as agricultural by-products or bark, enzymatic hydrolysis can be coupled with the extraction process, and GHs (sometimes also termed glycosidases) are commonly used for these purposes. These enzymes are generally easy to handle, as they do not require cofactors and they can be used at an early stage on the readily available material found in the forest and agricultural sectors (Turner et al. 2007). GHs are hydrolases responsible for the transfer of glycosyl moieties from a donor sugar to an acceptor and have either an inverting or retaining (Fig. 2.3) reaction mechanism, and in hydrolysis the acceptor is water (Ly and Withers 1999). The hydrolysed glycosidic bond can be located between two or more carbohydrates (e.g. polysaccharides) but also between a carbohydrate and a non-carbohydrate moiety (e.g. glycosylated antioxidants). In these types of applications, enzymes can (dependent on their specificity) thus be used both in pretreatment of the raw materials – acting on the

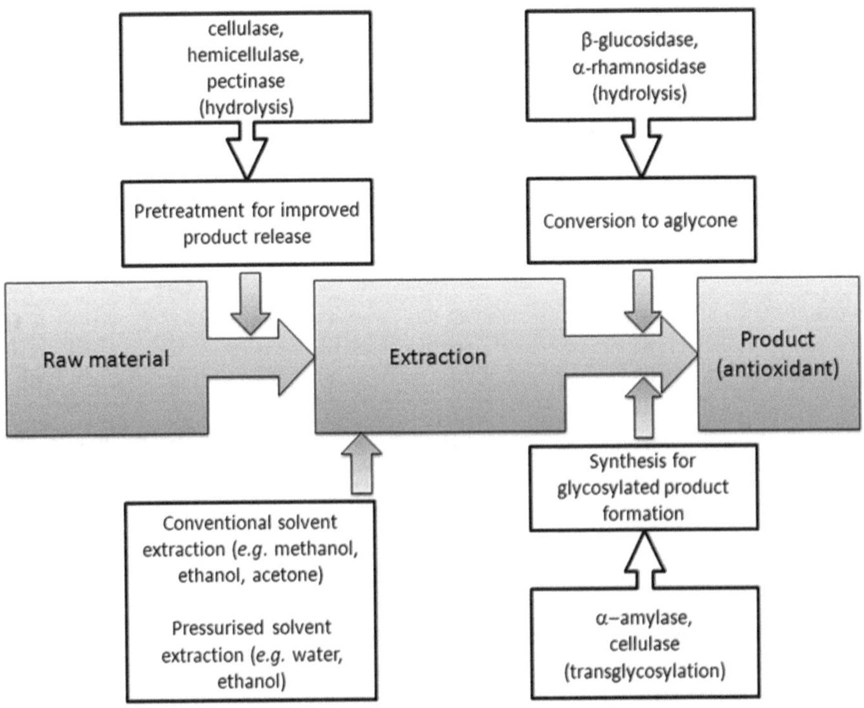

Fig. 2.2 Schematic overview of an extraction process to obtain antioxidants with desired glycosylation patterns. The possibilities to use glycoside hydrolases in pretreatment and in conversions to modify the glycosylation are indicated

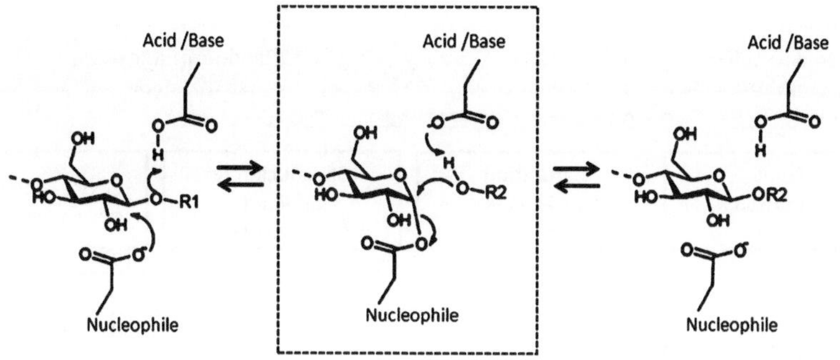

Fig. 2.3 The double displacement mechanism of retaining glycoside hydrolases. HO-R1 represents the group cleaved from the donor substrate, while HO-R2 represents the acceptor molecule. The covalent glycosyl-enzyme intermediate is boxed. For hydrolysis reactions HO-R2 is a water molecule and R2=H

polysaccharide fibres to simplify release of the secondary metabolites (the antioxidants) in the following extraction (Fig. 2.2) but also to change the glycosylation pattern (described in more detail in section "Glycoside Hydrolases in Flavonoid Conversions") on the polyphenolic products. Pretreatment with different types of polysaccharide-degrading glycoside hydrolases [cellulases, hemicellulases (e.g. xylanases and mannanases) and pectinases] before the extraction has, for example, been reported to promote release of the desired secondary metabolite flavonoids from matrices of different sources containing complex polysaccharides (Fu et al. 2008; Kapasakalidis et al. 2009; Landbo and Meyer 2001; Lin et al. 2009; Maier et al. 2008; Zheng et al. 2009). Sources investigated include fruits and berries, e.g. apples (Zheng et al. 2009), blackcurrants (Landbo and Meyer 2001) and grapes (Maier et al. 2008), but also agricultural products such as pigeon peas (Fu et al. 2008) or products from forestry, such as pine (Lin et al. 2009).

Development of Thermostability: A Means to Improve GHs as Extraction Aids

Thermostable GHs have been well documented for use in low-value, high-volume applications, such as starch degradation and the conversion of lignocellulosics (Turner et al. 2007), but they are still relatively rarely used in extractions/conversions of glycosylated flavonoids. Extractions from biomass often benefit from high-temperature processing, as this aids in loosening recalcitrant polysaccharide fibre structures. A step in this direction is also taken in flavonoid extractions, in which novel technologies striving to increase the environmental performance have been used that replace traditionally used extraction solvents (e.g. methanol and where deglycosylation is made by acid) with pressurised hot water where deglycosylation is made in an enzymatic step (Turner et al. 2006; Lindahl et al. 2010). The high-temperature extraction method puts in a need of a thermostable enzyme, which in this case was obtained from a thermophilic microorganism (*Thermotoga neapolitana*) but which also can be developed from enzymes originally active at ambient temperatures by mutagenesis. In the latter case, both rational and random methods have been utilised, but due to relatively straight forward screening possibilities (often relying on incubations and activity assays at the desired temperature), different random strategies are frequently utilised (Fig. 2.4). Successful combinatorial designs for enhanced (thermal) stability development have, for instance, been reviewed by Bommarius et al. (2006).

Glycoside Hydrolases in Flavonoid Conversions

Use of GHs as specific catalysts to modify the substituents on the target product is currently also gaining attention. Taking advantage of the

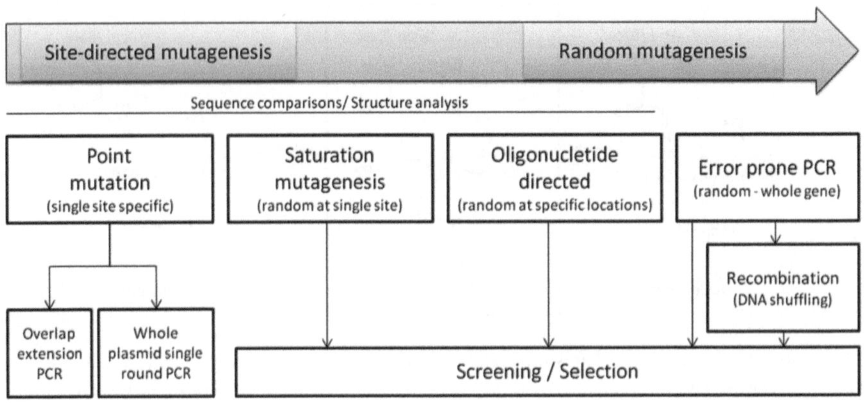

Fig. 2.4 Strategies for mutagenesis of enzymes by rational and random methodologies

possibilities to utilise retaining enzymes for both synthesis and hydrolysis, GHs can be used to either remove glycoside substituents (by hydrolysis using water as acceptor) or to add substituents (using in this case flavonoid acceptor molecules) (Fig. 2.3). Hydrolases express catalytic activity also in media with low water content such as organic solvents (resulting in less competition with water as acceptor molecule) and may under these conditions catalyse new reactions (Klibanov 2001). GHs however work rather poorly in organic media (compared to other hydrolytic enzymes, e.g. lipases) due to the requirement of higher thermodynamic water activity (Ljunger et al. 1994). The reasons for this are largely unknown, but indicate that water molecules have a role in interactions between substrate and enzyme. Use of thermostable GHs, when organic media are used, may again be advantageous as these enzymes are often resistant to denaturation by organic solvents, especially when run below their temperature optima for activity.

Enzymatic hydrolysis of flavonoid glycosides is dependent on the aglycone moiety, type of sugar and linkage, and is, e.g., used to obtain uniform flavonoid molecules with often higher antioxidising power than their glycosylated counterparts (Turner et al. 2006; Lindahl et al. 2010). On the other hand, glycosylation of flavonoids is one of the predominant approaches by which the biological activity of these natural compounds is regulated in living organisms (Yang et al. 2007) and will also increase water solubility of the molecule. Many well-designed chemical glycosylation methods are available, but due to limitation of acceptor, it is not possible to obtain regioselective glycosylation by using those methods (Davis 2000; Kong 2003). The delicate selectivity of biocatalysts can instead be used for this purpose, and as stated above, GHs provide versatile tools for both glycosylation and deglycosylation. Below, a few examples of GHs used (i) for hydrolysis of glycosidic groups and developed to improve deglycosylation of flavonoids and (ii) for synthesis (introducing new glycosidic groups) and developed to increase glycosylation on flavonoid backbones are given. For hydrolysis, the examples focus on β-glucosidases, while for the synthesis the examples shown mainly concerns endoglucanases but also mention use of α-amylase.

Deglycosylation of Flavonoids Using β-Glucosidases

Glycosylated flavonoids are the favoured forms for uptake in the human intestine but are in the body converted to the aglycone and free carbohydrates in hydrolysis reactions. The hydrolysis reactions are mainly catalysed by β-glucosidases (Walle 2004). The β-glucosidases are also helpful catalysts in analysis of flavonoid content in food and act by releasing the nonreducing terminal glucosyl residue leaving a uniform aglycone that is easier to quantify. The β-glucosidases (EC 3.2.1.21) are widely distributed in nature and found in all domains of living organisms and are classified under six GH families (GH1, GH3, GH5, GH9, GH30 and GH116), of which GH1, GH5 and GH30 display related $(\beta/\alpha)_8$ barrel structures and are classified as GH clan A (Ketudat Cairns and Esen 2010). Five of the families host enzymes with retaining reaction mechanism (only GH9 holds inverting enzymes), but examples of β-glucosidases used for deglucosylation of flavonoids mainly include enzymes from GH1 and GH3.

From GH3 a few examples of enzymes used to deglycosylate flavonoids are published [including thermostable enzymes from *T. neapolitana* (termed *Tn*Bgl3B) and *Dictyoglomus turgidum* (Turner et al. 2006; Kim et al. 2011)], showing that glycosidic groups at position 7 and 4′ on the flavonoid backbone (Fig. 2.1) were readily hydrolysable.

From GH1 which includes the largest number of characterised β-glucosidases, examples include enzymes from different domains of life, such as another enzyme from the prokaryotic thermophile *T. neapolitana* (*Tn*Bgl1A) (Turner et al. 2006) but also eukaryotic enzymes of fungal as well as human origin. *Tn*Bgl1A is a GH1 enzyme, which (like the GH3 candidates) efficiently hydrolyses glucosylations at the 4′- and 7-positions, but in this case hydrolysis of glucosides at the 3-position was also recognisable, although less efficient (Lindahl et al. 2010). The two intracellular GH1 β-glucosidases from the fungus *Penicillium decumbens* named G_I and G_{II} also displayed low activity towards the 3-glucoside and were most active towards flavonoids glycosylated at the 7-position (Mamma et al. 2004). The human β-glucosidase (hCBG) also preferred deglycosylation at the 7-position, but in this case glucosides located at the 3-position were not hydrolysed (Berrin et al. 2003). Comparison of the 3D structures (Fig. 2.5) showed that this difference between *Tn*Bgl1A and hCBG can be attributed to the wider opening of the active site cleft in *Tn*Bgl1A, making it possible for the 3-glucoside to fit (Khan et al. 2011).

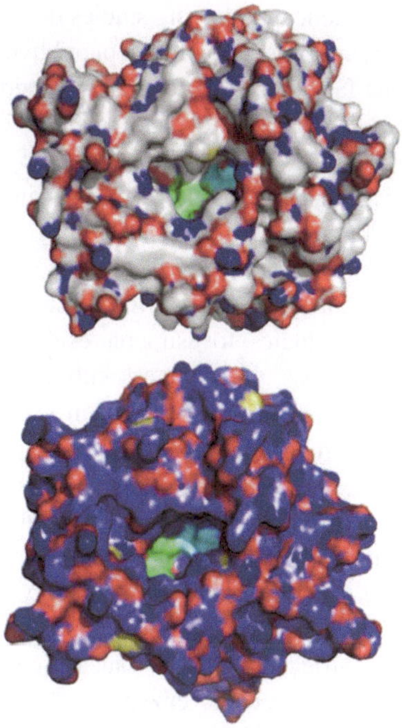

Fig. 2.5 Spacefill representations of the human cytosolic β-glucosidase (*top*) and the thermostable β-glucosidase from *Thermotoga neapolitana* (*bottom*) facing the active site. The wider active site opening of the flavonoid 3-glucoside hydrolysing *T. neapolitana* enzyme is clearly visible

Increased Flavonoid Hydrolysis in GH1 by Structure-Based Site-Directed Mutagenesis

The catalytic domain regions in GH1 are well conserved, but the enzymes in the family have varying substrate specificities, with some enzymes very specific for only one sugar (e.g. true cellobiases) or one aglycone (i.e. aryl-β-glucosidases), while others have a broad range of specificities for the glycones, the aglycones or both and are broad substrate specificity enzymes (e.g. using cellulose and β-glucan as well as flavonoids as substrates) (Bhatia et al. 2002). These differences make enzymes from GH1 interesting models for studies of the relationship between structure and activity (Chuenchor et al. 2008).

Most genetic engineering studies done on the GH1 enzymes in relation to flavonoid hydrolysis have thus far focused on identifying residues of importance for activity and mainly involve *Tn*Bgl1A (Khan et al. 2011) and the human cytosolic β-glucosidase hCBG (Berrin et al. 2003), while a multitude of GH1 enzymes have been studied and mutated for other purposes. The studies on *Tn*Bgl1A and hCBG are site-directed mutagenesis studies, focusing on residues identified by analysis of enzymes with known 3D structure. In our laboratory, the aim was to also improve the hydrolysis of quercetin-3-glucosides using *Tn*Bgl1A as model enzyme. To start this work, bioinformatic analysis of *Tn*Bgl1A was made, comparing this broad specificity enzyme with GH1 enzymes active on other bulky phenol-containing substrates (e.g. isoflavonoids and the alkaloids strictosidine and raucaffricine) and oligosaccharide-specific enzymes. The analysis was made to identify differences between specificity groups in regions around the +1 and +2 sugar-binding subsites in GH1 enzymes. To locate these subsites, 3D structure-determined enzymes were included in the analysis and the structure of *Tn*Bgl1A was determined (Khan et al. 2011; Kulkarni et al. unpublished). From the comparison, nonconserved amino acid residues located in β-strand 5 (spanning the +1 and +2 subsites) were mutated to residues occurring in the enzymes identified to use polyphenolic substrates. Different flavonoid glucosides like quercetin-3-glucoside, quercetin-3,4′-diglucoside and quercetin-4′-glucoside were also docked in the active site *Tn*Bgl1A in order to identify putative interactions of the amino acid residues chosen for mutagenesis. For example, the GH1 sequences showed variability at positions 219, 221 and 222 (*Tn*Bgl1A numbering), and the mutations F219L, N221S, G222Q and G222M were made on the basis of residues found in enzymes hydrolysing the bulky phenol-containing substrates. The site-directed mutagenesis methodology used was a ligation-independent whole plasmid amplification methodology (Fig. 2.4) utilising a methylated template that was selectively degraded by a methylation-specific restriction enzyme after PCR amplification. After introducing the mutations, it was revealed that mutant N221S led to a significant increase in conversion of quercetin-3-glucosides to quercetin, while no effect was observed for F219L and a limited increase was seen after the G222 mutations. The effect of the N221S mutation was suggested to occur via a loss in backbone carbonyl interactions that resulted in an increased flexibility of the parallel β-sheets, which could be a reason for the observed increase in catalytic efficiency towards quercetin-3-glucosides (Khan et al. 2011).

Mutation of a neighbouring residue in the human enzyme (hCBG, F225S) resulted in almost complete loss of activity of the enzyme (Berrin et al. 2003; Tribolo et al. 2007). To elucidate the role of the corresponding residue (N220 in *Tn*Bgl1A), this was also selected for mutagenesis and was mutated to S as well as to F (which was originally present in the human enzyme). In case of *Tn*Bgl1A, the N220S mutation increased the catalytic efficiency towards quercetin-3-glucosides compared to the wild-type enzyme (a result of a combined drop in K_M value together with threefold increase in the turnover number). Moreover, replacement of the hydrophilic amino acid residue N220 by the aromatic hydrophobic residue F resulted in drastic drop in the hydrolysis of both flavonoid glycosides as well as smaller model substrates like *para*-nitrophenyl-β-D-glucopyranoside (Kulkarni et al. unpublished). A similar effect was also seen at position 221. The mutation N221F, which introduced small local changes in the range of 0.4–0.7 Å, led to loss of catalytic activity compared to the wild type. This confirmed that also N220 plays a role in hydrolysis but that the interactions vary between specific GH1 enzymes and that the neighbouring residues N220 and N221 in *Tn*Bgl1A may display a similar rearrangement upon a single mutation of the respective residue. The corresponding N245 (at the +2 sugar-binding site) in the homologous rice β-glucosidase Os3BGlu7 was mutated to M and resulted in 6.5-fold loss of catalytic efficiency towards laminaribiose and 17–30-fold loss of catalytic efficiency for cellooligosaccharides with degree of polymerisation >2. On the other hand, the corresponding mutation M251N in Os3BGlu6 led to 15-fold increase in the catalytic efficiency for laminaribiose and 9–24-fold increase in catalytic efficiency for cellooligosaccharides with degree of polymerisation >2 (Sansenya et al. 2012). These observations show this position to be important for substrate interactions in GH1 enzymes. Presence of a bulkier hydrophobic group in the local area around the amino acids 220 and 221 in *Tn*Bgl1A was not favoured neither for the hydrolysis of long-chain cellooligosaccharides nor for the bulkier flavonoid glucosides.

Glycosylation of Flavonoids Using Cellulase and Amylase

Glycosylation of flavonoids is one of the predominant approaches by which the biological activity of these natural compounds is regulated in living organisms (Yang et al. 2007). Many well-designed chemical glycosylation methods are available, but due to limitation of acceptor, it is not possible to obtain regioselective glycosylation by using those methods (Davis 2000; Kong 2003). However, to overcome this problem, the delicate selectivity of a biocatalyst can be used, and glycosyltransferases and GHs can assist as useful tools for synthesis of defined glycosylated flavonoids (Hancock et al. 2006). In terms of substrate specificity, a glycosyltransferase is usually strict and requires a complex sugar nucleotide as the donor for the catalysis reaction (Wang and Huang 2009). On the other hand, GHs can also catalyse transglycosylation reactions. If the water molecule in the hydrolysis reaction is replaced by another acceptor molecule (e.g. a sugar molecule or a flavonoid), the double displacement mechanism of retaining GHs will result in transfer of the covalently bound glycosyl group from the donor substrate to the acceptor molecule. Transglycosylation is kinetically controlled, and during the reaction it is assumed that there is competition between the nucleophilic water and the acceptor substrate at the glycosyl-enzyme intermediate (Nakatani 2001; Park et al. 2005; Hancock et al. 2006).

Cellulases are GHs that catalyse the hydrolysis of the 1,4-β-D-glycosidic linkages in cellulose (and other related substrates, e.g. lichenan and cereal β-glucans). The name cellulase can refer to different types of enzymes acting on cellulose but is most commonly used for endoglucanases (endo-1,4-β-D-glucanases, EC 3.2.1.4), which can be found in at least 17 different GH families (see http://www.cazy.org/Glycoside-Hydrolases.html).

For the synthesis of flavonoid glycosides, a cellulase from *Aspergillus niger* has been used to add a fucose sugar moiety to a catechin backbone (Fig. 2.6). Catechin is an antioxidant that is, for instance, found in bark, a renewable resource of significant volume. During the reaction *para*-nitrophenyl-*β*-D-fucopyranoside was acting as a donor, while catechin monohydrate acted as acceptor resulting in a 26 % yield of catechin-β-d-fucopyranoside (Gao et al. 2000). In another study, an α-amylase from *Trichoderma viride* was reported to show transglycosylation activity towards both catechin and epigallocatechin gallate using dextrin (α-1,4-linked oligosaccharides resulting from starch degradation) as donor substrate (Noguchi et al. 2008). α-Amylases (EC 3.2.1.1) are enzymes classified under the large glycoside hydrolase family 13, and their main activity is hydrolysis of α-1,4-bonds of starch and glycogen.

Glycosynthases: Application of a Nucleophile-Mutated Cellulase in Flavonoid Glycosylation

The application of GHs in synthesis of carbohydrate or non-carbohydrate substrates has two major limitations: mainly low transglycosylation yield and secondary hydrolysis of the product. In order to overcome this problem, the invention of glycosynthases was a major development (Ly and Withers 1999). Glycosynthases is a class of unique GH mutants (mutated in the catalytic nucleophile) that can promote glycosidic bond formation in the presence of an activated glycosyl donor, and there is no further hydrolysis of the newly formed glycosidic linkage (Ly and Withers 1999; Wang and Huang 2009). Drawbacks with this methodology are however the necessity to

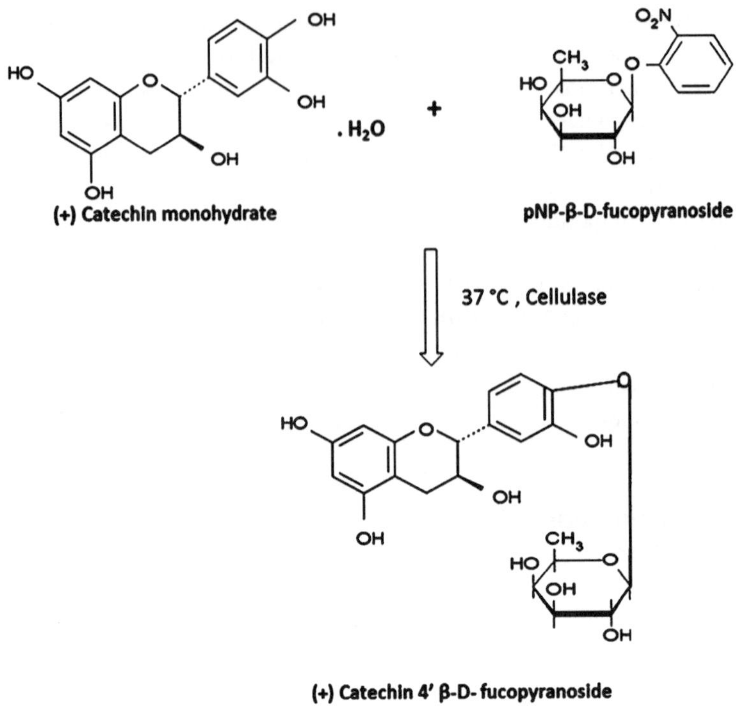

Fig. 2.6 Glycosylation of (+) catechin monohydrate. Synthesis of (+) catechin-β-D-fucopyranoside in the presence of cellulase at 37 °C while using pNP-β-D-fucopyranoside as donor (Modified from Gao et al. 2000)

produce activated glycosyl donors, with sometimes limited stability. Glycosynthases have been studied for the synthesis of complex glycoconjugates, and in a recent report it was shown that a glycosynthase mutant E197S of the *Humicola insolens* cellulase Cel7B was able to glycosylate flavonoids (Yang et al. 2007). Cel7B is a GH7 retaining endoglucanase, and the predominant activity of the wild-type enzyme is hydrolysis of the α-1, 4-linked glucosidic bond in cellulose, which in the mutant is circumvented by replacing the catalytic nucleophile (E197) with S. In this study a high-throughput MS-based method was used to screen 80 different acceptors and more than 20 glycosyl donors for substrate activity. In this screening process, a subclass of flavonoids was identified as an acting acceptor for Cel7B-E197S during transglycosylation reaction (Fig. 2.7). According to kinetic studies, the rate of glycosylation by Cel7B-E197S was comparable to glycosyltransferases which is very promising for the synthesis of glycosylated flavonoids (Yang et al. 2007).

Conclusions

To increase extraction yields and allow modification of antioxidants, new biocatalysts have the potential to adapt the compounds to different applications. Biocatalysis also has the potential of being a sustainable technology – something that is given increased attention today with current concerns about climate change and scarcity of fossil resources. In this chapter, we have reviewed emerging attempts to use and develop GHs as biocatalysts for modification of polyphenolic antioxidants classified under the flavonoid group. One aspect is the use of the GHs as extraction aids, and for this purpose an important property of the enzyme is its stability. Concerning the flavonoid product, both the antioxidative power and stability are affected by glycosylation. Currently studies to specifically modulate this (both remove and add glycosidic groups) have started. Although work in this field is just emerging, positive results have been published,

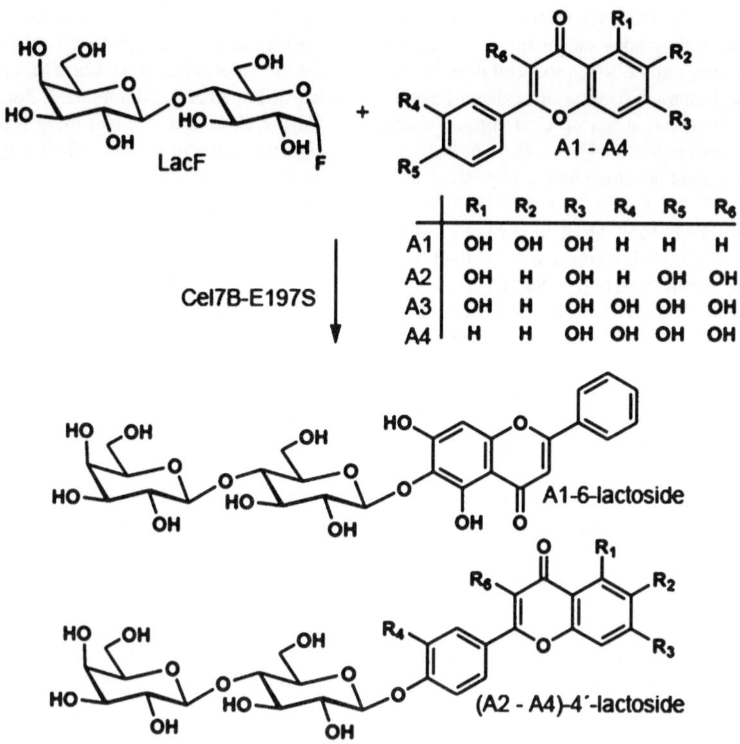

	R_1	R_2	R_3	R_4	R_5	R_6
A1	OH	OH	OH	H	H	H
A2	OH	H	OH	H	OH	OH
A3	OH	H	OH	OH	OH	OH
A4	H	H	OH	OH	OH	OH

Fig. 2.7 Glycosylation of flavonoids by Cel7B-E197S glycosynthase (Adapted from Yang et al. 2007; Wang and Huang 2009). The nucleophile (E197) of the GH7 cellulase from *Humicola insolens* is mutated to S. The lactosyl fluoride (LacF) was the disaccharide donor, and transfer of lactosyl from LacF to a number of flavonoids was catalysed by the Cel7B-E197S mutant in yields of 72–95 %. The synthesis was stereoselective (only β-glycosides) and regioselective for the glycosylation using the hydroxyl group at 4′ (*as in A2–A4*), while in absence of the hydroxyl group at this position (*as in A1*), the 6-position was glycosylated instead. A1 = baicalein, A2 = luteolin, A3 = quercetin, A4 = fisetin

where the action of GHs allows modification at uncommon positions or with new sugar moieties. Further development in this field is expected, allowing increased development and use of novel ingredients and bioactive compounds from poylphenolics.

Acknowledgements The authors wish to thank Formas 2009-1527 (SuReTech), the European project AMYLOMICS and the Antidiabetic Food Centre, a VINNOVA VINN Excellence Centre at Lund University.

References

Berrin JG, Czjzek M, Kroon PA, McLauchlan WR, Puigserver A, Williamson G, Juge N (2003) Substrate (aglycone) specificity of human cytosolic β-glucosidase. Biochem J 373:41–48

Bhatia Y, Mishra S, Bisaria VS (2002) Microbial β-glucosidases: cloning, properties, and applications. Crit Rev Biotechnol 22:375–407

Bommarius AS, Broering JM, Chaparro-Riggers JF, Polizzi KM (2006) High-throughput screening for enhanced protein stability. Curr Opin Biotechnol 17:606–610

Boudet AM (2007) Evolution and current status of research in phenolic compounds. Phytochemistry 68:2722–2735

Chuenchor W, Pengthaisong S, Robinson RC, Yuvaniyama J, Oonanant W, Bevan DR, Esen A, Chen C-J, Opassiri R, Svasti J, Cairns JRK (2008) Structural insights into rice bglu1 β-glucosidase oligosaccharide hydrolysis and transglycosylation. J Mol Biol 377:1200–1215

Corradini E, Foglia P, Giansanti P, Gubbiotti R, Samperi R, Lagana A (2011) Flavonoids: chemical properties and analytical methodologies of identification and quantitation in foods and plants. Nat Prod Res 25:469–495

Cuyckens F, Shahat AA, Van den Heuvel H, Abdel-Shafeek KA, El-Messiry MM, Seif-El Nasr MM,

Pieters L, Vlietinck AJ, Claeys M (2003) The application of liquid chromatography-electrospray ionization mass spectrometry and collision-induced dissociation in the structural characterization of acylated flavonol O-glycosides from the seeds of *Carrichtera annua*. Eur J Mass Spectrom 9:409–420

Davis BG (2000) Recent developments in oligosaccharide synthesis. J Chem Soc Perkin Trans 1:2137–2160

Fu Y-J, Liu W, Zu Y-G, Tong M-H, Li S-M, Yan M-M, Efferth T, Luo H (2008) Enzyme assisted extraction of luteolin and apigenin from pigeonpea [*Cajanus cajan* (L.) Millsp.] leaves. Food Chem 111:508–512

Gao C, Mayon P, MacManus DA, Vulfson EN (2000) Novel enzymatic approach to the synthesis of flavonoid glycosides and their esters. Biotechnol Bioeng 71:235–243

Griffiths G, Trueman L, Crowther T, Thomas B, Smith B (2002) Onions—a global benefit to health. Phytother Res 16(7):603–615. doi:10.1002/ptr.1222

Haddad AQ, Venkateswaran V, Viswanathan L, Teahan SJ, Fleshner NE, Klotz LH (2005) Novel antiproliferative flavonoids induce cell cycle arrest in human prostate cancer cell lines. Prostate Cancer Prostatic Dis 9:68–76

Hancock SM, Vaughan MD, Withers SG (2006) Engineering of glycosidases and glycosyltransferases. Curr Opin Chem Biol 10:509–519

Havsteen BH (2002) The biochemistry and medical significance of the flavonoids. Pharmacol Ther 96:67–202

Hollman PCH, Arts ICW (2000) Flavonols, flavones and flavanols – nature, occurrence and dietary burden. J Sci Food Agric 80:1081–1093

Hughes RJ, Croley TR, Metcalfe CD, March RE (2001) A tandem mass spectrometric study of selected characteristic flavonoids. Int J Mass Spec 210/211:371–385

Ishihara K, Nakajima N (2003) Structural aspects of acylated plant pigments: stabilization of flavonoid glucosides and interpretation of their functions. J Mol Catal B Enzym 23:411–417

Iwashina T (2000) The structure and distribution of the flavonoids in plants. J Plant Res 113:287–299

Kapasakalidis PG, Rastall RA, Gordon MH (2009) Effect of a cellulase treatment on extraction of antioxidant phenols from black currant (*Ribes nigrum* L.) pomace. J Agric Food Chem 57:4342–4351

Ketudat Cairns J, Esen A (2010) β-Glucosidases. Cell Mol Life Sci 67:3389–3405

Khan S, Pozzo T, Megyeri M, Lindahl S, Sundin A, Turner C, Nordberg Karlsson E (2011) Aglycone specificity of *Thermotoga neapolitana* β-glucosidase 1A modified by mutagenesis, leading to increased catalytic efficiency in quercetin-3-glucoside hydrolysis. BMC Biochem 12:11

Kim Y-S, Yeom S-J, Oh D-K (2011) Characterization of a GH3 family β-glucosidase from *Dictyoglomus turgidum* and its application to the hydrolysis of isoflavone glycosides in spent coffee grounds. J Agric Food Chem 59:11812–11818

Klibanov AM (2001) Improving enzymes by using them in organic solvents. Nature 409:241–246

Kong F (2003) Regio- and stereoselective synthesis of oligosaccharides with unprotected or lightly protected sugars as glycosyl acceptors. Curr Org Chem 7:841–865

Kulkarni TS, Khan S, Mahmood T, Sundin A, Lindahl S, Turner C, Logan DT, Nordberg Karlsson E (unpublished) Structure of *Thermotoga neapolitana* β-glucosidase 1A and comparison of active site mutants in hydrolysis of pNPGlc and quercetin-3 glucosides

Landbo A-K, Meyer AS (2001) Enzyme-assisted extraction of antioxidative phenols from black currant juice press residues (*Ribes nigrum*). J Agric Food Chem 49:3169–3177

Lee KW, Lee HJ (2006) The roles of polyphenols in cancer chemoprevention. Biofactors 26:105–121

Lin JK, Weng MS (2006) Flavonoids as nutraceuticals. In: Grotewold E (ed) The science of flavonoids. Springer, New York, pp 213–238

Lin S-C, Chang C-MJ, Deng T-S (2009) Enzymatic hot pressurized fluids extraction of polyphenolics from *Pinus taiwanensis* and *Pinus morrisonicola*. J Taiwan Inst Chem Eng 40:136–142

Lindahl S, Ekman A, Khan S, Wennerberg C, Borjesson P, Sjoberg PJR, Nordberg Karlsson E, Turner C (2010) Exploring the possibility of using a thermostable mutant of β-glucosidase for rapid hydrolysis of quercetin glucosides in hot water. Green Chem 12: 159–168

Ljunger G, Adlercreutz P, Mattiasson B (1994) Enzymatic synthesis of octyl-β-glucoside in octanol at controlled water activity. Enzyme Microb Technol 16:751–755

Ly HD, Withers SG (1999) Mutagenesis of glycosidases. Ann Rev Biochem 68:487–522

Maier T, Göppert A, Kammerer D, Schieber A, Carle R (2008) Optimization of a process for enzyme-assisted pigment extraction from grape (*Vitis vinifera L.*) pomace. Eur Food Res Technol 227:267–275

Mamma D, Hatzinikolaou DG, Christakopoulos P (2004) Biochemical and catalytic properties of two intracellular β-glucosidases from the fungus *Penicillium decumbens* active on flavonoid glucosides. J Mol Catal B Enzym 27:183–190

Nakatani H (2001) Analysis of glycosidase-catalyzed transglycosylation reaction using probabilistic model. Arch Biochem Biophys 385:387–391

Nijveldt RJ, van Nood E, van Hoorn DE, Boelens PG, van Norren K, van Leeuwen PA (2001) Flavonoids: a review of probable mechanisms of action and potential applications. Am J Clin Nutr 74:418–425

Noguchi A, Inohara-Ochiai M, Ishibashi N, Fukami H, Nakayama T, Nakao M (2008) A novel glucosylation enzyme: molecular cloning, expression, and characterization of *Trichoderma viride* Jcm22452 α-amylase and enzymatic synthesis of some flavonoid monoglucosides and oligoglucosides. J Agric Food Chem 56:12016–12024

Park T-H, Choi K-W, Park C-S, Lee S-B, Kang H-Y, Shon K-J, Park J-S, Cha J (2005) Substrate specificity and transglycosylation catalyzed by a thermostable β-glucosidase from marine hyperthermophile

Thermotoga neapolitana. Appl Microbiol Biotechnol 69:411–422

Robards K, Li X, Antolovich M, Boyd S (1997) Characterisation of citrus by chromatographic analysis of flavonoids. J Sci Food Agric 75:87–101

Sansenya S, Maneesan J, Ketudat Cairns JR (2012) Exchanging a single amino acid residue generates or weakens a +2 cellooligosaccharide binding subsite in rice beta-glucosidases. Carbohydr Res 351:130–133

Stobiecki M, Malosse C, Kerhoas L, Wojlaszek P, Einhorn J (1999) Detection of isoflavonoids and their glycosides by liquid chromatography/electrospray ionization mass spectrometry in root extracts of lupin (*Lupinus albus*). Phytochem Anal 10:198–207

Tribolo S, Berrin JG, Kroon PA, Czjzek M, Juge N (2007) The crystal structure of human cytosolic β-glucosidase unravels the substrate aglycone specificity of a family 1 glycoside hydrolase. J Mol Biol 370:964–975

Turner P, Mamo G, Nordberg Karlsson E (2007) Potential and utilization of thermophiles and thermostable enzymes in biorefining. Microb Cell Fact 6:9

Turner C, Turner P, Jacobson G, Almgren K, Waldeback M, Sjoberg P, Nordberg Karlsson E, Markides KE (2006) Subcritical water extraction and β-glucosidase-catalyzed hydrolysis of quercetin glycosides in onion waste. Green Chem 8:949–959

Walle T (2004) Absorption and metabolism of flavonoids. Free Radic Biol Med 36(7):829–837

Wang L-X, Huang W (2009) Enzymatic transglycosylation for glycoconjugate synthesis. Curr Opin Chem Biol 13:592–600

Yang M, Davies GJ, Davis BG (2007) A glycosynthase catalyst for the synthesis of flavonoid glycosides. Angew Chem 46:3885–3888

Zheng H-Z, Hwang I-W, Chung S-K (2009) Enhancing polyphenol extraction from unripe apples by carbohydrate-hydrolyzing enzymes. J Zhejiang Univrsity Sci B 10:912–919

On the Enzyme Specificity for the Synthesis of Prebiotic Galactooligosaccharides

3

Barbara Rodriguez-Colinas, Lucia Fernandez-Arrojo, Miguel de Abreu, Paulina Urrutia, Maria Fernandez-Lobato, Antonio O. Ballesteros, and Francisco J. Plou

Abstract

β-Galactosidases catalyze transgalactosylation reactions in which lactose as well as the glucose and galactose released by hydrolysis serve as galactosyl acceptors yielding a series of galactooligosaccharides (GOS). GOS constitute the major part of oligosaccharides in human milk and are responsible of the formation of a *Bifidus* microbiota in the intestine of milk-fed babies. The bioactive properties of GOS depend on their chemical composition, structure, and polymerization degree. We have analyzed the product specificity of various β-galactosidases, namely, those from *Kluyveromyces lactis*, *Bacillus circulans*, and *Aspergillus oryzae*. The major products synthesized by *B. circulans* β-galactosidase contained only β-(1→4) bonds, whereas the enzyme from *K. lactis* synthesized GOS with major presence of β-(1→6) linkages. The *A. oryzae* β-galactosidase formed preferentially β-(1→6) bonds, with minor proportion of β-(1→3). *B. circulans* and *K. lactis* β-galactosidases produce nearly 45–50 % (w/w) GOS, whereas the *A. oryzae* enzyme produces less than 30 % (w/w). Another difference between the three enzymes was the polymerization degree of products; in particular, for a GOS mixture enriched in disaccharides, *K. lactis* and *A. oryzae* β-galactosidases are the best choices. In contrast, the *B. circulans* enzyme would be preferable for a GOS product with a high trisaccharides and tetrasaccharides content.

B. Rodriguez-Colinas • L. Fernandez-Arrojo • A.O. Ballesteros • F.J. Plou (✉)
Instituto de Catálisis y Petroleoquímica, CSIC, Cantoblanco, Marie Curie 2, Madrid 28049, Spain
e-mail: fplou@icp.csic.es

M. de Abreu • M. Fernandez-Lobato
Centro de Biología Molecular Severo Ochoa (CSIC-UAM), Universidad Autónoma de Madrid, Madrid 28049, Spain

P. Urrutia
Instituto de Catálisis y Petroleoquímica, CSIC, Cantoblanco, Marie Curie 2, Madrid 28049, Spain

School of Biochemical Engineering, Pontificia Universidad Católica de Valparaíso, Valparaíso, Chile

P. Shukla and B.I. Pletschke (eds.), *Advances in Enzyme Biotechnology*, DOI 10.1007/978-81-322-1094-8_3, © Springer India 2013

Keywords
Glycosidase • Galactooligosaccharides • Prebiotics • Transglycosylation • Beta-galactosidase • Oligosaccharides • Microbiota

Introduction

β-galactosidases (β-D-galactoside galactohydrolases, EC 3.2.1.23), also called lactases, catalyze the hydrolysis of the galactosyl moiety from the nonreducing end of various oligosaccharides. β-galactosidases are retaining glycosidases, resulting in the net retention of the anomeric configuration (Plou et al. 2007). The β-galactosidases are members of GH1 and GH2 families of the glycoside hydrolases (Cantarel et al. 2009).

β-galactosidases have attracted attention from industry as they can be used in different applications. Due to the deficiency of human lactase, many people in the world are intolerant to lactose present in dairy products; the problem is currently solved by means of treatment of lactose-rich products with microbial lactases (Adam et al. 2004). On the other hand, several companies have recently established projects for the transformation of whey (one of the most important by-products of the agrofood industry, with a high lactose content), into bioethanol by means of a pretreatment with β-galactosidases to hydrolyze lactose. Another application of β-galactosidases is related to transgalactosylation reactions in which lactose (as well as the released glucose and galactose) serve as galactosyl acceptors yielding a series of disaccharides, trisaccharides and higher oligosaccharides called galactooligosaccharides (GOS) (Park and Oh 2010; Torres et al. 2010). GOS constitute the major part of oligosaccharides in human milk (Gosling et al. 2010; Rastall et al. 2005; Shadid et al. 2007). The formation of a beneficial *Bifidus* microflora in the intestine of milk-fed babies seems to be related with the prebiotic effect of GOS present in the maternal milk.

In the context of functional foods, a prebiotic is a nondigestible food ingredient that beneficially affects the host by selectively stimulating the growth and/or the activity of certain types of bacteria in the colon, basically of the genera *Bifidobacterium* and *Lactobacillus* (Gibson and Ottaway 2000). Prebiotics produce positive effects on human health as the metabolism of these bacteria releases short-chain fatty acids (acetate, propionate and butyrate) and L-lactate (Roberfroid 2007). Among them, they exert protective effects against colorectal cancer and bowel infectious diseases by inhibiting putrefactive and pathogen bacteria, improve the bioavailability of essential minerals, reduce the level of cholesterol in serum or enhance the glucid and lipid metabolism (Tuohy et al. 2005). GOS, fructooligosaccharides (FOS), isomaltooligosaccharides (IMOS), lactulose, soybean oligosaccharides, lactosucrose, gentiooligosaccharides, and xylooligosaccharides are among the most consumed prebiotics (Playne et al. 2003).

It is well reported that the chemical structure of the obtained oligosaccharides (composition, number of hexose units, and types of linkages between them) may affect their fermentation pattern by probiotic bacteria in the gut (Cardelle-Cobas et al. 2011; Martinez-Villaluenga et al. 2008). Varying the source of β-galactosidase, the yield and composition of GOS can be modified (Iqbal et al. 2010; Maischberger et al. 2010; Rodriguez-Colinas et al. 2011; Splechtna et al. 2006). In general, complex mixtures of GOS of various chain lengths and glycosidic bonds are commonly obtained. The most studied β-galactosidases are those from *Kluyveromyces lactis* (Chockchaisawasdee et al. 2005; Martinez-Villaluenga et al. 2008; Maugard et al. 2003), *Aspergillus oryzae* (Albayrak and Yang 2002c; Guerrero et al. 2011; Iwasaki et al. 1996), *Bacillus circulans* (Gosling et al. 2011) and *Bifidobacterium* sp. (Hsu et al. 2007).

In our laboratory, we have studied the transgalactosylation activity of several β-galactosidases. In this chapter, we will summarize the structural features of the GOS synthesized by the different enzymes. In addition, the kinetics of transgalactosylation and the GOS yields will be comparatively discussed.

Bacillus circulans β-Galactosidase

Different isoforms of the β-galactosidase from *Bacillus circulans* have been reported in the commercial preparation Biolacta (Daiwa Kasei). At least three isoforms with distinctive behavior in GOS production have been characterized: β-galactosidase-1 showed very low transglycosylation activity (Mozaffar et al. 1984), β-galactosidase-2 contributed most significantly to GOS synthesis (Mozaffar et al. 1984, 1986), and β-galactosidase-3 was able to produce GOS with β(1 → 3) bonds (Fujimoto et al. 1998). More recently, Song et al. (2011a) described in Biolacta four isoforms differing in their molecular size: β-gal-A (189 kDa), β-gal-B (154 kDa), β-gal-C (134 kDa), and β-gal-D (91 kDa). The transferase activity of β-galactosidase from *B. circulans* has been also applied to the synthesis of lactosucrose (Wei et al. 2009), N-acetyl-lactosamine (Bridiau et al. 2010), and other galactosylated derivatives (Farkas et al. 2003). Interestingly, the enzyme is also able to catalyze the galactosylation of different acceptors in the presence of high percentages of organic cosolvents, up to 50 % v/v (Usui et al. 1993). The *B. circulans* β-galactosidase has been immobilized on different supports (Mozaffar et al. 1986; Torres and Batista-Viera 2012). However, only partial analysis of the GOS formed in the transglycosylation reaction with lactose has been described (Mozaffar et al. 1986), probably owing to the complexity of the reaction mixture derived from the presence of several isoforms with different regiospecificity.

A novel commercial preparation of β-galactosidase from *B. circulans* (Biolactase) was recently studied in our laboratory. Its volumetric activity towards *o*-nitrophenyl-β-D-galactopyranoside (ONPG) was 2,740 U/ml at pH 5.5 and 40 °C. We analyzed in detail the synthesis of galactooligosaccharides (GOS) catalyzed by Biolactase using 400 g/l lactose and 1.5 U/ml (β-galactosidase activity towards ONPG), in particular the selectivity of the bonds formed.

Specificity of *B. circulans* β-Galactosidase

The complete identification of the GOS synthesized by a particular enzyme is a difficult task. In fact, considering the formation of Gal-β(1 → 2), Gal-β(1 → 3), Gal-β(1 → 4), and Gal-β(1 → 6) bonds, the theoretical number of synthesized GOS accounts for 7 disaccharides, 32 trisaccharides, 128 tetrasaccharides, and so on.

Figure 3.1 shows the chromatogram obtained by high-performance anion-exchange chromatography coupled with pulsed amperometric detection (HPAEC-PAD) of the reaction mixture with *B. circulans* β-galactosidase close to the time of maximum GOS concentration. Peaks *1*, *2*, and *4* corresponded to galactose, glucose, and lactose, respectively. As illustrated in the chromatogram, the two main products present in the reaction mixture were peaks *11* and *17*. Using a commercial standard, peak *11* was identified as the trisaccharide 4-galactosyl-lactose [Gal-β(1 → 4)-Gal-β(1 → 4)-Glc]. Peak *17* was purified by semipreparative chromatography using an amino column, as the sugar concentration in samples for HPAEC-PAD analysis was too low for an efficient scaling up. The mass spectrum of peak *17* showed that it was a tetrasaccharide. The NMR data for peak *17* was consistent with a molecule that presented two galactosyl moieties β-(1 → 4)-linked to the O-4 of the galactose unit of lactose, resulting in the tetrasaccharide Gal-β(1 → 4)-Gal-β(1 → 4)-Gal-β(1 → 4)-Glc. Commercially available standards and other GOS purified in our laboratory allowed us to identify in the chromatograms the disaccharides allolactose [Gal-β(1 → 6)-Glc] (peak *3*), 3-galactobiose [Gal-β(1 → 3)-Gal] (peak *5*), 4-galactobiose [Gal-β(1 → 4)-Gal] (peak *6*) and Gal-β(1 → 3)-Glc (peak *8*), as well as the

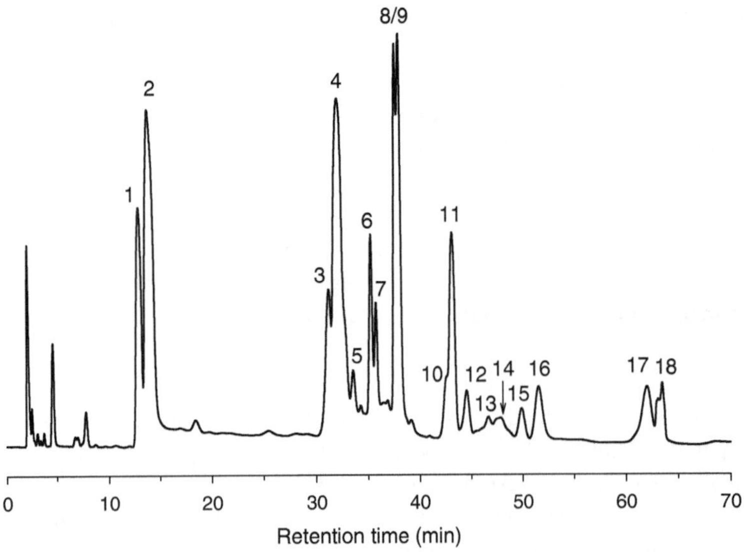

Fig. 3.1 HPAEC-PAD analysis of the reaction of lactose with *B. circulans* β-galactosidase (Biolactase). The peaks correspond to: *1* galactose, *2* glucose, *3* allolactose, *4* lactose, *5* 3-galactobiose, *6* 4-galactobiose, *7* 6-galactosyl-lactose, *8* 3-galactosyl-glucose, *11* 4-galactosyl-lactose, *15* Gal-β(1→4)-Gal-β(1→3)-Glc, *17* Gal-β(1→4)-Gal-β(1→4)-Gal-β(1→4)-Glc, and *9*, *10*, *12*, *13*, *14*, *16* other GOS (unknown). The chromatogram corresponds to the reaction mixture after 6.5 h with Biolactase. Conditions: 400 g/l lactose, 0.1 M sodium acetate buffer (pH 5.5), 40 °C

trisaccharide 6-galactosyl-lactose [Gal-β(1→6)-Gal-β(1→4)-Glc] (peak *7*). We also purified peak *15* by semipreparative hydrophilic interaction chromatography (HPLC-HILIC), whose mass and NMR spectra indicated that it was the trisaccharide Gal-β(1→4)-Gal-β(1→3)-Glc. Peaks *9*, *10*, *12*, *13*, *14*, and *16* remained unknown.

It is worth noting that two of the major products synthesized by *B. circulans* β-galactosidase (peaks *11* and *17*) contained only β-(1→4) bonds (Fig. 3.2). Yanahira et al. (1995) were the first in performing structural analysis of the GOS formed by *B. circulans* β-galactosidase employing Biolacta from Daiwa Kasei; they reported that the main product was 4-galactosyl-lactose, but the formation of tetrasaccharides was not mentioned. In that paper, several disaccharides and trisaccharides were purified and characterized; the authors reported the presence of various GOS with β-(1→2) bonds (Yanahira et al. 1995), which may correspond to some of the unidentified peaks in our study. Recently, Song et al. (2011b) analyzed the GOS production by the different isoforms of *B. circulans* β-galactosidase; although the authors found significant differences in total GOS yield, the structural analysis of the synthesized compounds was not reported.

GOS Production with *B. circulans* β-Galactosidase

In the presence of lactose, most of the β-galactosidases catalyze both GOS synthesis and lactose hydrolysis. The transferase/hydrolase ratio, which determines the maximum yield of GOS, depends basically on two parameters: (a) the concentration of lactose and (b) the intrinsic enzyme properties, that is, the ability of the enzyme to bind the nucleophile (to which a galactosyl moiety is transferred) and to exclude H_2O from the acceptor binding site. Maximum GOS production for any particular enzyme depends basically on the relative rates of the transgalactosylation and hydrolysis reactions. As the lactose is consumed, the concentration of GOS increases until it reaches a maximum. At this point, the rate of synthesis of GOS products equals its rate of

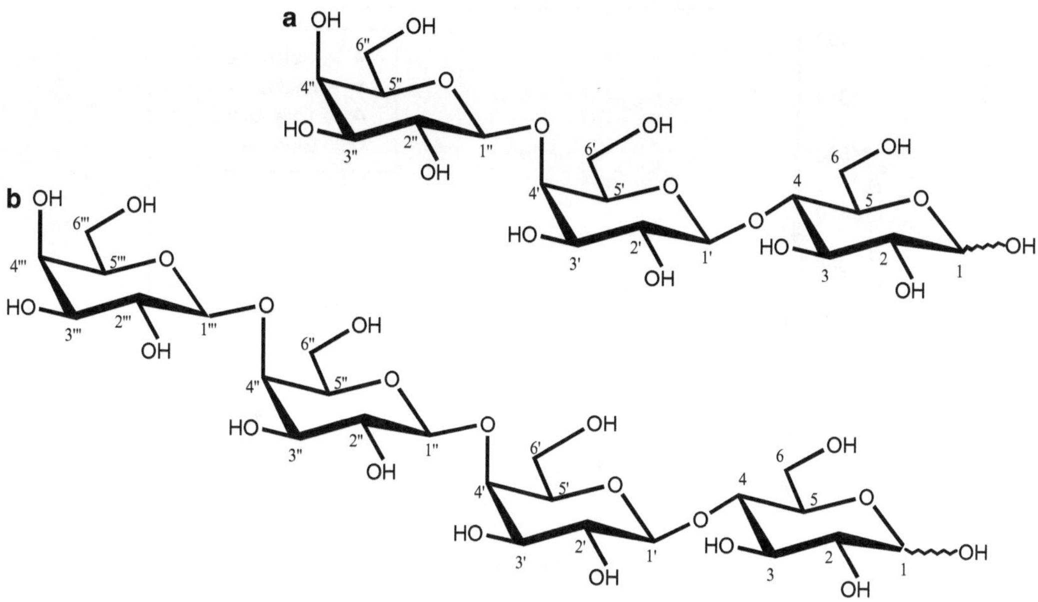

Fig. 3.2 Galactooligosaccharides synthesized by β-galactosidase from *B. circulans* containing only β(1→4) bonds: (**a**) Gal-β(1→4)-Gal-β(1→4)-Glc; (**b**) Gal-β(1→4)-Gal-β(1→4)-Gal-β(1→4)-Glc

hydrolysis. Subsequently, kinetic control is lost, and the reaction must be stopped quickly before product hydrolysis becomes the major process and a thermodynamic equilibrium is reached. The existence of this maximum explains why transglycosylation results in higher yields of condensation products compared with equilibrium-controlled processes.

We performed the GOS synthesis at 400 g/l lactose with a biocatalyst concentration of 15 U/ml. It has been widely reported that, working under kinetic control conditions, enzyme concentration has no effect on the maximum GOS yield as long as no enzyme inactivation takes place and it only exerts a marked influence on the reaction time at which the maximum oligosaccharide concentration is achieved (Buchholz et al. 2005; Chockchaisawasdee et al. 2005). Figure 3.3 shows that the maximum GOS production at 15 U/ml was achieved in 6.5 h, with a yield of 198 g/l. This value corresponds to 49.4 % (w/w) of total sugars, which was higher than the value obtained at 1.5 U/ml (165 g/l, 41.3 %). Interestingly enough, the remaining lactose at the equilibrium (10 g/l, 2.5 % of total carbohydrates) is significantly lower than the obtained at 1.5 U/ml. This effect indicates that the stability of *B. circulans* β-galactosidase is only moderate under typical GOS formation conditions; at 1.5 U/ml, the reaction is stopped before reaching the final composition.

The upper range of GOS yield reported is close to 40–45 % (Hansson and Adlercreutz 2001; Rabiu et al. 2001; Splechtna et al. 2006), which is lower than that obtained in the case of other prebiotics such as fructooligosaccharides (approx. 65 %) (Alvaro-Benito et al. 2007; Ghazi et al. 2006). In consequence, the GOS production with *B. circulans* β-galactosidase is one of the highest values reported to date.

Kluyveromyces lactis β-Galactosidase

The major commercial source of β-galactosidase by far is the mesophile yeast *Kluyveromyces lactis* (Chockchaisawasdee et al. 2005; Martinez-Villaluenga et al. 2008; Maugard

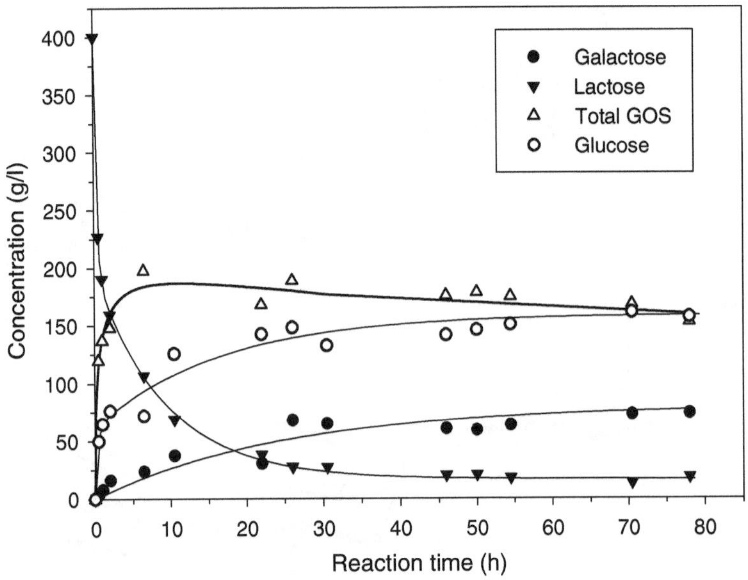

Fig. 3.3 Kinetics of GOS formation at 15 U/ml using 400 g/l lactose catalyzed by β-galactosidase from *B. circulans* (Biolactase). Reaction conditions: 0.1 M sodium acetate buffer (pH 5.5), 40 °C

et al. 2003; Pal et al. 2009). Due to its intracellular nature, production of cell-free *K. lactis* β-galactosidase is limited by the high cost associated to enzyme extraction and downstream processing as well as the low stability of the enzyme (Park and Oh 2010; Pinho and Passos 2011). The production of GOS in batch and continuous bioreactors of *K. lactis* β-galactosidase has been described (Chockchaisawasdee et al. 2005; Martinez-Villaluenga et al. 2008). Several approaches for immobilization of this enzyme have been proposed using different carriers (Maugard et al. 2003; Zhou et al. 2003; Zhou and Chen 2001). Recently, the crystallization and resolution of its three-dimensional structure has been reported (Pereira-Rodriguez et al. 2010).

We studied in detail the transgalactosylation activity of *K. lactis* β-galactosidase. Different enzyme preparations were assayed: (1) ethanol-permeabilized cells of a strain of *K. lactis* and (2) two soluble β-galactosidases from *K. lactis* commercially available (Lactozym 3000L and Maxilact LGX 5000). Structural characterization of the synthesized galactooligosaccharides was carried out.

Specificity of *K. lactis* β-Galactosidase

Figure 3.4 shows the HPAEC-PAD chromatogram of the reaction mixture with *K. lactis* β-galactosidase (Lactozym 3000L) after 7 h. Peaks *1*, *2*, and *5* were assigned to galactose, glucose, and lactose, respectively. We found that the most abundant GOS synthesized by this enzyme corresponded to peak *3* (the disaccharide 6-galactobiose [Gal-β(1 → 6)-Gal]), peak *4* (allolactose, Gal-β(1 → 6)-Glc), and peak *8*. Purification of peak *8* was performed by semipreparative HPLC-HILIC. Its mass spectrum indicated that it was a trisaccharide, and its 2D-NMR spectra showed a galactosyl moiety β-(1 → 6) linked to the galactose ring of lactose (Rodriguez-Colinas et al. 2011).

In consequence, the three major products synthesized by *K. lactis* cells contained a β-(1 → 6) bond between two galactoses or between one galactose and one glucose (Fig. 3.5). This product selectivity may have significant implications, as it was reported that β(1 → 6) linkages are cleaved very fast by β-galactosidases from Bifidobacteria (Martinez-Villaluenga et al. 2008), a key factor in the prebiotic properties.

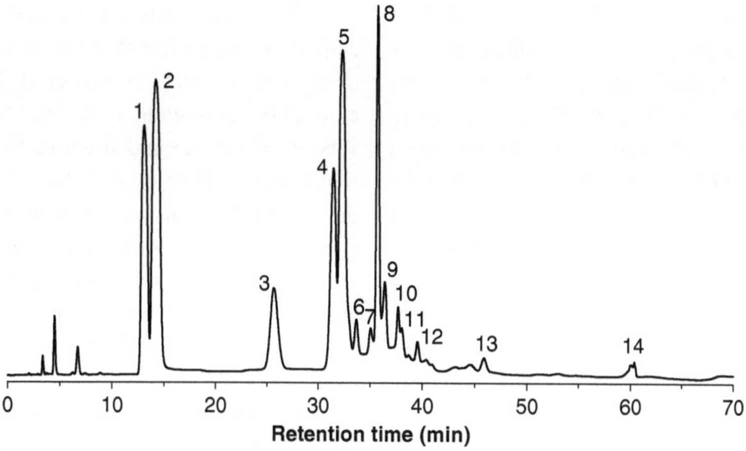

Fig. 3.4 HPAEC-PAD analysis of the reaction of lactose with *K. lactis* β-galactosidase (Lactozym 3000L): *1* galactose, *2* glucose, *3* 6-galactobiose, *4* allolactose, *5* lactose, *6* 3-galactobiose, *7* 4-galactobiose, *8* 6-galactosyl-lactose, *10* 3-galactosyl-glucos, and *9*, *11*, *12*, *13*, *14* other GOS (unknown). The chromatogram corresponds to the reaction mixture after 7 h. Conditions: 400 g/l lactose, 0.1 M sodium phosphate buffer (pH 6.8), 40 °C

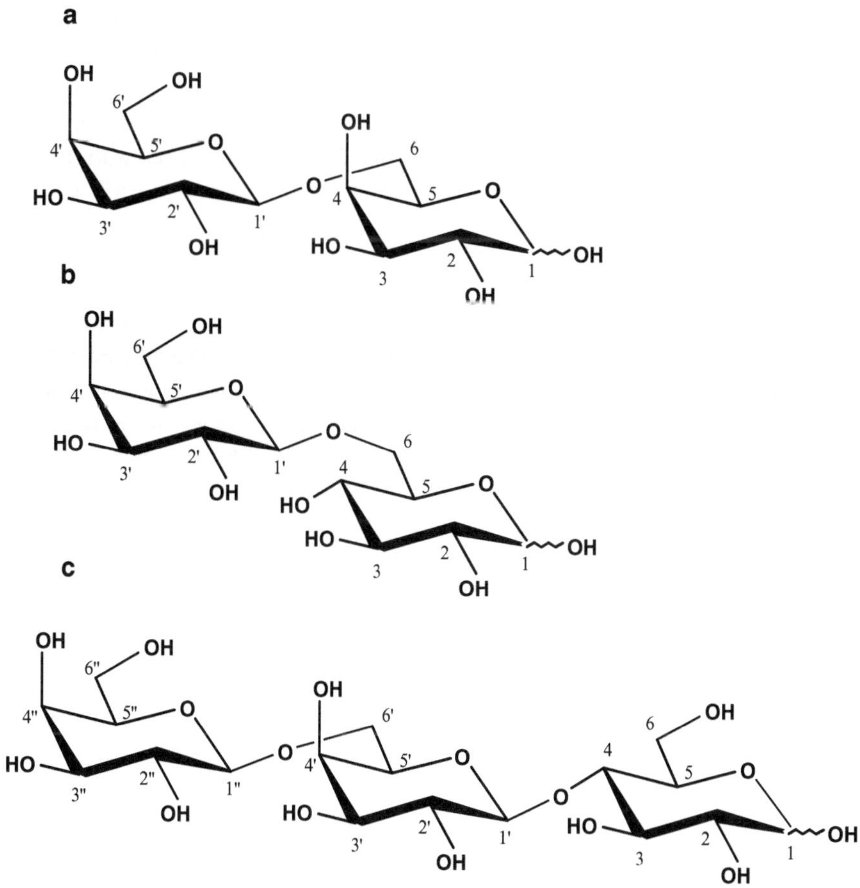

Fig. 3.5 Major galactooligosaccharides synthesized by β-galactosidase from *K. lactis*: (**a**) 6-galactobiose [Gal-β(1 → 6)-Gal], (**b**) allolactose [Gal-β(1 → 6)-Glc], (**c**) 6-galactosyl-lactose [Gal-β(1 → 6)-Gal-β(1 → 4)-Glc]

Chockchaisawasdee et al. (2005) were the first in performing a preliminary structural analysis of the GOS formed by soluble *K. lactis* β-galactosidase (Maxilact L2000), concluding that the major bonds were β-(1→6). Maugard et al. (2003) and Cheng et al. (2006) detected the formation of disaccharides by *K. lactis*, but no structural identification was done. Martinez-Villaluenga et al. (2008) reported the formation of products with β-(1→6) bonds with *K. lactis* β-galactosidase, in accordance with our findings, but the formation of other GOS was not considered.

The product specificity of *K. lactis* β-galactosidase contrasts with that of its *B. circulans* counterpart. The former exhibits a tendency to synthesize β-(1→6) bonds (Martinez-Villaluenga et al. 2008; Rodriguez-Colinas et al. 2011), whereas the latter prefers the formation of β-(1→4) bonds. Another difference between both enzymes deals with the formation of disaccharides, because *B. circulans* β-galactosidase yields a moderate amount of allolactose, 4-galactobiose, and 3-galactosyl-glucose, whereas the *K. lactis* enzyme is able to use efficiently free galactose and glucose as acceptors yielding 6-galactobiose and allolactose, respectively, with notable yields (Rodriguez-Colinas et al. 2011).

GOS Production with Permeabilized Cells of *K. lactis*

Biochemical reactions using whole cells have advantages over soluble enzymes in many industrial bioconversion processes, allowing the reuse of the biocatalyst and continuous processing. However, the permeability barrier of the cell envelope for substrates and products often causes very low reaction rates, especially in yeasts. In order to increase the volumetric activity, permeabilization of yeast cells is an economical, simple, and safe process that usually facilitates substrate access to the intracellular enzymes (Fontes et al. 2001; Kondo et al. 2000; Siso et al. 1992). The permeabilizing agent may decrease the phospholipids content in the membrane thus allowing the transit of low molecular weight compounds in and out of the cells (Manera et al. 2010).

In this context, ethanol-permeabilized *K. lactis* cells have been evaluated for lactose hydrolysis (Fontes et al. 2001; Genari et al. 2003; Panesar et al. 2007; Siso et al. 1992) and for the bioconversion of lactose and fructose to the disaccharide lactulose (Lee et al. 2004). We investigated the use of *K. lactis* permeabilized cells for the synthesis of GOS. Although permeabilized cells of *Kluyveromyces marxianus* were also recently employed for GOS production, the structure of the synthesized products was not reported (Manera et al. 2010).

Different methods of permeabilizing cells that do not significantly affect enzymatic activity have been described, including drying, treatment with solvents or surfactants, lyophilization, ultrasonic treatment, and mechanical disruption (Panesar et al. 2007, 2011). In our work, a double treatment (ethanol incubation and lyophilization) was applied to harvested *K. lactis* CECT 1,931 cells. Although cells treated with short-chain alcohols may die in many cases, intracellular enzymes may resist such treatment and, in consequence, are not inactivated. In addition, the permeability barriers of cell walls are lowered, so the cells themselves can be utilized as a biocatalyst repeatedly. A scanning electron micrograph (SEM) of the permeabilized cells obtained in our experiments is shown in Fig. 3.6.

Figure 3.7 shows the reaction progress with permeabilized cells. The maximum GOS concentration (177 g/l, which represents 44 % (w/w) of the total carbohydrates in the reaction mixture) was observed at 6 h. After that, the amount of GOS diminished by the hydrolytic action of β-galactosidase until reaching an equilibrium GOS concentration of 105 g/l. Interestingly, lactose disappeared almost completely at the end of the process. As illustrated in Fig. 3.7, GOS containing β-(1→6) bonds contributed substantially to total GOS throughout the process. At the point of maximal GOS concentration (6 h), the weigh composition of the mixture was: monosaccharides (32 %), lactose (24 %), 6-galactobiose (6 %), allolactose (10 %), 6-galactosyl-lactose (16 %), and other GOS (12 %). Martinez-Villaluenga et al. (2008), using a soluble *K. lactis* β-galactosidase, reported a mixture formed by monosaccharides (49 %), residual lactose (21 %), the disaccharides

Fig. 3.6 SEM micrograph of permeabilized *K. lactis* cells at 1,000x

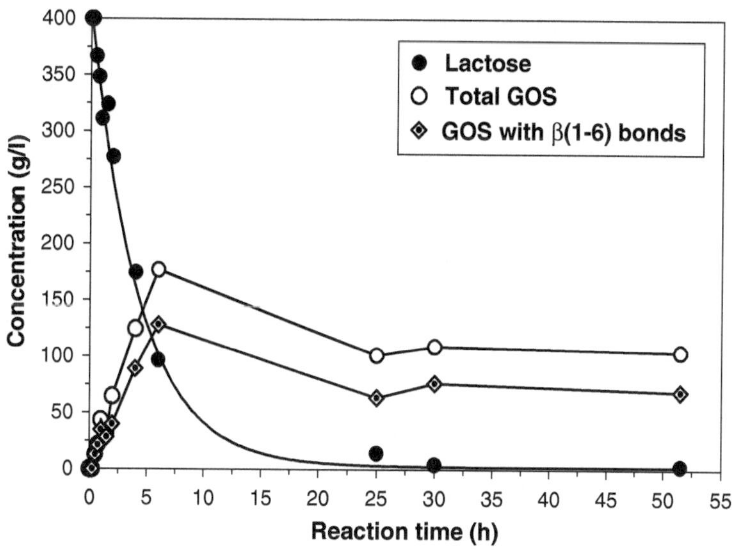

Fig. 3.7 Galactooligosaccharides production from lactose by permeabilized *K. lactis* cells. Experimental conditions: 400 g/l lactose in 0.1 M sodium phosphate buffer (pH 6.8), 1.2–1.5 U/ml, 40 °C

6-galactobiose and allolactose (13 %), and 6-galactosyl-lactose (17 %).

In our analyzes with *K. lactis* β-galactosidase, GOS yield was slightly higher than the reported in previous studies with the same enzyme (Chockchaisawasdee et al. 2005; Martinez-Villaluenga et al. 2008; Maugard et al. 2003), which could be the consequence of: (1) the

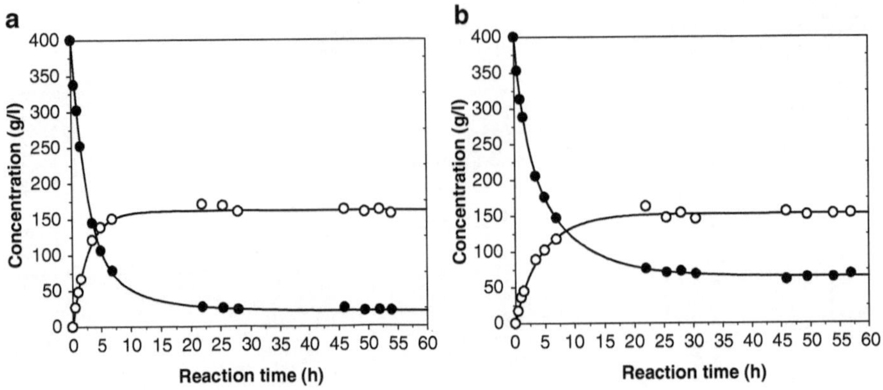

Fig. 3.8 Galactooligosaccharides production from lactose by (**a**) Lactozym 3000L and (**b**) Maxilact LGX 5000. Experimental conditions: 400 g/l lactose in 0.1 M sodium phosphate buffer (pH 6.8), 1.2–1.5 U/ml, 40 °C

overestimation of lactose in previous works due to its tendency to coelute with allolactose, or (2) the contribution of minor GOS to the total yield is not considered in such works.

GOS Production with Soluble *K. lactis* β-Galactosidase

We studied the behavior of two commercially available soluble β-galactosidases (Lactozym 3000L and Maxilact LGX 5000). The experiments were carried out at 1.2–1.5 U/ml, which is a lower enzyme concentration than the typically used (3–12 U/ml) in similar experiments with *K. lactis* β-galactosidase (Chockchaisawasdee et al. 2005; Lee et al. 2004; Martinez-Villaluenga et al. 2008). At higher enzyme concentrations, reactions are so fast (less than 240 min) that is difficult to observe any effect of enzyme stability on reaction progress. The rest of the conditions used in our experiments were similar to those employed for soluble *K. lactis* β-galactosidase (400 g/l lactose, pH 6.8, 40 °C).

The activity of Lactozym and Maxilact towards ONPG was 645 and 2,145 U/ml, respectively. The reaction profile of both enzymes in the GOS synthesis is shown in Fig. 3.8. Several differences were found in the behavior of soluble β-galactosidases compared with permeabilized cells. Firstly, Fig. 3.8 does not show the typical pattern in which a maximum GOS yield is followed by the hydrolysis of the synthesized products. Secondly, lactose is not completely consumed at the equilibrium, with a remaining concentration of 21 and 65 g/l for Lactozym and Maxilact, respectively, whereas only 3 g/l was determined for permeabilized cells. Maximum GOS yield was slightly lower for soluble β-galactosidases (160 g/l for Lactozym and 154 g/l for Maxilact, which represent 42 % and 41 % w/w of the total carbohydrates present in the mixture) compared with permeabilized cells (177 g/l, 44 % w/w). The above results seem to indicate that, due to the low stability of soluble β-galactosidases, the reaction finishes before reaching the final equilibrium. In contrast, the higher stability of permeabilized cells is demonstrated by the characteristic reaction profile – with a maximum – when a competition exists between hydrolysis and transglycosylation (kinetic control) (Rodriguez-Alegria et al. 2010).

Figure 3.9 illustrates the GOS synthesis vs. lactose conversion with the three *K. lactis* biocatalysts. The maximum GOS concentration with permeabilized cells was obtained when lactose conversion was 76 %. Lactozym and Maxilact reached 95 % and 83 % of lactose conversion, without displaying a maximum in GOS concentration. This could be indicating that the microenvironment of the permeabilized cells also exerts an influence on the transglycosylation/hydrolysis ratio.

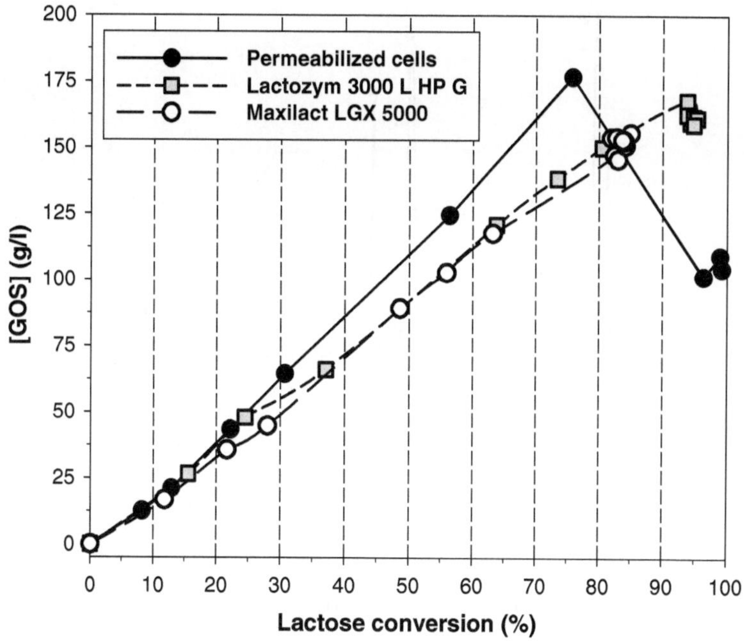

Fig. 3.9 Analysis of GOS formation as a function of lactose conversion catalyzed by β-galactosidase from *K. lactis* in soluble form or in permeabilized cells. Reaction conditions: 400 g/l lactose in 0.1 M sodium phosphate buffer (pH 6.8), 1.2–1.5 U/ml, 40 °C

Aspergillus oryzae β-Galactosidase

The β-galactosidase from *Aspergillus oryzae* is a monomeric enzyme whose molecular mass and isoelectric point are 105 kDa (Tanaka et al. 1975) and 4.6 (Ansari and Husain 2010; Yang et al. 1994), respectively. The biocatalyst optimum temperature is in the range 45–55 °C (Guidini et al. 2010; Guleç et al. 2010) and shows an optimum pH of 4.5 with ONPG as substrate and 4.8 towards lactose (Tanaka et al. 1975).

The β-galactosidase from *A. oryzae* has been applied to the synthesis of different transgalactosylated products such as GOS (Albayrak and Yang 2002b; Iwasaki et al. 1996; Vera et al. 2012), lactulose (Guerrero et al. 2011; Mayer et al. 2004), and galactosyl-polyhydroxyalcohols (Irazoqui et al. 2009; Klewicki 2007), using free and immobilized enzyme. The β-galactosidase from *A. oryzae* has been immobilized by different strategies including entrapment in alginate (Freitas et al. 2011), covalent attachment onto various carriers (Gaur et al. 2006; Huerta et al. 2011; Neri et al. 2011; Sheu et al. 1998) or combined ionic adsorption and cross-linking (Guidini et al. 2011).

Specificity of *A. oryzae* β-Galactosidase

We have studied the synthesis of GOS catalyzed by a preparation of β-galactosidase from *A. oryzae* commercially available (Enzeco® Fungal Lactase) using 400 g/l lactose and 15 U/ml (β-galactosidase activity towards 30 mM ONPG at 25 °C and pH 4.5 in citrate-phosphate buffer 0.1 M). We analyzed by HPAEC-PAD the product specificity of *A. oryzae* β-galactosidase (Urrutia et al. 2013). Figure 3.10 shows the chromatogram obtained with this enzyme after 4 h of reaction. The main galactooligosaccharide synthesized by this enzyme was the trisaccharide 6-galactosyl-lactose (peak *8*). Several disaccharides containing different bonds were also identified: Gal-β(1→6)-Gal (peak *3*), Gal-β(1→6)-Glc (peak *4*), Gal-β(1→3)-Gal (peak *7*), and

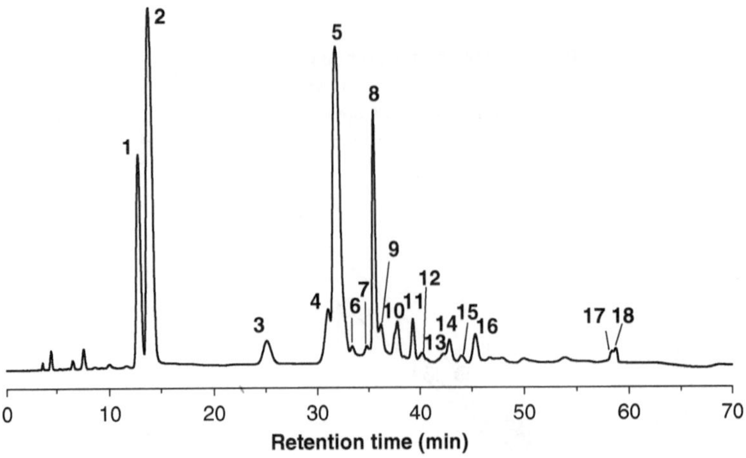

Fig. 3.10 HPAEC-PAD analysis of the reaction of lactose with *A. oryzae* β-galactosidase. The peaks correspond to the following: *1* galactose, *2* glucose, *3* 6-galactobiose, *4* allolactose, *5* lactose, *6* 3-galactobiose, *7* n.d., *8* Gal-β(1→6)-Gal-β(1→4)-Glc, *9* n.d., *10* Gal-β(1→3)-Glc, *11* n.d., *12* n.d., *13* n.d., *14* Gal-β(1→4)-Gal-β(1→4)-Glc, *15* n.d., *16* n.d., *17* n.d., and *18* n.d. The chromatogram corresponds to the reaction mixture after 4 h. Reaction conditions: 400 g/l lactose, 0.1 M citrate-phosphate buffer (pH 4.5), 40 °C. n.d. not detected

Gal-β(1→3)-Glc (peak *10*). The enzyme was also able to synthesize tetrasaccharides (at least peaks *17–18*, considering their retention times).

Toba et al. (1985) were the first to report the characterization of some of the tri-, tetra-, and pentasaccharides synthesized by *A. oryzae* β-galactosidase. In particular, they identified three trisaccharides, 6-galactosyl-lactose (peak *8* in Fig. 3.10), 3-galactosyl-lactose, and Gal-β(1→4)-Gal-β(1→6)-Glc (the two later not identified by us in the chromatograms); two tetrasaccharides, Gal-β(1→6)-Gal-β(1→6)-Gal-β(1→4)-Glc and Gal-β(1→6)-Gal-β(1→3)-Gal-β(1→4)-Glc; and one pentasaccharide, Gal-β(1→6)-Gal-β(1→6)-Gal-β(1→6)-Gal-β(1→4)-Glc. Neri et al. (2011) characterized several of the GOS obtained with a preparation of *A. oryzae* β-galactosidase covalently immobilized onto a hydrazide-Dacron-magnetite composite. In particular, they identified three of the GOS previously characterized by Toba et al. (1985): the trisaccharides 6-galactosyl-lactose and Gal-β(1→4)-Gal-β(1→6)-Glc and the tetrasaccharide Gal-β(1→6)-Gal-β(1→6)-Gal-β(1→4)-Glc. The authors also mentioned the presence of a disaccharide containing a β(1→6) bond, but no information on its composition and structure was given.

One of the main contributions of our work is the identification of at least four disaccharides in the reaction mixture, two of them containing a β(1→6) bond and the other two with a β(1→3) bond. The specific features of this enzyme indicate a tendency to form β(1→6) bonds followed by β(1→3) linkages, with a minor contribution of β(1→4) bonds.

GOS Production with *A. oryzae* β-Galactosidase

Figure 3.11 illustrates the kinetics of GOS production using 400 g/l lactose and an enzyme concentration of 15 U/ml. GOS reached a maximum concentration of 107 g/l at 8 h of reaction, which corresponded to 26.8 % of total carbohydrates in the mixture. From that point of maximum yield, the GOS concentration slowly decreased down to values of approximately 60 g/l.

It has been proved that the increase of lactose concentration [up to 30–40 % (w/v)] has a strong positive effect on the maximum GOS obtained

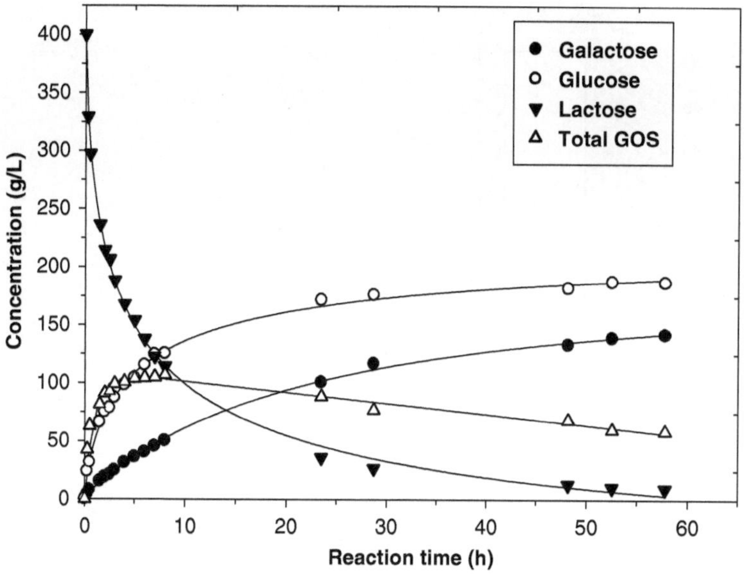

Fig. 3.11 Kinetics of GOS formation at 15 U/ml using 400 g/l lactose catalyzed by β-galactosidase from *A. oryzae*. Reaction conditions: 0.1 M citrate-phosphate buffer (pH 4.5), 40 °C

(Iwasaki et al. 1996; Matella et al. 2006); in consequence, initial lactose concentration should be considered for an adequate comparison with other studies. Albayrak and Yang (2002a) used an initial lactose concentration of 500 g/l obtaining a maximum GOS concentration of 27 % (w/w), of which more than 70 % was formed by trisaccharides. In agreement with these results, Neri et al. (2011) obtained 20.2 % trisaccharides and 5.9 % tetrasaccharides using 500 g/l lactose and 40 °C, which accounted for a total yield of 26.1 % (130 g/l). No significant differences were observed when employing the free or immobilized enzyme. Gaur et al. (2006), starting with 200 g/l lactose, reported that *A. oryzae* β-galactosidase formed only trisaccharides, with a maximum concentration of 22.6 % (w/w) for soluble enzyme and 25.5 % (w/w) for chitosan-immobilized β-galactosidase. Guleç et al. (2010) reported the production of mainly trisaccharides, with a maximum of GOS concentration of 20.8 % (w/w) using 320 g/l lactose and 55 °C.

In our study, we noted that the contribution of disaccharides to total GOS was very significant, a finding that had not been previously described for *A. oryzae* β-galactosidase. In fact, at the point of maximum GOS concentration (7 h, Fig. 3.11), the amounts of di-, tri-, and tetrasaccharides (expressed in weight percentage referred to the total sugars) were 9.9 %, 15.2 % and 1.8 %, respectively.

Regarding the effect of pH and temperature on GOS synthesis, it has been found that even though both parameters affect reaction rate, they do not modify maximum GOS concentration (Albayrak and Yang 2002b; Neri et al. 2011).

Conclusions

Significative differences have been found in the behavior of β-galactosidases from *Bacillus circulans*, *Kluyveromyces lactis*, and *Aspergillus oryzae* for the synthesis of GOS. First of all, the product specificity was very dependent on the enzyme origin. Thus, *K. lactis* β-galactosidase exhibits a tendency to synthesize β-(1→6) bonds, whereas the *B. circulans* counterpart prefers the formation of β-(1→4) bonds. The β-galactosidase from *A. oryzae* displays a marked preference to form β-(1→6) linkages followed by β-(1→3).

Maximum GOS concentration is also dependent on biocatalyst. The GOS production with the three enzymes – at the point of maximum

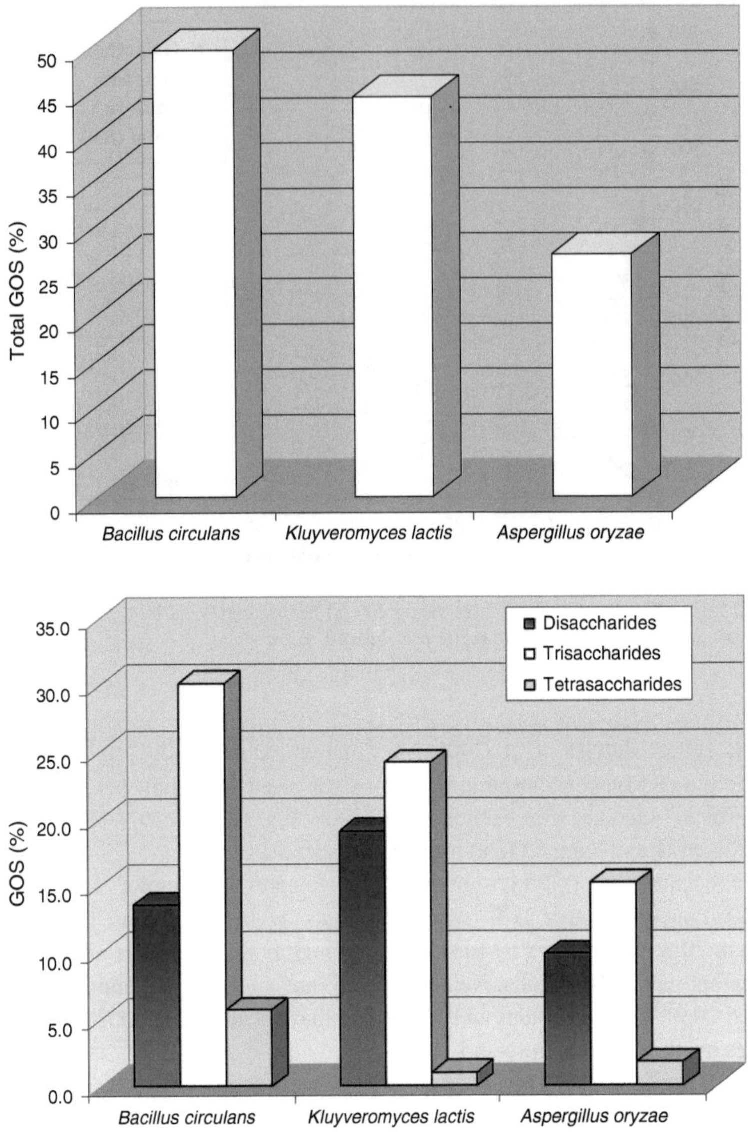

Fig. 3.12 Comparison of the three β-galactosidases studied in this work: (*top*) maximum yield of GOS, (*bottom*) distribution of di-, tri-, and tetrasaccharides at the point of maximum GOS production

concentration – is represented in Fig. 3.12 (top). *B. circulans* and *K. lactis* β-galactosidases produce nearly 45–50 % (w/w) GOS referred to the total amount of carbohydrates in the system. In contrast, *A. oryzae* β-galactosidase produces less than 30 % (w/w) GOS. Another difference between the three enzymes deals with the product distribution, as illustrated in Fig. 3.12 (bottom). For a product enriched in disaccharides, *K. lactis* β-galactosidase would be the best choice, as the amount of disaccharides almost equals that of trisaccharides. However, the *B. circulans* enzyme would be preferable if a GOS with a high tri- and tetrasaccharides content is desirable.

As the properties of GOS may depend significantly on their degree of polymerization and chemical structure, the selection of the appropriate enzyme could have a considerable effect on the bioactivity of the resulting product. However, more structure-function studies (in vitro and in vivo) assaying different GOS are required to determine the target molecules to be synthesized.

Acknowledgements We thank Ramiro Martínez (Novozymes A/S, Madrid, Spain) for supplying Lactozym and for useful suggestions. We thank DSM Food Specialties for supplying Maxilact LGX 5000. Projects BIO2010-20508-C04-01 and BIO2010-20508-C04-04 from Spanish Ministry of Science and Innovation supported this research. B.R.C and M.A. were supported by fellowships from the Spanish Ministries of Science and Innovation (FPI program) and Education and Culture (FPU program), respectively. P. U was supported by a PhD Fellowship and international internship from Conicyt-Chile.

References

Adam AC, Rubio-Texeira M, Polaina J (2004) Lactose: the milk sugar from a biotechnological perspective. Crit Rev Food Sci Nutr 44:553–557

Albayrak N, Yang ST (2002a) Immobilization of *Aspergillus oryzae* beta-galactosidase on tosylated cotton cloth. Enzyme Microb Technol 31:371–383

Albayrak N, Yang ST (2002b) Immobilization of beta-galactosidase on fibrous matrix by polyethyleneimine for production of galacto-oligosaccharides from lactose. Biotechnol Prog 18:240–251

Albayrak N, Yang ST (2002c) Production of galacto-oligosaccharides from lactose by *Aspergillus oryzae* beta-galactosidase immobilized on cotton cloth. Biotechnol Bioeng 77:8–19

Alvaro-Benito M, De Abreu M, Fernandez-Arrojo L, Plou FJ, Jimenez-Barbero J, Ballesteros A, Polaina J, Fernandez-Lobato M (2007) Characterization of a b-fructofuranosidase from *Schwanniomyces occidentalis* with transfructosylating activity yielding the prebiotic 6-kestose. J Biotechnol 132:75–81

Ansari SA, Husain Q (2010) Lactose hydrolysis by β-galactosidase immobilized on concanavalin A-cellulose in batch and continuous mode. J Mol Catal B: Enzym 63:68–74

Bridiau N, Issaoui N, Maugard T (2010) The effects of organic solvents on the efficiency and regioselectivity of N-acetyl-lactosamine synthesis, using the β-galactosidase from *Bacillus circulans* in hydro-organic media. Biotechnol Prog 26:1278–1289

Buchholz K, Kasche V, Bornscheuer UT (2005) Biocatalysts and enzyme technology. Wiley-VCH, Weinheim

Cantarel BL, Coutinho PM, Rancurel C, Bernard T, Lombard V, Henrissat B (2009) The Carbohydrate-Active EnZymes database (CAZy): an expert resource for glycogenomics. Nucleic Acids Res 37:D233–D238

Cardelle-Cobas A, Corzo N, Olano A, Pelaez C, Requena T, Avila M (2011) Galactooligosaccharides derived from lactose and lactulose: influence of structure on *Lactobacillus, Streptococcus* and *Bifidobacterium* growth. Int J Food Microbiol 149:81–87

Cheng CC, Yu MC, Cheng TC, Sheu DC, Duan KJ, Tai WL (2006) Production of high-content galacto-oligosaccharide by enzyme catalysis and fermentation with *Kluyveromyces marxianus*. Biotechnol Lett 28:793–797

Chockchaisawasdee S, Athanasopoulos VI, Niranjan K, Rastall RA (2005) Synthesis of galacto-oligosaccharide from lactose using beta-galactosidase from *Kluyveromyces lactis*: studies on batch and continuous UF membrane-fitted bioreactors. Biotechnol Bioeng 89:434–443

Farkas E, Schmidt U, Thiem J, Kowalczyk J, Kunz M, Vogel M (2003) Regioselective synthesis of galactosylated tri- and tetrasaccharides by use of β-galactosidase from *Bacillus circulans*. Synthesis 5:699–706

Fontes EAF, Passos FML, Passos FJV (2001) A mechanistical mathematical model to predict lactose hydrolysis by β-galactosidase in a permeabilized cell mass of *Kluyveromyces lactis*: validity and sensitivity analysis. Process Biochem 37:267–274

Freitas FF, Marquez LDS, Ribeiro GP, Brandao GC, Cardoso VL, Ribeiro EJ (2011) A comparison of the kinetic properties of free and immobilized *Aspergillus oryzae* β-galactosidase. Biochem Eng J 58–59:33–38

Fujimoto H, Miyasato M, Ito Y, Sasaki T, Ajisaka K (1998) Purification and properties of recombinant β-galactosidase from *Bacillus circulans*. Glycoconj J 15:155–160

Gaur R, Pant H, Jain R, Khare SK (2006) Galacto-oligosaccharide synthesis by immobilized *Aspergillus oryzae* beta-galactosidase. Food Chem 97:426–430

Genari AN, Passos FV, Passos FML (2003) Configuration of a bioreactor for milk lactose hydrolysis. J Dairy Sci 86:2783–2789

Ghazi I, Fernandez-Arrojo L, Gomez de Segura AG, Alcalde M, Plou FJ, Ballesteros A (2006) Beet sugar syrup and molasses as low-cost feedstock for the enzymatic production of fructo-oligosaccharides. J Agric Food Chem 54:2964–2968

Gibson GR, Ottaway RA (2000) Prebiotics: new developments in functional foods. Crit Rev Foof Sci. Chandos, Oxford

Gosling A, Stevens GW, Barber AR, Kentish SE, Gras SL (2010) Recent advances refining galactooligosaccharide production from lactose. Food Chem 121:307–318

Gosling A, Stevens GW, Barber AR, Kentish SE, Gras SL (2011) Effect of the substrate concentration and water activity on the yield and rate of the transfer reaction of β-galactosidase from *Bacillus circulans*. J Agric Food Chem 59:3366–3372

Guerrero C, Vera C, Plou F, Illanes A (2011) Influence of reaction conditions on the selectivity of the synthesis of lactulose with microbial beta-galactosidases. J Mol Catal B: Enzym 72:206–212

Guidini CZ, Fischer J, Resende MMD, Cardoso VL, Ribeiro EJ (2011) β-Galactosidase of Aspergillus oryzae immobilized in an ion exchange resin combining the ionic-binding and crosslinking methods: kinetics and stability during the hydrolysis of lactose. J Mol Catal B: Enzym 71:139–145

Guidini CZ, Fischer J, Santana LNS, Cardoso VL, Ribeiro EJ (2010) Immobilization of *Aspergillus oryzae* β-galactosidase in ion exchange resins by combined ionic-binding method and cross-linking. Biochem Eng J 52:137–143

Guleç HA, Gürdaç S, Albayrak N, Mutlu M (2010) Immobilization of *Aspergillus oryzae* β-galactosidase on low-pressure plasma-modified cellulose acetate membrane using polyethyleneimine for production of galactooligosaccharide. Biotechnol Bioproc Eng 15:1006–1015

Hansson T, Adlercreutz P (2001) Optimization of galactooligosaccharide production from lactose using beta-glycosidases from hyperthermophiles. Food Biotechnol 15:79–97

Hsu CA, Lee SL, Chou CC (2007) Enzymatic production of galactooligosaccharides by beta-galactosidase from *Bifidobacterium longum* BCRC 15708. J Agric Food Chem 55:2225–2230

Huerta LM, Vera C, Guerrero C, Wilson L, Illanes A (2011) Synthesis of galacto-oligosaccharides at very high lactose concentrations with immobilized β-galactosidases from *Aspergillus oryzae*. Process Biochem 46:245–252

Iqbal S, Nguyen TH, Nguyen TT, Maischberger T, Haltrich D (2010) β-Galactosidase from *Lactobacillus plantarum* WCFS1: biochemical characterization and formation of prebiotic galacto-oligosaccharides. Carbohydr Res 345:1408–1416

Irazoqui G, Giacomini C, Batista-Viera F, Brena BM, Cardelle-Cobas A, Corzo N, Jimeno ML (2009) Characterization of galactosyl derivatives obtained by transgalactosylation of lactose and different polyols using immobilized β-galactosidase from *Aspergillus oryzae*. J Agric Food Chem 57:11302–11307

Iwasaki KI, Nakajima M, Nakao SI (1996) Galacto-oligosaccharide production from lactose by an enzymic batch reaction using β-galactosidase. Process Biochem 31:69–76

Klewicki R (2007) The stability of gal-polyols and oligosaccharides during pasteurization at a low pH. LWT- Food Sci Technol 40:1259–1265

Kondo A, Liu Y, Furuta M, Fujita Y, Matsumoto T, Fukuda H (2000) Preparation of high activity whole cell biocatalyst by permeabilization of recombinant flocculent yeast with alcohol. Enzyme Microb Technol 27:806–811

Lee YJ, Kim CS, Oh DK (2004) Lactulose production by beta-galactosidase in permeabilized cells of *Kluyveromyces lactis*. Appl Microbiol Biotechnol 64:787–793

Maischberger T, Leitner E, Nitisinprasert S, Juajun O, Yamabhai M, Nguyen TH, Haltrich D (2010) β-Galactosidase from *Lactobacillus pentosus*: purification, characterization and formation of galacto-oligosaccharides. Biotechnol J 5:838–847

Manera AP, De Almeida Costa FA, Rodrigues MI, Kalil SJ, Maugeri Filho F (2010) Galacto-oligosaccharides production using permeabilized cells of *Kluyveromyces marxianus*. Int J Food Eng 6, article 4

Martinez-Villaluenga C, Cardelle-Cobas A, Corzo N, Lano A, Villamiel M (2008) Optimization of conditions for galactooligosaccharide synthesis during lactose hydrolysis by beta-galactosidase from *Kluyveromyces lactis* (Lactozym 3000 L HP G). Food Chem 107:258–264

Matella NJ, Dolan KD, Lee YS (2006) Comparison of galactooligosaccharide production in free-enzyme ultrafiltration and in immobilized-enzyme systems. J Food Sci 71:C363–C368

Maugard T, Gaunt D, Legoy MD, Besson T (2003) Microwave-assisted synthesis of galacto-oligosaccharides from lactose with immobilized beta-galactosidase from *Kluyveromyces lactis*. Biotechnol Lett 25:623–629

Mayer J, Conrad J, Klaiber I, Lutz-Wahl S, Beifuss U, Fischer L (2004) Enzymatic production and complete nuclear magnetic resonance assignment of the sugar lactulose. J Agric Food Chem 52:6983–6990

Mozaffar Z, Nakanishi K, Matsuno R (1984) Purification and properties of β-galactosidases from *Bacillus circulans*. Agric Biol Chem 48:3053–3061

Mozaffar Z, Nakanishi K, Matsuno R (1986) Continuous production of galacto-oligosaccharides from lactose using immobilized beta-galactosidase from *Bacillus circulans*. Appl Microbiol Biotechnol 25:224–228

Neri DFM, Balcao VM, Cardoso SM, Silva AMS, Domingues MDRM, Torres DPM, Rodrigues LRM, Carvalho LB, Teixeira JAC (2011) Characterization of galactooligosaccharides produced by β-galactosidase immobilized onto magnetized Dacron. Int Dairy J 21:172–178

Pal A, Pal V, Ramana KV, Bawa AS (2009) Biochemical studies of β-galactosidase from *Kluyveromyces lactis*. J Food Sci Technol 46:217–220

Panesar R, Panesar PS, Singh RS, Kennedy JF, Bera MB (2007) Production of lactose-hydrolyzed milk using ethanol permeabilized yeast cells. Food Chem 101:786–790

Panesar R, Panesar PS, Singh RS, Kennedy JF (2011) Hydrolysis of milk lactose in a packed bed reactor system using immobilized yeast cells. J Chem Technol Biotechnol 86:42–46

Park AR, Oh DK (2010) Galacto-oligosaccharide production using microbial β-galactosidase: current state and perspectives. Appl Microbiol Biotechnol 85:1279–1286

Pereira-Rodriguez A, Fernández-Leiro R, Cerdán E, Becerra M, Sanz-Aparicio J (2010) Crystallization and preliminary X-ray crystallographic analysis of β-galactosidase from *Kluyveromyces lactis*. Acta Crystallogr Sect F F66:297–300

Pinho JMR, Passos FML (2011) Solvent extraction of β-galactosidase from *Kluyveromyces lactis* yields a stable and highly active enzyme preparation. J Food Biochem 35:323–336

Playne MJ, Bennett LE, Smithers GW (2003) Functional dairy foods and ingredients. Aust J Dairy Technol 58:242–264

Plou FJ, de Gómez Segura A, Ballesteros A (2007) Application of glycosidases and transglycosidases for the synthesis of oligosaccharides. In: Polaina J, MacCabe AP (eds) Industrial enzymes: structure, function and application. Springer, New York

Rabiu BA, Jay AJ, Gibson GR, Rastall RA (2001) Synthesis and fermentation properties of novel galacto-oligosaccharides by beta-galactosidases from *Bifidobacterium* species. Appl Environ Microbiol 67:2526–2530

Rastall RA, Gibson GR, Gill HS, Guarner F, Klaenhammer TR, Pot B, Reid G, Rowland IR, Sanders ME (2005) Modulation of the microbial ecology of the human colon by probiotics, prebiotics and synbiotics to enhance human health: an overview of enabling science and potential applications. FEMS Microbiol Ecol 52:145–152

Roberfroid M (2007) Prebiotics: the concept revisited. J Nutr 137:830S–837S

Rodriguez-Alegria ME, Enciso-Rodriguez A, Ortiz-Soto ME, Cassani J, Olvera C, Munguia AL (2010) Fructooligosaccharide production by a truncated *Leuconostoc citreum* inulosucrase mutant. Biocatal Biotransform 28:51–59

Rodriguez-Colinas B, De Abreu MA, Fernandez-Arrojo L, De Beer R, Poveda A, Jimenez-Barbero J, Haltrich D, Ballesteros AO, Fernandez-Lobato M, Plou FJ (2011) Production of galacto-oligosaccharides by the β-galactosidase from Kluyveromyces lactis: comparative analysis of permeabilized cells versus soluble enzyme. J Agric Food Chem 59:10477–10484

Shadid R, Haarman M, Knol J, Theis W, Beermann C, Rjosk-Dendorfer D, Schendel DJ, Koletzko BV, Krauss-Etschmann S (2007) Effects of galactooligosaccharide and long-chain fructooligosaccharide supplementation during pregnancy on maternal and neonatal microbiota and immunity – a randomized, double-blind, placebo-controlled study. Am J Clin Nutr 86:1426–1437

Sheu DC, Li SY, Duan KJ, Chen CW (1998) Production of galactooligosaccharides by beta-galactosidase immobilized on glutaraldehyde-treated chitosan beads. Biotechnol Tech 12:273–276

Siso MIG, Cerdán E, Picos MAF, Ramil E, Belmonte ER, Torres AR (1992) Permeabilization of *Kluyveromyces lactis* cells for milk whey saccharification: a comparison of different treatments. Biotechnol Tech 6:289–292

Song J, Abe K, Imanaka H, Imamura K, Minoda M, Yamaguchi S, Nakanishi K (2011a) Causes of the production of multiple forms of β-galactosidase by *Bacillus circulans*. Biosci Biotechnol Biochem 75:268–278

Song J, Imanaka H, Imamura K, Minoda M, Katase T, Hoshi Y, Yamaguchi S, Nakanishi K (2011b) Cloning and expression of a β-galactosidase gene of *Bacillus circulans*. Biosci Biotechnol Biochem 75:1194–1197

Splechtna B, Nguyen TH, Steinbock M, Kulbe KD, Lorenz W, Haltrich D (2006) Production of prebiotic galacto-oligosaccharides from lactose using beta-galactosidases from *Lactobacillus reuteri*. J Agric Food Chem 54:4999–5006

Tanaka Y, Kagamiishi A, Kiuchi A, Horiuchi T (1975) Purification and properties of β-galactosidase from *Aspergillus oryzae*. J Biochem 77:241–247

Toba T, Yokota A, Adachi S (1985) Oligosaccharide structures formed during the hydrolysis of lactose by *Aspergillus oryzae* β-galactosidase. Food Chem 16:147–162

Torres DP, Goncalves M, Teixeira JA, Rodrigues LR (2010) Galacto-oligosaccharides: production, properties, applications, and significance as prebiotics. Compr Rev Food Sci Food Saf 9:438–454

Torres P, Batista-Viera F (2012) Immobilization of β-galactosidase from *Bacillus circulans* onto epoxy-activated acrylic supports. J Mol Catal B: Enzym 74:230–235

Tuohy KM, Rouzaud GCM, Bruck WM, Gibson GR (2005) Modulation of the human gut microflora towards improved health using prebiotics – assessment of efficacy. Curr Pharm Des 11:75–90

Urrutia P, Rodriguez-Colinas B, Fernandez-Arrojo L, Ballesteros AO, Wilson L, Illanes A, Plou FJ (2013) Detailed analysis of galactooligosaccharides synthesis with β-galactosidase from *Aspergillus oryzae*. J Agric Food Chem 61:1081–1087

Usui T, Kubota S, Ohi H (1993) A convenient synthesis of β-D-galactosyl disaccharide derivatives using the β-d-galactosidase from *Bacillus circulans*. Carbohydr Res 244:315–323

Vera C, Guerrero C, Conejeros R, Illanes A (2012) Synthesis of galacto-oligosaccharides by β-galactosidase from *Aspergillus oryzae* using partially dissolved and supersaturated solution of lactose. Enzyme Microb Technol 50:188–194

Wei L, Xiaoli X, Shufen T, Bing H, Lin T, Yi S, Hong Y, Xiaoxiong Z (2009) Effective enzymatic synthesis of lactosucrose and its analogues by β-galactosidase from *Bacillus circulans*. J Agric Food Chem 57:3927–3933

Yanahira S, Kobayashi T, Suguri T, Nakakoshi M, Miura S, Ishikawa H, Nakajima I (1995) Formation of oligosaccharides from lactose by *Bacillus circulans* β-galactosidase. Biosci Biotechnol Biochem 59:1021–1026

Yang ST, Marchio JL, Yen JW (1994) A dynamic light scattering study of β-galactosidase: environmental effects on protein conformation and enzyme activity. Biotechnol Prog 10:525–531

Zhou QZ, Chen XD, Li X (2003) Kinetics of lactose hydrolysis by β-galactosidase of *Kluyveromyces lactis* immobilized on cotton fabric. Biotechnol Bioeng 81:127–133

Zhou QZK, Chen XD (2001) Effects of temperature and pH on the catalytic activity of the immobilized β-galactosidase from *Kluyveromyces lactis*. Biochem Eng J 9:33–40

Microbial Mannanases: Properties and Applications

4

Hemant Soni and Naveen Kango

Abstract

Mannans are a major constituent of the hemicellulose fraction of lignocelluloses. Mannans perform distinct functions as structural components in cell walls of softwoods and storage functions in seeds. Enzymatic hydrolysis of mannan involves the backbone hydrolyzing endo-β-mannanases and β-mannosidases. Mannans are heteropolymeric and their hydrolysis also requires the action of β-glucosidases and side-chain cleaving α-galactosidases and acetyl mannan esterases. Microorganisms are therefore explored for the production of such repertoire of enzymes so that effective mannan hydrolysis can be achieved. The present chapter discusses the occurrence and structural properties of mannans in plant materials and its hydrolysis using enzymes sourced from various fungi and other microorganisms. The production and properties of mannanolytic enzymes, their cloning and expression in heterologous hosts, and their application have also been discussed.

Keywords
Hemicellulose • Mannans • β-Mannanase • β-Mannosidase • Locust bean gum

Introduction

Hemicelluloses are structural polysaccharides of the plant cell wall. Hemicellulose is associated with cellulose and lignin and forms about 30 % of the lignocellulosic biomass. Lignocellulose is abundant and represents one of the major natural renewable resources and a dominating waste material from agriculture. This renewable resource can be used in several industries, including the pharmaceutical, biofuel, and pulp and paper industries, and many more (Kango et al. 2003; Kango 2007). The generation of feedstock is possible by hydrolysis of lignocellulosic biomass using various microorganisms and their enzymes. The hydrolysis of lignocellulose has become the

H. Soni • N. Kango (✉)
Department of Applied Microbiology and Biotechnology, Dr. Hari Singh Gour Vishwavidyalaya, Sagar 470003, MP, India
e-mail: nkango@gmail.com

P. Shukla and B.I. Pletschke (eds.), *Advances in Enzyme Biotechnology*,
DOI 10.1007/978-81-322-1094-8_4,

"hot spot" and a crucial part of lignocellulose biotechnology. According to Chaikumpollert et al. (2004), hemicelluloses form about one third of all the components available in plants and are the second major heteropolymer present in nature. Distribution of hemicellulose in gymnosperms and angiosperms varies. Hemicelluloses consist of different heterogeneous polymers of sugars such as xylose, arabinose, mannose, glucose, galactose, and sugar acids. These hemicelluloses are named according to their main sugar component (80–90 %) which is present in its backbone, e.g., mannose is present in mannan hemicelluloses. Enzyme-based hydrolysis of hemicelluloses, especially mannan and xylan, significantly affects the prospects of biobleaching and saccharification of lignocellulosic biomass (Viikari et al. 1993).

Mannan: Occurrence and Structure

Most of the main-chain sugars in hemicellulose are linked together by β-1,4-glycosidic bonds. Mannans are one of the most important constituents of hemicelluloses in the wall of higher plants. Mannan is composed of repeating units of mannose (a second carbon epimer of glucose) linked by β-1,4-glycosidic linkages. Besides D-mannose, other sugars like glucose, galactose, and acetyl groups can be present in various mannans. Mannans are further classified on the basis of other sugars present in the structure, e.g., glucose-containing mannan is called glucomannan. Similarly, when galactose is present as a side chain linked to the main chain, the polymer is called galactomannan, and when both glucose and galactose are linked to the mannose sugars of the main-chain backbone, it is called galactoglucomannan. Mannan exists in both linear and branched forms with a β-1,4-linked backbone. Mannans are an important part of the hemicellulose family, which are further classified as linear mannan, glucomannan, galactomannan, and galactoglucomannan. Linear mannans are homopolysaccharides which have a main chain composed of 1,4-linked β-D-mannopyranosyl (mannose-mannose) residues. The percentage of galactose in linear mannans is 5 % or less. In

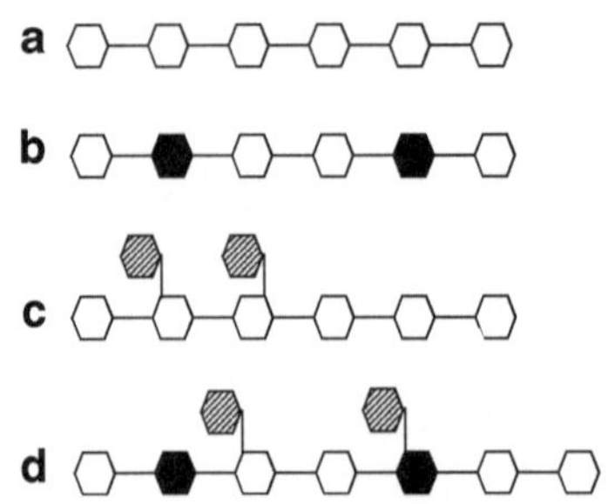

Fig. 4.1 Structure of different types of mannans found in nature: (**a**) Linear mannan (**b**) Glucomannan (**c**) Galactomannan (**d**) Galactoglucomannan. ○ Mannose ● Glucose ◍ Galactose

glucomannan, the main chain consists of randomly β-1,4-linked D-mannose and D-glucose residues, while in galactomannan, the galactose sugars are present as single side chains substituted on the main-chain sugar, mannose (Fig. 4.1). In galactoglucomannan, galactose sugars are present as single side chains in α-1,6-linkage with the main chain, which consists of both mannose and glucose.

The ratio of sugars present in mannan varies with respect to the different sources from which it is obtained, which indicates the polydiversity of the polymer. In linear mannans, mannose is predominantly present, while in galactomannan the ratio of galactose to mannose is 1:3. True galactomannan is considered to contain more than 5 % of galactose residues in side chains. In galactoglucomannan, the ratio of galactose/glucose/mannose is observed to be 1:1:3. Mannans are found in nature as part of the hemicellulose fraction of hardwood and softwood. It is predominantly found in the endosperm of copra, locust bean, guar beans, seeds of other leguminous plants, coffee beans, roots of the konjac tree, and ivory nuts. Locust bean gum (LBG) is a galactomannan, while mannan from konjac trees is a glucomannan. Linear mannans are the major structural units in woods, in seeds of ivory nut, and in green coffee beans. Petkowicz et al. (2007) separated mannans from ivory nuts into two components, mannan I and mannan II. Mannan I, extracted with alkali, displayed a crystalline structure, while mannan II was not amenable to direct extraction and displayed a less crystalline

Table 4.1 Mannan content of some plants

Source of mannan	Plant part	Type of mannan	Ratio of sugars
Ceratonia siliqua (carob or locust bean)	Endosperm of seed	Galactomannan	~1:4 (Gal:Man)
Phytelephas macrocarpa (ivory nut)	Endosperm of seed	Linear mannan	Homopolymer (mannose)
Schizolobium amazonicum	(a) Seed coat side or exterior section	Linear mannan	Homopolymer (mannose)
	(b) An intermediate section of seed endosperm	Rich in galactomannan	~1:3 (Gal:Man)
Schizolobium parahybum	Endosperm of seed	Galactomannan	~1:3 (Gal:Man)
Carum carvi	Endosperm of seed	Linear mannan	Homopolymer (mannose)
Cyamopsis tetragonolobus (guar seed)	Endosperm of seed	Galactomannan	~1:2 (Gal:Man)
Amorphophallus konjac	Roots	Glucomannan	~3:4 (Glu:Man)
Coffea arabica (coffee bean)	Endosperm	Galactomannan	~1:2 and ~1:7 (Gal:Man)[a]
Aloe barbadensis (acemannan)	Leaves	Linear mannan[b]	Homopolymer (mannose)
Cesalpinia spinosa (tara tree)	Endosperm of seed	Galactomannan	~1:3 (Gal:Man)

[a]Developmentally regulated
[b]Associated with acetyl group

structure. Molecular size also varied as mannan I was smaller compared to mannan II. Linear mannans are present in the seed coat or exterior section of leguminous plants, while galactomannan occurs in the intermediate section of the seed endosperm. Various gum extracts from plants are conspicuous sources for galactomannan, for example, locust bean gum, tara gum, fenugreek gum, and guar gum. In these sources, the main chain of galactomannan contains 1,4-linked β-d-mannopyranosyl residues with side chains of single 1,6-linked α-D-galactopyranosyl groups attached along the main chain (Fig. 4.1). The distribution of galactose in galactomannan varies between mannans obtained from different sources. It is observed that all types of galactomannans have more than 5 % of galactose residues as side chains. Galactomannan obtained from the endosperm of locust bean or *Ceratonia siliqua* (carob) has a ratio of 1:4 (galactose/mannose). The galactomannan of the intermediate section of the seed endosperm of *Schizolobium amazonicum* and *Cesalpinia spinosa* has a sugar ratio of 1:3 (galactose/mannose) (Table 4.1). The function of galactomannan in seeds, in addition to the retention of water by solvation, is to prevent complete drying of seeds in high atmospheric temperatures so that the enzymes, which are crucial for seed germination, remain active. Liepman et al. (2007) have showed some evidence that mannan also functions as a signaling molecule in plant growth and development. Three-dimensional structure studies of guaran or guar fibers (*Cyamopsis tetragonolobus*) were done by Chandrasekaran et al. (1998) using x-ray diffraction, and they revealed that the hydrogen of the galactosyl side chain interacted with the mannan backbone and provided structural stability. The structure showed a flat twofold helix with a pitch of 10.38 Å. Glucomannans are the principal components of softwood hemicelluloses and consist of β-1,4-linked D-mannose and D-glucose residues with a 3:1 ratio. Hongshu et al. (2002) obtained glucomannan from ramie (*Boehmeria nivea*) which contained 95–99 % of D-glucose and D-mannose residues with a ratio of 1.3–1.7:1. The presence of D-galactose residues in glucomannan is very rare, but Puls and Schuseil (1993), working with softwood, observed D-galactose residues attached to the main-chain mannose residues with α-1,6-linked terminal units and observed a ratio of mannose/glucose/galactose as 3:1:0.1. The glucomannan from *Amorphophallus* (konjac) showed an association with starch-like α-glucan, comprised of 1,4-linked β-d-mannopyranose and D-glucopyranose in 70 %

and 30 %, respectively (Aspinall 1959). Kenne et al. (1975) studied distribution of the *O*-acetyl groups in glucomannan from pine and observed that acetyl groups are irregularly distributed in pine glucomannan. In galactoglucomannans, galactose residues are attached to both D-glucosyl and D-mannosyl units with a α-1,6-linkage, and mannosyl units also have partial substitution by *O*-acetyl groups. Several reports showed that about 60–65 % of mannose residues in galactoglucomannan from native Norway spruce wood and pulp were acetylated at either the C-2 or C-3 position (Aspinall et al. 1962; Popa and Spiridon 1998; Timell 1967; Willfor et al. 2003). Lundqvist et al. (2002, 2003) extracted galactoglucomannan from spruce (*Picea abies*) by heat fractionation at different temperatures and characterized it. Galactoglucomannan from spruce contained about one third of D-mannosyl units substituted by *O*-acetyl groups with an equal distribution between C-2 and C-3 and a molar ratio of 0.1:1:4 (galactose/glucose/mannose). Various types of mannans with their sources and sugar ratios are listed in Table 4.1.

Solubility among mannans towards water varies due to the presence of D-galactose side chains. The solubility of galactomannan and galactoglucomannan is higher in comparison to linear and glucomannan homopolymers. The D-galactose side chains prevent alignment of macromolecules and lead to formation of strong hydrogen bonds (Timell 1965). In addition to aforesaid structures, mannans also display a range of curious structures and configurations. For instance, Ishurd et al. (2004) observed and isolated galactomannan from *Retama raetam* (Fabaceae). Its backbone consisted mostly of 1-3-linked β-D-mannopyranosyl residues with attachment of galactopyranosyl residues observed at C-6. Nunes et al. (2005) observed arabinosyl and glucosyl residues in galactomannan from green and roasted coffee infusions. The acetyl groups were present in the main chain of mannan at the O-2 position of mannose residues, while arabinose residues were at O-6 of mannose residues as side chains. Omarsdottir et al. (2006) isolated galactomannan from a number of lichen species like foliose lichen (*Peltigera canina*). The backbone of these mannans displayed odd structures, being composed of α-1,6-linked mannopyranosyl residues with a difference in the side-chain pattern at O-2 and O-4 instead of O-6, which is observed in various galactomannan structures. Singh and Malviya (2006) observed D-glucopyranosyl units in glucomannan from seeds of a medicinal plant, *Bryonia laciniosa*, which displayed α-1-6-linkages in the main chain with a 1:1.01 ratio of glucose and mannose.

The degree of polymerization (DP) of any macromolecule is a manifestation of the approximate number of monomer units present in polymer. The DP of any polymer influences its various properties such as colligative properties, boiling point, freezing point, solubility, viscosity, toughness, and somatic pressure. The DP also helps in calculating the average molecular weight of the polymer. Petkowicz et al. (2007) isolated mannan from ivory nuts and observed two types, viz., mannan I and II in which mannan I has a lower molecular weight and a DP of ~15, while mannan II had a DP of about ~80 with higher molecular weight. Softwood glucomannans with a 3:1 ratio of mannose:glucose exhibit higher DPs of more than 200. Softwood galactoglucomannans with a 1:1:3 ratio for galactose/glucose/mannose exhibit a DP between 100 and 150. Enzymes play a crucial role in the modification of polymers and its structural analysis. Analysis or sequencing of mannan requires several mannanases from legume seeds and microorganisms to act on the various mannans. Selection of enzymes is important because their differential activity towards substrates reveals the structural difference. A mannanase from *Trichoderma reesei* was able to hydrolyze fiber-bound galactoglucomannan from pine kraft pulp, while an enzyme from *Bacillus subtilis* was not effective for its hydrolysis (Ratto et al. 1993). Tenkanen et al. (1997) studied the action of a mannanase from *T. reesei* on galactoglucomannan in pine kraft pulp and analyzed the hydrolysate by ^{1}H NMR spectroscopy and high-performance anion-exchange chromatography (HPAEC-PAD). The relative amount of sugar residues in the hydrolysate of pine kraft pulp, after extensive hydrolysis by a mannanase from *T. reesei*, was analyzed. The molar percentage of mannose, glucose, and galactose was 73.4, 20.4, and 5.8,

Table 4.2 Some microbial sources of mannanases

	Microorganisms		
Major groups	Genus	Species	References
Fungi	*Malbranchea*	*M. cinnamomea*	Maijala et al. (2012)
	Myceliophthora	*M. fergusii*	Maijala et al. (2012)
	Aspergillus	*A. niger*	Benech et al. (2007)
		A. fumigatus	Puchart et al. (2004)
		A. aculeatus	Setati et al. (2001)
		A. niger	Ademark et al. (1998)
	Trichoderma	*T. harzianum*	Ferreira and Filho (2004)
		T. reesei	Stalbrand et al. (1993)
	Penicillium	*P. oxalicum*	Kurakake et al. (2006)
		P. citrinum	Yoshida et al. (1993)
	Sclerotium	*S. coffeicola*	Groβwidnhager et al. (1999)
		S. rolfsii	Gubitz et al. (1996)
Yeast	*Saccharomyces*	*S. cerevisiae*	Setati et al. (2001)
	Candida	*C. albicans*	Reyna et al. (1999)
Bacteria	*Bacillus*	*B. subtilis* WY34	Jiang et al. (2006)
	Enterococcus	*E. casseliflavus*	Oda et al. (1993)
Actinomycetes	*Streptomyces*	*Streptomyces* sp. S27	Shi et al. (2011)
		S. galbus NR	Kansoh and Nagieb (2004)
		S. lividans	Arcand et al. (1993)
	Thermomonospora	*T. fusca*	Hilge et al. (1998)
	Cellulomonas	*Cellulomonas* sp.	Takegawa et al. (1989)

respectively, as determined by ^{1}H NMR spectroscopy, and 71.8, 20.3, and 6.9 by the HPAEC-PAD method. LBG hydrolysis products of the recombinant man*5S27* enzyme were analyzed using HPAEC. The approximate percentages of mannose, mannobiose, mannotriose, mannotetraose, mannopentaose, and other sugar oligosaccharides were 3.23, 0.74, 22.14, 2.21, 6.89, and 64.79 (Shi et al. 2011). Analysis of the sequence and percentage of sugars in mannan requires a particular enzyme or enzymes with their homo- and hetero-synergistic actions leading to the hydrolysis of hemicelluloses. The analysis of hydrolysate needs a suitable method or a combination of methods like TLC, HPLC, and NMR spectroscopy.

Mannan-Degrading Enzymes and Their Sources

Mannans can be present as linear and branched, homo- as well as heteropolymers. In general, enzymes involved in the hydrolysis of mannan are called as mannanases. Complete biodegradation of mannans necessitates the use of various enzymes. Enzymes that actively participate in mannan hydrolysis include β-mannanase (1,4-β-D-mannan mannohydrolase, EC 3.2.1.78), β-mannosidase (1,4-β-D-mannopyranoside hydrolase, EC 3.2.1.25), β-glucosidase (1,4-β-D-glucoside glucohydrolase, EC 3.2.1.21), α-galactosidase (1,4-α-D-galactoside galactohydrolase, EC 3.2.1.22), and acetyl esterase (EC 3.1.1.6). A number of microorganisms, including fungi, yeast, bacteria, and actinomycetes, produce β-mannanases and other accessory enzymes. Among these, fungi have been investigated by various workers for mannanase production (Dhawan and Kaur 2007; Moreira and Filho 2008; Van Zyl et al. 2010). Some of the prominent mannanase producers are listed in Table 4.2.

β-Mannanase distribution is also observed in plants. Seeds are the most preferred sources for isolation of mannanases. However, other plant organs like fruits also displayed the presence of mannanases (Bourgault et al. 2001). Schroder et al. (2006) obtained endo-β-mannanase from ripe tomato fruit. For mannanase production, microorganisms which are selected from various sources (soil, compost, water, agriculture waste) are grown on basal media containing mannan (LBG) as sole carbon source (Ratto and Poutanen 1988; Puchart et al. 2004; Maijala et al. 2012).

Mode of Action of Mannanases

Polysaccharides like mannans can exist in linear, homo, hetero, or branched form. β-Mannanases find application in the extraction of vegetable oil, in the manufacture of instant coffee as a viscosity reducer agent for coffee extract, nutraceuticals such as the production of MOS (mannose oligosaccharides), pharmaceuticals, food and feed, production of second-generation biofuels, paper and pulp, and various other industries (Sachslehner et al. 2000; Van Zyl et al. 2010). At least one main-chain hydrolyzing enzyme, like β-mannanase, and one side-chain hydrolyzing enzyme, like α-galactosidase, are required for the breakdown of branched mannan (LBG). β-Mannanase cleaves internal β-1,4-linked residues of mannose/glucose in the mannan backbone. This enzyme mainly produces oligomannan/oligoglucomannan and is very effective on linear mannan and glucomannan (homopolymer), although the hydrolysis action of this enzyme is affected in galactomannan due to the presence of side chains. β-Mannosidase helps to remove mannose from the nonreducing end of mannan and cleaves β-1,4-linked mannose residues. Similarly, β-glucosidase removes glucose residue from the nonreducing end of the oligoglucomannan and cleaves 1,4-β-D-glucopyranose. Besides these main-chain enzymes, two side-chain cleaving enzymes are very important for complete biodegradation of mannan, viz., α-galactosidase cleaves the α-1,6 glycosidic bonds between galactose and the main-chain sugars (mannose/glucose) and leads to hydrolysis of D-galactopyranosyl side chains of galactomannan and galactoglucomannan. Acetyl mannan esterase removes acetyl groups from galactoglucomannan. The delineation of various mannanases action is shown in Fig. 4.2.

Mannan composition also affects the action of enzymes, and to achieve complete degradation of heteromannan like locust bean gum, fungi and bacteria have to produce three enzymes, namely, β-mannanase, β-mannosidase, and α-galactosidase (Hilge et al. 1998). These hemicellulases also show synergistic action. When β-mannanase and β-mannosidase (main-chain cleaving enzymes) or α-galactosidase and acetyl mannan esterase (side-chain cleaving enzymes) cooperate, it is called homosynergistic action. Heterosynergy refers to the interaction of main- and side-chain cleaving enzymes working together (Fig. 4.3). Homosynergy between β-mannosidase and two β-mannanases obtained from the enzyme extract of *Sclerotium rolfsii* (Gubitz et al. 1996; Moreira and Filho 2008) and heterosynergy between β-mannanase, β-mannosidase, and α-galactosidase have been observed in enzyme extractions of *Thermotoga neapolitana* 68 on galactomannan (Duffaud et al. 1997).

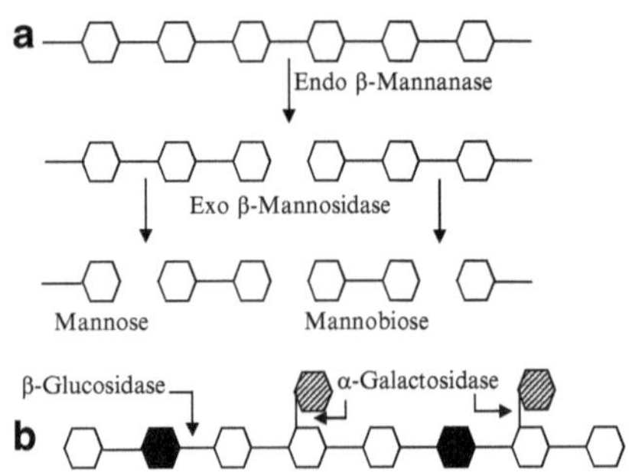

Fig. 4.2 (**a**) Action of β-mannanase and β-mannosidase on linear mannan. (**b**) Action of debranching enzyme α-galactosidase and β-glucosidase on galactoglucomannan. ○ Mannose ● Glucose ◍ Galactose

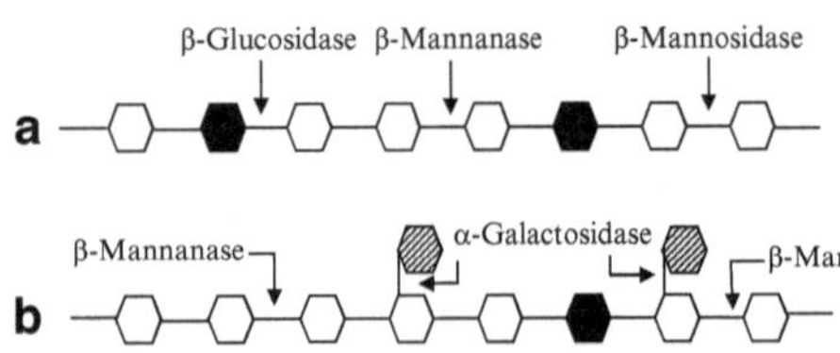

Fig. 4.3 (**a**) Homosynergetic actions of β-mannanase, β-mannosidase, and β-glucosidase on glucomannan. (**b**) Heterosynergetic action of β-mannanase and β-galactosidase on galactoglucomannan. ○ Mannose ● Glucose ◍ Galactose

Microbial Production of Mannanases

The best and richest sources of enzymes are microorganisms (Kirk et al. 2002). For mannanase production, mainly fungi and some bacteria are used at a commercial level and their enzyme systems are reported to be inducible. Hemicelluloses like xylan are not able to cross cell walls; therefore, small oligosaccharides formed as a result of xylan degradation act as an inducer and also play an important role as a regulation factor for xylanase biosynthesis (Singh et al. 2003). Both submerged and solid-state fermentation (SSF) have been examined for mannanase production. The cost of enzymes remains a bottleneck in realizing their application on a large scale. Use of inexpensive substrates, such as by-products of agro-industries and forestry waste, can effectively subsidize the recurring cost of enzyme production. Ratto and Poutanen (1988) have used wheat bran with locust bean galactomannan for mannanase production by bacteria and fungi. Mannanase activities were found to be 256, 34, and 24 nkat ml^{-1} with *Bacillus subtilis*, *Aspergillus awamori*, and *T. reesei*, respectively. Abdeshahian et al. (2010) used palm kernel cake (PKC) as a substrate for β-mannanase production by *Aspergillus niger* FTCC5003 through solid-state fermentation. Production was evaluated by response surface methodology on the basis of a central composite face-centered (CCF) design with three independent variables, namely, incubation, temperature, initial moisture content of substrate, and airflow rate. The highest level of β-mannanase (2,117.89 U/g) was obtained when the incubation temperature, initial moisture level, and aeration rate were 32.5 °C, 60 %, and 0.5 l/min, respectively, during SSF. There are many species of fungi reported to produce significantly high mannanase activity. For instance, Lin and Chen (2004) observed 27.4 U/ml mannanase activity in a submerged culture of *Aspergillus niger* NCH 189. Similarly, Hossain et al. (1996) obtained about 90 U/ml mannanase activity in submerged conditions using *Bacillus* sp. KK01. Production of enzymes is affected by temperature, pH, agitation, and aeration. The overall production process gives a good outcome in terms of enzyme activity at optimum temperature, pH, and other factors. The effect of the agitation speed, aeration rate, and temperature on the production of β-mannanase by *Bacillus licheniformis* NK 27 in a batch fermenter was studied by Feng et al. (2003). They concluded that temperature was the most significant factor in β-mannanase production. Feng et al. (2003) obtained a maximum activity of 212 U/ml in 36 h at an aeration rate of 0.75 vvm, agitation of 600 rpm, and a constant temperature of 30 °C. Mannanase production by microorganisms is influenced by the media composition, mostly carbon and nitrogen (Kataoka and Tokiwa 1998; Dhawan and Kaur 2007). Groβwindhager et al. (1999) used glucose and cellulose for *S. rolfsii*, while Ademark et al. (1998) and Gomes et al. (2007) used locust bean gum (LBG) for *A. niger* and a thermophilic fungus, *Thermoascus aurantiacus*. Ferreira and Filho (2004) have used wheat bran as the carbon source for the production of β-mannanase from mesophilic fungus *Trichoderma harzianum* strain T4. Besides the carbon source, various organic or inorganic nitrogen sources play an important role in mannanase production. Organic nitrogen sources like peptone, yeast autolysate, corn steep liquor (CSL), and beef extract are preferred (Puchart et al. 2004; Zhang et al. 2006; Cui et al. 1999; Kataoka and Tokiwa 1998), while inorganic nitrogen sources like ammonium sulfate, diammonium hydrogen phosphate, ammonium dihydrogen phosphate, and sodium nitrate have been found to play an effective role (Zakaria et al. 1998; Perret et al. 2004). Gomes et al. (2007) achieved the highest β-mannanase and β-mannosidase activity by the thermophilic fungus (*Thermoascus aurantiacus*) with soya meal as nitrogen source, supplemented with LBG as carbon source. Recently, Mohamad et al. (2011) performed a comparison study of different carbon and nitrogen sources for their effect on mannan-degrading enzyme production by *Aspergillus niger*. They revealed in their result that guar gum (GG) and bacteriological peptone supported the highest β-mannanase activity. They achieved β-mannanase activities equivalent to 1,495, 1,148, 10.7, 8.8, and 4.6 nkat ml^{-1}

with guar gum (GG), LBG, α-cellulose, glucose, and carboxymethyl cellulose as carbon sources, respectively. Activity levels equivalent to 1,744, 1,168, 817, 241, 113, and 99 nkat ml^{-1} were achieved with bacteriological peptone, yeast extract, ammonium sulfate, ammonium nitrate, and ammonium chloride as nitrogen sources, respectively. The above results showed that mannanase production by *A. niger* can be enhanced with GG and LBG. Inorganic nitrogen sources reduced β-mannanase production greatly, while organic nitrogen sources enhanced β-mannanase production. In contrast, Kalogeris et al. (2003) have obtained better production of cellulases by *Thermoascus aurantiacus* using inorganic nitrogen sources. Various industries like paper and pulp and detergent industries need enzymes that function well at a high pH. Alkaline β-mannanase was obtained for the first time from alkaliphilic *Bacillus* sp. AM001 by Akino et al. (1987). This mannanase showed a pH optimum between 7.0 and 9.0. Mudau and Setati (2006) have studied endo-mannanase-producing molds from hypersaline environments and observed the effect of salt (NaCl) on growth and enzyme production. All four isolates, *Scopulariopsis brevicaulis* LMK002, *S. candida* LMK004, *S. candida* LMK008, and *Verticillium dahliae* LMK006, showed growth on NaCl concentrations of up to 10 %. Endo-mannanase production by *Scopulariopsis* isolates was found to increase with NaCl concentration. Groβwindhager et al. (1999) have shown efficient β-mannanase production by *Sclerotium rolfsii* and *S. coffeicola* under derepressed condition by using cellulose- and glucose-based media. They have concluded that cellulose is the best inducer for both *S. rolfsii* and *S. coffeicola* strains for mannanase production with maximum activities of 677 and 461 Uml^{-1}, respectively. In a glucose-based medium, activities were 96.6 and 67.7 Uml^{-1}. Glucose is an easily metabolizable substrate, and in the presence of this substrate, glycosyl hydrolase systems get repressed (Ronne 1995; Ruijter and Visser 1997). However, both the strains *S. rolfsii and S. coffeicola* were observed to produce mannanase activity when a typical repressing substrate, glucose, was used as sole carbon source in batch cultivation. Mannan-degrading enzyme production started only when the glucose concentration in the medium dropped low. High mannanase activity (240 Uml^{-1}) by *S. rolfsii* CB5191.62 was achieved in a glucose fed-batch system in which glucose concentration in the media was maintained low (Groβwindhager et al. 1999). Table 4.3 displays an overview of production and properties of mannanases from various microorganisms.

Heterologous Production

Higher yield, ease of operational conditions, simple recovery, and downstream processing have prompted several workers towards cloning and heterologous production of mannanases. The recombinant DNA technique provides enormous opportunity to make genetically modified microbial strains. More than 50 % of mannanase-producing microorganisms, which are being used at industrial level, are genetically engineered (Dhawan and Kaur 2007).

S. cerevisiae is not known for production of mannanase by itself, but the heterologous production of endo-β-1,4-mannanase has been done using *S. cerevisiae* as a genetically modified host by Setati et al. (2001). Similarly, Qiao et al. (2008) have used *Pichia pastoris* as a host for expression of MAN gene of *Bacillus subtilis*. It has been observed that, if the same gene encoding mannanase is expressed in different hosts, the resultant recombinant enzymes show somewhat different properties. For instance, MAN1 gene of *Aspergillus aculeatus* MRC 11624 was cloned and expressed in *S. cerevisiae*, *A. niger*, and *Y. lipolytica*. Besides higher enzyme activity, the resultant recombinant enzymes showed different temperature and pH optima as compared to the native enzymes. Isolation and cloning of genes encoding mannanases and their expression in a suitable host play an important role in the molecular and structural studies of enzyme proteins and protein engineering thereof. Eight essentially conserved active site residues of β-mannanases, viz., Arg-83, His-119, Asn-157, Glu-158, His-224, Tyr-226, Gly-254, and Trp-283, are reported in *Bacillus* N16-5 mannanase (Ma et al. 2004).

Table 4.3 Production and properties of mannanases from various microorganisms

Microorganism	Production	Optimum pH	Optimum T °C	Substrate	Mr (kDa)	Incubation	Rev/min	PI	Reference
Bacillus subtilis WY 34	1,105 U/ml	–	50	Konjac powder	39.6	4 days	200	–	Jiang et al. (2006)
A. niger NCH-189	28 U/ml	5	30	Defatted copra	–	7 days	120	–	Lin and Chen (2004)
Aspergillus fumigatus Man I/Man II	40 U/ml	–	45	Glucose/LBG	60/63	75 h	200	4.9/4.7	Puchart et al. (2004)
Aspergillus awamori K4									Kurakake and Komaki (2001)
β-Mannanase	50 U/g	–	30	CW + WB		88 h			
β-Mannosidase	1.4 U/g	–	30	CW + WB		88 h			
Sclerotium rolfsii CBS147082									Großwindhager et al. (1999)
Mannanase	209 U/ml	5	30	Cellulose		13 days	150		
Mannanase	8.2 U/ml	5	30	Glucose		13 days	150		
Sclerotium rolfsii CBS 151.31									Großwindhager et al. (1999)
Mannanase	348 U/ml	5	30	Cellulose		13 days	150		
Mannanase	95.8 U/ml	5	30	Glucose		13 days	150		
Sclerotium coffeicola CBS 667.85									Großwindhager et al. (1999)
Mannanase	461 U/ml	5	30	Cellulose		13 days	150		
Mannanase	67.7 U/ml	5	30	Glucose		13 days	150		
Trichoderma harzianum E 58	0.60/0.66 IU/ml	–	28	LBG/Avicel	–	6–8 days	100	6.5	Torrie et al. (1990)
Thielavia terrestris									Araujo and Ward (1990)
Mannanase	6.7 U/ml	5	45	LBG	M1-52 M2-30 M3-55 M4-89	48	200		
β-D-Mannosidase	0.001 U/ml	5	45	LBG	72	48	200		
Aspergillus awamori									Ratto and Poutanen (1988)
Mannanase	34 nkat/ml	5.3	28	WB + LBG					
Trichoderma reesei									Ratto and Poutanen (1988)
Mannanase	24 nkat/ml	5.3	28	WB + LBG					

CW coffee waste, *WB* wheat bran

A 1,345 bp gene encoding mannanase (ManN) from *Aspergillus sulphureus* was expressed in *Pichia pastoris* (Chen et al. 2007). Alkaline β-mannanase (ManA) was cloned by Ma et al. (2004) from *Bacillus* sp. N165, and its overproduction and optimization have been studied by Lin et al. (2007). They achieved a maximum yield of 310 U/ml after optimization. Recently, Pan et al. (2011) demonstrated heterologous expression of alkaline β-mannanase by a yeast expression system. Pan et al. (2011) have used *Kluyveromyces cicerisporus* Y179U and *Pichia pastoris* GS115 for expression of MAN 330 (truncated β-mannanase) and MAN 493. MAN330 and MAN 493 genes were amplified and alkaline mannanase was successfully expressed using Y179U/pUKD-S-MAN 330 and GS115/pPIC-9 k MAN 493 (vectors), and high yields of 1,378 and 1,114 U/ml in shake flasks were obtained, respectively. Both enzymes had a maximum activity at pH 9.5 and 70 °C. β-Mannanase from *Bacillus subtilis* has been purified and characterized (Jiang et al. 2006). Recently, a thermostable β-mannanase from *Bacillus subtilis* BCC41051 was expressed in *E. coli* and *Bacillus megaterium* (Summpunn et al. 2011).

The open reading frame of the gene coding β-mannanase was amplified by PCR using Man-CHF (5′-GTACGCCATATGTTTAAGAAACATAC GATCTCTTTGC-3′) and Man-CHR (5′-GTACG CCTCGAGTTCAACGATTGGCGTT AAAGAATC-3′) primers, and the recombinant vector pEManAhis was transfected into *E. coli* BL21. The gene was expressed and induced by IPTG (isopropyl-β-D-1thiogalactopyranoside), and the highest activity of 415.18 U/ml was obtained. For expression in *B. megaterium*, *E. coli*, *Bacillus* shuttle, and expression vector, Pxb was used. The gene coding for β-mannanase was amplified with the primer Man-F1 (5′-GTACGCG GATCCGACAAATGTTTAAGAAACATA CGATC-3′) and Man-R1 (5′-CTGATTCATT CAACGATTGG-3′) and transformed into *B. megaterium* with the help of a pXManA plasmid. The expression of the cloned gene was induced by xylose to obtain 359 U/ml enzyme activity. Various examples of heterologous production of mannanases, vectors employed, and the properties of recombinant enzyme are presented in Table 4.4. Heterologous expression allows a simpler and cheaper means of production using desired hosts, while induction of β-mannanases by native organisms needs mostly expensive and complex medium components (Kote et al. 2009; Viniegra et al. 2003; Van Zyl et al. 2010).

Applications of Mannanases

Mannan-degrading enzymes find various uses in different industries. More recently, mannanases have been used for the production of manno-oligosaccharides (MOS), feed upgradation, biobleaching, and detergents. The details of these applications are detailed below.

Production of Manno-Oligosaccharide

Mannanases degrade mannans and produce manno-oligosaccharides (MOS) and mannose. The MOS contribute to human health and are considered to confer prebiotic benefits. Fan et al. (2009a, b) have showed that glucomannan enhanced fecal probiotics. Hydrolyzed glucomannan can be used as a prebiotic to augment growth of fecal probiotics. MOS is demonstrated to confer a similar result as oligofructose prebiotics. Kobayashi et al. (1987) have noticed that oligosaccharides, which are used as prebiotics to enhance growth of human intestinal microflora, including manno-oligosaccharide. MOS also used as functional food ingredients.

Coffee and Coconut Oil Extraction

In coffee extract, mannan is present as the main polysaccharide and this mannan increases the viscosity of the coffee extract, which unfavorably affects instant coffee preparation. β-Mannanase is used for reducing the viscosity of coffee extract, because it hydrolyzes mannan into simple

Table 4.4 Heterologous production and properties of expressed mannanase

Microorganism (gene source)	Host for cloning	Host for expression	Promoter/terminator	Vector/plasmid	Optima for enzymes pH	Optima for enzymes Temp °c	Stability for enzyme pH	Stability for enzyme Temp °c	K_M mg/ml	Vmax µmolmin^{-1} mg^{-1}	Activity (flask) nkatml^{-1}	Reference
B. subtilis BCC41051	–	*E. coli* BL21 (DE3)	–	pET24b(+)	–		–		–	–	~6,930	Summpunn et al. (2011)
		B. megaterium	–	pXb	–		–		–	–	~5,995	
Bacillus sp. N16-5 (MAN330)	–	*Kluyveromyces cicerisporus*	–	pUKD-S-MAN330	9.5	70	5–11	60–70			~23,012	Pan et al. (2011)
Bacillus sp. N16-5 (MAN493)	–	*Pichia pastoris*	–	pPIC-9 K-MAN493	9.5	70	5–11	60–70			~18,603	
Streptomyces sp. *S27* (man5)	*E. coli* JM109	*E. coli* BL21 (DE3)	–	pET-30a(+)	7	65	5–9	50	0.16	3,739	~447.2	Shi et al. (2011)
Paenibacillus sp. *BME-14* (man26B)	*E. coli* DH5α	*E. coli* BL21 (DE3)	–	pGEX-6P-1	4.5	60	4–8.5	35–65	3.80	91.70	–	Xiaoyu et al. (2010)
A. aculeatus MRC11624 (man 1)	*E. coli* JM109	*A. niger*	gpdp/glaAT	pGT-man1	3.8	~80	–		–	–	16,596	Van Zyl et al. (2009)
Aspergillus aculeatus MRC11624 man1	*E. coli* DH5α	*Yarrowia lipolytica*	hp4dp/-	pYL-man1-HmA	–	–	5.5	28	–	–	13,073	Roth et al. (2009)
A. sulphureus (MANN)	*E. coli* Top10	*Pichia pastoris*	–	pPICZαA	2.4	50	2.2–8.0	<40	0.93	~344U/mg	~1,603	Chen et al. (2007)
A. aculeatus MRC11624	*S. cerevisiae*	*S. cerevisiae*	*ADH2P/ADH2$_T$*	pMES1	3	60	–		–	–	521	Setati et al. (2001)
			PGK1P/PGK1	pMES2							379	

sugars. The hydrolyzed mannan is also beneficial to consumer health as it decreases fat utilization. Endo-mannanase is also applicable in coconut extraction. Mannan is present as the main component in the cell wall of coconuts, so the use of mannanase helps to achieve a higher yield of oil and improves the refining properties of oil. Commercially, endo-β-mannanase from *Aspergillus niger* is marketed as "GAMANASE" for coconut oil extraction (Novo Nordisk, Denmark).

Feed Upgradation

Mannan is commonly present in feed ingredients such as soybean meal, copra meal, and palm kernel meal. Among these meals, soybean meal is mostly used as a protein source in poultry feed. This meal is the product remaining after complete oil extraction from soybean seeds. Due to high levels of mannan, this meal negatively affects growth performance of animals because of its indigestive nature. This mannan component can be hydrolyzed by addition of endo-β-mannanase in the meal. These additions of β-mannanase improve digestibility by hydrolysis of mannan and thereby enhance the performance of poultry. Mannanase produced from *Trichoderma longibrachiatum* and *B. lentus* is marketed as "Hemicell" by ChemGen, a US-based company, and is potentially used as an animal feed supplement.

Biobleaching

The use of xylanase in biobleaching of pulp is well known (Viikari et al. 1993). Mannanase is also used in enzymatic bleaching of softwood pulp. The extraction of lignin from softwood is important. For this purpose, alkaline treatment is performed for hydrolyzing hemicelluloses and removal of lignin. This alkaline treatment of wood pulps creates environment pollution, especially water pollution. Application of mannanases has significantly reduced the use of alkaline chemicals in treatment of wood pulps (Gubitz et al. 1997; Puchart et al. 2004). Mannanases also help in removal of lignin from pulps by biobleaching without affecting the quality of fibers. Here are so many thermophilic microorganisms like *Thermomyces lanuginosus*, *Malbranchea cinnamomea*, *Myceliophthora fergusii*, and *Bacillus subtilis* which are able to produce thermostable mannanases (Maijala et al. 2012; Summpunn et al. 2011). The thermostable β-mannanases and β-mannosidases offer a significant advantage for biobleaching at elevated temperatures. A thermostable extracellular β-mannanase from *A. niger* was studied by Naganagouda et al. (2009). They have shown that this β-mannanase was active over a wide pH and temperature range. Due to these properties of this enzyme, it is potentially useful in biobleaching and food processing. Tenkanen et al. (1997) have shown that mannanases can be used as an alternative in place of hydrogen peroxide in biobleaching.

Detergent and Textile Industries

Amylases and cellulases are well-known common enzymes which are used for many decades in detergents. Gums are used as stabilizing and thickening agents worldwide in many foods and in various household products. These gums contain mannans like galactomannan, guar gum, and tara gum. Mannanases can be added in detergents to remove this gummy matter from clothes. Detergents or cleaners, which are used to clean contact lenses, and hard surface cleaners also contain mannanases. It is observed that detergents containing mannanase can improve the whiteness of cellulosic material. But utilization of mannanase has been limited because most of the mannanases have their optima around neutral or somewhat acidic pH (4.0–6.0). Some fungal β-mannanases can tolerate an alkaline pH up to eight. β-Mannanase from alkaliphilic *Bacillus* sp. N16-5, cloned and expressed in *Kluyveromyces cicerisporus* and *Pichia pastoris,* had a pH optimum of 9.5 and was stable over a pH range of 5.0–11.0 (Pan et al. 2011). A mannanase preparation, "MANNAWAY," used in washing detergents is marketed by Novozymes, a US-based company. Galactomannans such as guar gum and LBG are widely employed as print paste in textile printing. Mannanase, used in enzyme consortia, helps to degrade mannan after the printing of cloth.

Other Applications

Mannanase is also used in oil drilling or gas oil well stimulation. There are many fungal and bacterial β-mannanases available which have high-temperature optima. Heterologous production is providing access to thermostable β-mannanases. To open crevices and cracks for oil and gas flow, mannans like guar gum and tara gum are used as polymer solutions with sand particles. Polymer solutions are hydrolyzed with a suitable thermostable mannanase for better oil and gas recovery. Organisms from hydrothermal vents are isolated and used as a source of endo- and exo-mannanase, and this preparation is marketed by companies, for example, pyrolase 160 and pyrolase 200 products are marketed by Diversa Company with a 37–93 °C recommended temperature. In the production of second-generation fuel (bioethanol, biodiesel), mannanases are crucial with other enzymes. Synergy between mannan-degrading enzymes and cellulases was demonstrated with a fivefold increase in glucose yield from lignocellulose polysaccharides (Jorgensen et al. 2010). Partially hydrolyzed guar gum (PHGG) is used as relief agent for irritable bowel syndrome (IBS).

References

Abdeshahian P, Samat N, Hamid AA, Yusoff WM (2010) Utilization of palm kernel cake for production of beta-mannanase by *Aspergillus niger* FTCC 5003 in solid substrate fermentation using an aerated column bioreactor. J Ind Microbiol Biotechnol 37(1):103–109

Ademark P, Varga A, Medve J, Harjunpää V, Drakenberg T, Tjerneld F, Stålbrand H (1998) Softwood hemicellulose-degrading enzymes from *Aspergillus niger*: purification and properties of a β-mannanase. J Biotechnol 63:199–210

Akino T, Nakamura N, Horikoshi K (1987) Production of β-mannosidase and β-mannanase by an alkaliphilic *Bacillus* sp. Appl Microbiol Biotechnol 26:323–327

Araujo A, Ward OP (1990) Hemicellulases of *Bacillus* species: preliminary comparative studies on production and properties of mannanases and galactanases. J Appl Bacteriol 68:253–261

Arcand N, Kluepfel D, Paradis FW, Morosoli R, Shareck F (1993) Beta-mannanase of *Streptomyces lividans* 66: cloning and DNA sequence of the manA gene and characterization of the enzyme. Biochem J 290:857–863

Aspinall GO (1959) Structural chemistry of the hemicelluloses. Adv Carbohydr Chem 14:429–468

Aspinall GO, Begbie R, McKay JE (1962) The hemicelluloses of European larch (Larix decidua). Part II. J Chem Soc 214:219

Benech RO, Li X, Patton D, Powlowski J, Storms R, Bourbonnais R, Paice M, Tsang A (2007) Recombinant expression, characterization, and pulp prebleaching property of a *Phanerochaete chrysosporium* endo-β-1,4-mannanase. Enzyme Microb Technol 41:740–747

Bourgault R, Bewley JD, Alberici A, Decker D (2001) Endo-β-mannanase activity in tomato and other ripening fruits. Hortscience 36:72–75

Chaikumpollert O, Methacanon P, Suchiva K (2004) Structural elucidation of hemicelluloses from Vetiver grass. Carbohydr Polym 57:191–196

Chandrasekaran R, Radha A, Okuyama K (1998) Morphology of galactomannans: an X-ray structure analysis of guaran. Carbohydr Res 306:243–255

Chen X, Cao Y, Ding Y, LuW LD (2007) Cloning, functional expression and characterization of *Aspergillus sulphureus* β-mannanase in *Pichia pastoris*. J Biotechnol 8:452–461

Cui F, Shi J, Lu Z (1999) Production of neutral beta-mannanase by *Bacillus subtilis* and its properties. Wei Sheng Wu Xue Bao 39(1):60–63

Dhawan S, Kaur J (2007) Microbial mannanases: an overview of production and applications. Crit Rev Biotechnol 27(4):197–216

Duffaud GD, McCutchen CM, Leduc P, Parker KL, Kelly RM (1997) Purification and characterization of extremely thermostable β-mannanase, β-mannosidase, and α-galactosidase from the hyperthermophilic Eubacterium *Thermotoga neapolitana* 5068. Appl Environ Microbiol 63:169–177

Fan Z, Wagschal K, Chen W, Montross MD, Lee CC, Yuan L (2009a) Multimeric hemicellulases facilitate biomass conversion. Appl Environ Microbiol 75:1754–1757

Fan Z, Werkman JR, Yuan L (2009b) Engineering of a multifunctional hemicellulose. Biotechnol Lett 31:751–757

Feng Y, He Z, Ong SL, Hu J, Zhang Z, Ng WJ (2003) Optimization of agitation, aeration and temperature conditions for maximum β-mannanase production. Enzyme Microb Technol 32:282–289

Ferreira HM, Filho EXF (2004) Purification and characterization of a β-mannanase from *Trichoderma harzianum* strain T4. Carbohydr Polym 57:23–29

Gomes J, Terler K, Kratzer R, Kainz E, Steiner W (2007) Production of thermostable β-mannosidase by a strain of *Thermoascus aurantiacus*: isolation, partial purification and characterization of the enzyme. Enzyme Microb Technol 40:969–975

Großwindhager C, Sachslehner A, Nidetzky B, Haltrich D (1999) Endo-*β*-1,4 mannanase is efficiently produced by *Sclerotium (Athelia) rolfsii* under derepressed conditions. J Biotechnol 67:189–203

Gubitz GM, Hayn M, Sommerauer M, Steiner W (1996) Mannan-degrading enzymes from *Sclerotium rolfsii*: characterization and synergism of two endo β-mannanases and β-mannosidase. Bioresour Technol 58:127–135

Gubitz GM, Haltrich D, Latal B, Steiner W (1997) Mode of depolymerisation of hemicellulose by various mannanases and xylanases in relation to their ability to bleach softwood pulp. Appl Microbiol Biotechnol 47:658–662

Hilge M, Gloor SM, Rypniewski W, Sauer O, Heightman TD, Zimmerman W et al (1998) High-resolution native and complex structures of thermostable *β*-mannanase from *Thermomonospora fusca*-substrate specificity in glycosyl hydrolase family 5. Structure 6:1433–1444

Hongshu Z, Jinggan Y, Yan Z (2002) The glucomannan from ramie. Carbohydr Polym 47:83–86

Hossain MZ, Abe JI, Hizukuri S (1996) Multiple forms of β-mannanase from *Bacillus* sp. KK01. Enzyme Microb Technol 18:95–98

Ishurd O, Kermagi A, Zgheel F, Flefla M, Elmabruk M, Yalin W, Kennedy JF, Yuanjiang P (2004) Structural aspects of water soluble galactomannan from the seeds of *Retama raetam*. Carbohydr Polym 58:41–44

Jiang Z, Wei Y, Li D, Chai P, Kusakabe I (2006) High-level production, purification and characterization of a thermostable β-mannanase from the newly isolated *Bacillus subtilis* WY34. Carbohydr Polym 66:88–96

Jorgensen H, Sanadi AR, Felby C, Lange NEK, Fischer M, Ernst S (2010) Production of ethanol and feed by high dry matter hydrolysis and fermentation of palm kernel press cake. Appl Biochem Biotechnol 161:318–332

Kalogeris E, Christakopoulos P, Katapodis P, Alexiou A, Vlachou S, Kekos D, Macris BJ (2003) Production and characterization of cellulolytic enzymes from the thermophilic fungus *Thermoascus aurantiacus* under solid state cultivation of agricultural wastes. Process Biochem 38:1099–1104

Kango N (2007) Lignocellulose biotechnology: harnessing renewable plant materials for energy and chemicals. Everyman's Sci XLII(4):194–197

Kango N, Agrawal SC, Jain PC (2003) Production of xylanase by *Emericella nidulans* NK-62 on low-value lignocellulosic substrates. World J Microbiol Biotechnol 19(7):691–694

Kansoh AL, Nagieb ZA (2004) Xylanase and mannanase enzymes from *Streptomyces galbus* NR and their use in biobleaching of softwood kraft pulp. Anton van Leeuwonhoek 85:103–114

Kataoka N, Tokiwa Y (1998) Isolation and characterization of an active mannanase-producing anaerobic bacterium, *C. tertium* KT-5A from lotus soil. J Appl Microbiol 84:357–367

Kenne L, Rossel K-G, Svensson S (1975) Studies on the distribution of the O-acetyl groups in pine glucomannan. Carbohydr Res 44:69–76

Kirk O, Borchet TV, Fuglsang CC (2002) Industrial enzyme applications. Curr Opin Biotechnol 13:345–351

Kobayashi Y, Echigen R, Mada M, Mutai M (1987) Effect of hydrolyzate of konjac mannan and soybean oligosaccharides on intestinal flora in man and rats. In: Mitsuoka T (ed) Intestinal flora and food factors. Gakkai Shuppan Centre, Tokyo, pp 79–97

Kote NV, Patil AGG, Mulimani VH (2009) Optimization of the production of thermostable endo-β-1,4 mannanases from a newly isolated *Aspergillus niger* gr and *Aspergillus flavus* gr. Appl Biochem Biotechnol 152(2):213–223

Kurakake M, Komaki T (2001) Production of β-Mannanase and β-Mannosidase from *Aspergillus awamori* K and their properties. Curr Microbiol 42:377–380

Kurakake M, Sumida T, Masuda D, Oonishi S, Komaki T (2006) Production of galacto-manno-oligosaccharides from guar gum by β-mannanase from *Penicillium oxalicum* SO. J Agric Food Chem 54:7885–7889

Liepman AH, Nairn CJ, Willats WGT, Sørensen I, Roberts AW, Keegstra K (2007) Functional genomic analysis supports conservation of function among cellulose synthase-like A gene family members and suggest diverse roles of mannans in plants. Plant Physiol 143:1881–1893

Lin TC, Chen C (2004) Enhanced mannanase production by submerged culture of *Aspergillus niger* NCH-189 using defatted copra based media. Process Biochem 39(9):1103–1109

Lin SS, Dou WF, Xu HY, Li HZ, Xu ZH, Ma YH (2007) Optimization of medium composition for the production of alkaline b-mannanase by alkaliphilic *Bacillus* sp.N16–5 using response surface methodology. Appl Microbiol Biotechnol 75:1015–1022

Lundqvist J, Teleman A, Junel L, Zacchi G, Dahlman O, Tjerneld F, Stalbrand H (2002) Isolation and characterization of galactoglucomannan from spruce (*Picea abies*). Carbohydr Polym 48:29–39

Lundqvist J, Jacobs A, Palm M, Zacchi G, Dahlman O, Stålbrand H (2003) Characterization of galactoglucomannan extracted from spruce (*Picea abies*) by heat-fractionation at different conditions. Carbohydr Polym 51:203–211

Ma YH, Xue YF, Dou YT, Xu ZH, Tao WY, Zhou PJ (2004) Characterization and gene cloning of a novel b-mannanase from alkaliphilic *Bacillus* sp. N16–5. Extremophiles 8:447–454

Maijala P, Kango N, Szijarto N, Viikari L (2012) Characterization of hemicellulases from thermophilic fungi. Anton van Leeuwonhoek 101(4):905–917

Mohamad SN, Nagasundara R, Mohamad RR, Ariff AB (2011) Improved mannan-degrading enzymes production by *Aspergillus niger* through medium optimization. New Biotechnol 28. doi:10.1016/j.nbt.2010.10.008

Moreira LRS, Filho EXF (2008) An overview of mannan structure and mannan-degrading enzyme systems. Appl Microbiol Biotechnol 79:165–178

Mudau MM, Setati ME (2006) Screening and identification of endomannanase-producing microfungi from hypersaline environments. Curr Microbiol 52:477–481

Naganagouda K, Salimath PV, Mulimani VH (2009) Purification and characterization of endo-β-1,4 mannanase from *Aspergillus niger* gr for application in food processing industry. J Microbiol Biotechnol 19(10):1184–1190

Nunes FM, Domingues MR, Comibra MA (2005) Arabinosyl and glucosyl residues as structural features of acetylated galactomannans from green and roasted coffee infusions. Carbohydr Res 340:1689–1698

Oda Y, Komaki T, Tonomura K (1993) Production of β-mannanase and β-mannosidase by *Enterococcus casseliflavus* FL2121 isolated from decayed konjac. Food Microbiol 10:352–358

Omarsdottir S, Petersen BO, Barsett H, Paulsen BS, Duus JØ, Olafsdottir ES (2006) Structural characterization of a highly branched galactomannan from the lichen *Peltigera canina* by methylation analysis and NMR-spectroscopy. Carbohydr Polym 63:54–60

Pan X, Zhou J, Tian A, Le K, Yuan H, Xue Y, Ma Y, Lu H (2011) High level expression of a truncated β-mannanase from alkaliphilic *Bacillus* sp. N16-5 in *Kluyveromyces cicerisporus*. Biotechnol Lett 33:565–570

Perret S, Belaich A, Fierobe HP, Belaich JP, Tardif C (2004) Towards designer cellulosomes in *Clostridia*: mannanase enrichment of the cellulosomes produced by *Clostridium cellulolyticum*. Biotechnol Appl Biochem 40:255–259

Petkowicz CLO, Schaefer S, Reicher F (2007) The mannan from *Schizolobium parahybae* endosperm is not a reserve polysaccharide. Carbohydr Polym 69:659–664

Popa VI, Spiridon J (1998) Hemicelluloses: structure and properties. In: Dumitriu S (ed) Polysaccharides: structural diversity and functional versatility. Marcel Dekker, New York, pp 297–311

Puchart V, Vrsanska M, Svoboda P, Pohl J, Biely P (2004) Purification and characterization of two forms of endo- β-1,4-mannanase from a thermotolerant fungus, *Aspergillus fumigatus* IMI 385708 (formerly *T. lanuginosus* IMI 158749). Biochim Biophys Acta 1674:239–250

Puls J, Schuseil J (1993) Chemistry of hemicellulose: relationship between hemicellulose structure and enzyme required for hydrolysis. In: Coughlan MP, Hazlewood GP (eds) Hemicellulose and hemicellulases. Portland Press, London, pp 1–27

Qiao Y, Chen X, Ding H, Yue M (2008) Heterologous expression and characterization of man gene from *Bacillus subtilis* in *Pichia pastoris*. Front Biol China 3:26–31

Ratto M, Poutanen K (1988) Production of mannan-degrading enzymes. Biotechnol Lett 10(9):661–664

Ratto M, Siika-aho M, Buchert J, Valkeajarvi A, Viikari L (1993) Enzymatic hydrolysis of isolated and fibre-bound galactoglucomannans from pine-wood and pine raft pulp. Appl Microbiol Biotechnol 40: 449–454

Reyna ABV, Noyola PP, Calvo-Méndez C, López-Romero E, Flores-Carreón A (1999) Purification and biochemical characterization of two soluble α-mannosidases from *Candida albicans*. Glycobiology 9:533–537

Ronne H (1995) Glucose repression in fungi. Trends Genet 11:12–17

Roth R, Moodley V, van Zyl P (2009) Heterologous expression and optimized production of an *Aspergillus aculeatus* endo-1,4-β-mannanase in *Yarrowia lipolytica*. Mol Biotechnol 43:112–120

Ruijter GJG, Visser J (1997) Carbon repression in *Aspergilli*. FEMS Microbiol Lett 151:103–114

Sachslehner A, Foidl G, Foidl N, Gubitz G, Haltrich D (2000) Hydrolysis of isolated coffee mannan and coffee extract by mannanases of *Sclerotium rolfsii*. J Biotechnol 80:127–134

Schroder R, Wegrzyn TF, Sharma NN, Atkinson RG (2006) LeMAN4 endo-β-mannanase from ripe tomato fruit can act as a mannan transglycosylase or hydrolase. Planta 224:1091–1102

Setati ME, Ademark P, VanZyl WH, Hahn-Hagerdal B, Stalbrand H (2001) Expression of the *Aspergillus aculeatus* endo-β-1,4-mannanase encoding gene (*man1*) in *Saccharomyces cerevisiae* and characterization of the recombinant enzyme. Protein Expr Purif 21:105–114

Shi P, Yuan T, Zhao J, Huang H, Luo H, Meng K, Wang Y, Yao B (2011) Genetic and biochemical characterization of a protease-resistant mesophilic β-mannanase from *Streptomyces* sp. S27. J Ind Microbiol Biotechnol 38:451–458

Singh V, Malviya T (2006) A non-ionic glucomannan from the seeds of an indigenous medicinal plant: *Bryonia laciniosa*. Carbohydr Polym 64:481–483

Singh S, Madlala AM, Prior BA (2003) *Thermomyces lanuginosus*: properties of strains and their hemicellulases. FEMS Microbiol Rev 27:3–16

Stalbrand H, Siika-aho M, Tenkanen M, Viikari L (1993) Purification and characterization of two β-mannanases from *Trichoderma reesei*. J Biotechnol 29:229–242

Summpunn P, Chaijan S, Isarangkul D, Wiyakrutta S, Meevootisom V (2011) Characterization, gene cloning and heterologous expression of β-mannanase from a thermophilic *bacillus subtilis*. J Microbiol 49:86–93

Takegawa K, Miki S, Jikibara T, Iwahara S (1989) Purification and characterization of exo-α-D-mannosidase from a *Cellulomonas* sp. Biochim Biophys Acta 991:431–437

Tenkanen M, Makkonen M, Perttula M, Viikari L, Teleman A (1997) Action of *Trichoderma reesei* mannanase on galactoglucomannan in pine kraft pulp. J Biotechnol 57:191–204

Timell TE (1965) Wood hemicelluloses: part II. Carbohydr Chem 20:409–483

Timell TE (1967) Recent progress in the chemistry of wood hemicelluloses. Wood Sci Technol 1:45–70

Torrie JP, Senior DJ, Saddler JN (1990) Production of β-mannanases by *Trichoderma harzianum* E58. Appl Microbiol Biotechnol 34:303–307

Van Zyl PJ, Moodley V, Rose SH, Roth RL, Van Zyl WH (2009) Production of the *Aspergillus aculeatus* endo-1,4-β-mannanase in *A. niger*. J Ind Microbiol Biotechnol 36:611–667

Van Zyl WH, Rosea SH, Trollopeb K, Gorgensb JF (2010) Fungal β-mannanases: mannan hydrolysis, heterologous production and biotechnological applications. Process Biochem 45:1203–1213

Viikari L, Tenkanen M, Buchert J, Ratto M, Bailey M, Siika-Aho M, Linko M (1993) Hemicellulases for industrial applications. In: Saddler JN (ed) Bioconversion of forest and agricultural plant residues. CAB International, Wallingford, pp 131–182

Viniegra-Gonzalez G, Favela-Torres E, Aguilar CN, Romero-Gomez SJ, Diaz-Godinez G, Augur C (2003) Advantages of fungal enzyme production in solid state over liquid fermentation systems. Biochem Eng J 13:157–167

Willfor S, Sjoholm R, Laine C, Roslund M, Hemming J, Holmbom B (2003) Characterisation of water-soluble galactoglucomannan from Norway spruce wood and thermomechanical pulp. Carbohydr Polym 52:175–187

Xiaoyu F, Huang X, Liu P, Lin L, Wu G, Li C, Feng C, Hong Y (2010) Cloning and characterization of a novel mannanase from *Paenibacillus* sp. BME-14. J Microbiol Biotechnol 20(3):518–524

Yoshida T, Inoue T, Ichishima E (1993) 1,2-i-D-Mannosidase from *Penicillium citrinum*: molecular and enzymic properties of two isoenzymes. Biochem J 290:349–354

Zakaria MM, Ashiuchi M, Yamamoto S, Yagi T (1998) Optimization for beta-mannanase production of a psychrophilic bacterium, *Flavobacterium* sp. Biosci Biotechnol Biochem 62:655–660

Zhang Q, Yan X, Zhang L, Tang W (2006) Cloning, sequence analysis, and heterologous expression of a *β*-mannanase gene from *Bacillus subtilis* Z-2. Mol Biol 40:368–374

5 Enzyme Synergy for Enhanced Degradation of Lignocellulosic Waste

J. Susan van Dyk and Brett I. Pletschke

Abstract

Lignocellulosic substrates are very complex with a number of different components associated within a three-dimensional network. The complexity of the substrate, as well as the close association between components, makes the substrate recalcitrant to degradation. Such substrates require a large number of hydrolytic enzymes working in synergy to achieve complete degradation. The use of synergy studies allows us to determine whether enzymes display cooperation in hydrolysis of lignocellulosic substrates. This allows us to evaluate which enzymes should possibly be included in designer enzyme cocktails for lignocellulose hydrolysis. It also provides us with insight into areas of substrate recalcitrance, including obstacles such as physical association between components which prevent access of enzymes to their substrate.

Keywords

Cellulose • Enzymes • Hemicellulase • Lignocellulose • Synergy

Introduction

Lignocellulose contains approximately 75 % polysaccharides in the form of cellulose and hemicellulose, and its degradation represents a sustainable option for obtaining large quantities of sugars for fermentation into bioethanol (Gomez et al. 2008).

J.S. van Dyk • B.I. Pletschke (✉)
Department of Biochemistry, Microbiology and Biotechnology, Rhodes University, PO Box 94, 6140 Grahamstown, South Africa
e-mail: B.Pletschke@ru.ac.za

Lignin is a polyphenolic compound that provides rigidity to plants and can make up to 20 % of the lignocellulose composition in secondary plant cell walls. It has a very complex structure and only a few microorganisms in nature (white-rot fungi) produce the enzymes required to degrade this component. These enzymes are listed in Table 5.1. Lignin represents the main obstacle to lignocellulose hydrolysis as it prevents hydrolytic enzymes from accessing the cellulose and hemicellulose polysaccharides (Varnai et al. 2010). Biological degradation of lignin in biofuel technologies is not commonly used, and generally, a physico-chemical pretreatment

P. Shukla and B.I. Pletschke (eds.), *Advances in Enzyme Biotechnology*,
DOI 10.1007/978-81-322-1094-8_5,

Table 5.1 Enzymes required to degrade lignocellulose components

Component	Enzymes required for degradation
Lignin	Laccase, manganese peroxidase, lignin peroxidase
Cellulose	Cellobiohydrolase, endoglucanase, β-glucosidase
Hemicellulose: xylan	Endo-xylanase, acetyl xylan esterase, β-xylosidase, α-L-arabinofuranosidase, α-glucuronidase, ferulic acid esterase, p-coumaric acid esterase
Hemicellulose: mannan	Endo-mannanase, β-mannosidase, α-galactosidase, β-glucosidase, acetyl mannan esterase
Hemicellulose: arabinan	Endo-arabinase, α-L-arabinofuranosidase
Pectin	Pectin methylesterase, pectin and pectate lyases, endo- and exopolygalacturonases, rhamnogalacturonan lyases and hydrolases

method is used to remove the lignin or to break the linkages between the lignin and polysaccharide components (see review by Hendriks and Zeeman 2009).

Cellulose consists of chains of glucose connected with β-1,4-linkages that are packed into microfibrils into the plant cells and is the main polysaccharide component in most plants. A minimum of three enzymes are required to degrade cellulose to glucose, namely, cellobiohydrolases (exoglucanases), endoglucanases and β-glucosidases. As glucose is easily converted into ethanol by *Saccharomyces cerevisiae*, cellulose hydrolysis is the main target in bioconversion. The hydrogen bonding between adjacent cellulose chains, the crystalline structure and the insolubility of the cellulose present great difficulty for the enzymes to hydrolyse this substrate.

Hemicellulose is defined as the component of the plant that hydrogen bonds to cellulose within the three-dimensional structure of the plant (Albersheim et al. 2011). It is a broad term that refers to various polysaccharides such as arabinoxylan, xyloglucan, galactoglucomannan, glucomannan and arabinan, each consisting of different combinations of sugars with a variety of linkages, including α-1,2, α-1,3, α-1,6 and β-1,4 linkages. Due to the variety of sugars and linkages, a large number of enzymes are required for the degradation of these substrates as enzymes may have specificity for particular sugars and/or particular linkages. Hemicellulose composition varies between primary and secondary cell walls and between sources of lignocellulose such as grasses, hardwoods and softwoods. Therefore, different hemicellulase enzymes may be required for each type of substrate. Table 5.1 lists some of the enzymes required to degrade the most common hemicelluloses. The hydrolysis of hemicellulose can release a number of sugars such as xylose, arabinose, mannose and galactose which could provide an additional source of substrate for ethanol production, although fermentation has to be carried out by other yeasts such as *Pichia stipitis*. This could potentially enhance the overall ethanol yield from lignocellulose substrates compared to using only the cellulose fraction.

An additional component that further increases the complexity of lignocellulose hydrolysis is the presence of pectin. Pectin may be present at high levels in the primary cell walls of plants, but becomes less than 10 % of secondary cell walls (Albersheim et al. 2011). However, in agricultural wastes such as citrus peel or apple pomace, high concentrations of pectin may be found which necessitates additional enzymes for their degradation. Three different types of pectin are found in plants, namely, homogalacturonan, rhamnogalacturonan I and rhamnogalacturonan II. Table 5.1 lists some of the main enzymes used to degrade pectins.

After removal of lignin, the main obstacle to the degradation of the polysaccharides in the substrate lies in the nature of the three-dimensional structure and association between the different components. Short sections of hemicellulose hydrogen bond with adjacent cellulose fibrils, masking the cellulose as well as forming cross-linkages between cellulose fibrils within the plant structure, resulting in a complex, interwoven structure (Albersheim et al. 2011). If pectin is further present, it will be found within this matrix between the cellulose and hemicellulose. The unravelling of this network requires that each enzyme obtains access to its particular substrate and bonds within the plant cell wall. This is achieved through cooperation between enzymes, referred to as synergistic interactions.

By studying synergistic interactions between enzymes, we are able to identify the physical barriers to access of enzymes or the types of bonds that contribute the most to recalcitrance. Studying the ratios of enzymes that provide the highest degree of synergy allows the optimisation of enzyme loadings which leads to reduced cost for the enzyme input. This forms the basis for the design of enzyme cocktails for the hydrolysis of different complex substrates.

Enzyme Synergy Studies

Enzyme synergy studies investigate the interaction between two or more enzymes on a substrate to determine whether the enzymes display cooperative behaviour in the degradation of the substrate. The degree of synergy/synergism is calculated as the observed activity of the enzyme mixture, divided by the theoretical sum of the individual activities of the same enzymes on the same substrate. The calculated value for degrees of synergy can be divided into three categories which provide different interpretations of the interactions taking place.

A degree of synergy of below 1 is an indication that no synergy took place and that the combined use of the enzymes resulted in a lower activity than the theoretical sum of their individual activities. This may indicate that the enzymes were competing for the same substrate sites and therefore did not enhance the degradation of the substrate. Thus, enzymes with similar or overlapping functions could display this degree of synergy, indicating that one of the enzymes is probably not required in a cocktail of multiple enzymes for degradation of the specific substrate.

A degree of synergy of 1 indicates that no synergy took place when the enzymes were used in combination and that the one enzyme did not contribute to the other enzyme's activity on the substrate. The enzymes most likely act independently on the substrate and the action of the one is not dependent on the other. It is not an indication that the one enzyme should be discarded from the eventual enzyme cocktail as it may still contribute to the overall degradation of the substrate. This could be evaluated in a different way by measuring the impact of the enzymes on increased yields of product.

The third category of synergy is where the degree of synergy is calculated to be above 1. This is a clear indication that the enzymes are cooperating in the degradation of the substrate which may take place in different ways. One enzyme may be hampered by substituents on the substrate which is removed by the other enzyme. Within the three-dimensional structure of lignocellulose, the one enzyme may be removing a portion of the substrate which masks the access of enzymes to other parts of the substrate.

The degree of synergy is an indication of the enzyme cooperation required for the degradation of the substrate. A high degree of synergy indicates the extent to which the enzymes are mutually dependent on each other for access to their substrate or bond.

The degree of synergy is dependent on the concentration and ratio of the enzymes, as well as the specific characteristics of the enzymes. Synergy is generally observed only at low concentrations of enzyme which retards the degradation in order to allow observation of the interaction between the enzymes (Andersen et al. 2008). At high concentrations of enzyme, synergy may not be observed.

The ratios in which enzymes are used have an impact on the degree of synergy observed. Synergistic interactions become more important when an enzyme is only present at a low ratio. Ratios are important to optimise enzyme loadings and also provide an indication of the level of recalcitrance associated with specific bonds or specific components of the substrate. The contribution of the enzymes is not necessarily related to the percentage composition of the substrate, and only direct biochemical assays can determine optimal ratios (Banerjee et al. 2010). A substrate may contain only a small percentage of xylan, yet the synergistic relationship may indicate a requirement for endo-xylanase supplementation at much greater levels. An optimal ratio between two or more enzymes may change if a further enzyme is added to the combination.

Substrate characteristics may also have an impact on the degree of synergy, for example, the crystallinity of the substrate or the degree of polymerisation (Andersen et al. 2008). The specific characteristics of enzymes have an impact on the degree of synergy observed. When studying synergistic interactions, it is therefore not sufficient to have a general classification of the hydrolytic enzymes involved. It is important that the family of the enzyme and its substrate specificity is known. If this information is not available, it may be difficult to draw clear conclusions from the study. For example, most xylanases are found in two families, namely, family 10 and 11, although some xylanases also fall within families 5, 8 and 43 (Kolenova et al. 2006). Endo-xylanases from the glycosyl hydrolase family 10 (GH10) have a smaller active site and a low substrate specificity and are able to cleave linkages in the xylan backbone close to any substituents present. On the other hand, endo-xylanases from glycosyl hydrolase family 11 (GH11) have a higher substrate specificity with a larger active site. Therefore, it is probable that substituents will pose a greater hindrance to GH11 endo-xylanases as cleavage of the backbone can only take place at a number of residues away from the substituent. Therefore, it is expected to find a high degree of synergy between the depolymerising (GH11) and the debranching enzymes (such as α-L-arabinofuranosidases and acetyl xylan esterases).

Studies in this regard have focused on arabinose and acetyl substituents on arabinoxylan. From these studies, it is apparent that it is not just the characteristics of the endo-xylanases but also the α-L-arabinofuranosidases and acetyl xylan esterases that affect synergistic behaviour. Shi et al. (2010) used a GH51 arabinofuranosidase with an endo-xylanase in synergy studies and found that the arabinofuranosidase was unable to cleave arabinose from xylan polymers unless the endo-xylanase had first cleaved the xylan backbone into shorter oligosaccharides. The arabinofuranosidase appeared to be unable to bind large polysaccharides. In a similar manner, synergy was found between endo-xylanases and acetyl xylan esterases. Raweesri et al. (2008) found that the release of arabinose and acetyl groups from substituted xylan before using a depolymerising endo-xylanase resulted in greater levels of synergy, indicating that the depolymerisation was restricted by the presence of substituents on the backbone.

Another enzyme characteristic that is important is the presence of carbohydrate-binding domains (CBMs), although their impact on synergistic relationships has received limited attention. These domains provide an interaction between the enzyme and the substrate that is separate from the active site and allows the enzyme to bind to the substrate. A large number of CBM families exist and enzymes may have no CBM, one CBM or multiple CBMs. While they may increase the activity of an enzyme on a substrate, they could have a specific impact on synergistic interactions between enzymes. If two enzymes have different substrate specificities, but similar CBMs, they could display competitive behaviour as they will both be binding to the same sites (Andersen et al. 2008). The rate of adsorption or desorption of enzymes could also result in competitive behaviour between enzymes (Andersen et al. 2008).

In addition, time can have an impact on the degree of synergy observed. The general pattern that has been found is that the highest degree of synergy is observed at the early stages of hydrolysis with the degree of synergy becoming lower during later stages (Andersen et al. 2008). It is argued that the reason for this pattern is that the greatest level of cooperation between enzymes is required at the early stages of deconstruction of the substrate. As the enzymes cooperate to unravel the substrate, more binding sites become available over time, requiring a lower degree of cooperation between the enzymes and a lower degree of synergy.

Synergistic cooperation may be different depending on whether enzymes are added in a sequential or simultaneous manner. Where one enzyme relies on the removal of a substituent before it is able to cleave the substrate, the synergy may be higher using the sequential method. The patterns observed in this regard are highly dependent on the specific characteristics of the enzymes involved. Shi et al. (2010) found similar degrees of synergy when an endo-xylanase or arabinofuranosidase was used simultaneously or sequentially, but only when the endo-xylanase was used first.

Synergy Studies in Our Research Group

Synergy studies were conducted in our research group using purified enzymes from *Clostridium cellulovorans*, namely, EngE, ManA, XynA and ArfA in combinations of two or three enzymes on different substrates. EngE is a GH5 cellulosomal enzyme from *C. cellulovorans* (Tamaru and Doi 1999). Purified EngE was able to hydrolyse CMC and lichenan, but had virtually no activity on acid swollen cellulose, Avicel, laminarin and xylan (Tamaru and Doi 1999). ManA is a GH5 cellulosomal enzyme from *C. cellulovorans* (Tamaru et al. 2000). ManA was examined for substrate specificity and only had activity on β-1, 4-mannosidic linkages and had no activity on CMC (Tamaru et al. 2000). XynA is an endo-1,4-β-xylanase with a family 11 (GH11) catalytic domain, a dockerin domain and an acetyl xylan esterase module (Kosugi et al. 2002b). The acetyl xylan esterase module was able to remove acetate from acetylated xylan and other acetylated substrates. The xylanase and acetyl xylan esterase module could act synergistically to degrade acetylated xylan (Kosugi et al. 2002b). ArfA is a non-cellulosomal α-L-arabinofuranosidase classified in glycosyl hydrolase family 51 (GH51), with specificity towards *p*-nitrophenol-arabinofuranoside and additional activity on sugar beet arabinan (arabinase activity) (Kosugi et al. 2002a).

Synergy studies were conducted on a variety of complex substrates such as corn stalk, grass, pineapple pulp, sugarcane bagasse (with various pretreatments) and lime pretreated sugar beet pulp (see Table 5.2). Many of these substrates are agricultural waste products produced in high quantities and have great potential as lignocellulosic substrates for biofuel production. The composition of the substrates, from our own analysis and from literature, is provided in Table 5.3 as there may be a relationship between composition and synergistic interactions.

By utilising two or three pure enzymes in combination in synergy studies, we were able to observe the interactions between these enzymes on complex substrates. From the results summarised in Table 5.2, it is apparent that most of the enzyme combinations displayed some degree of synergy, indicating that cooperation between the enzymes took place. The degrees of synergy varied with different enzyme combinations, indicating that different levels of cooperation took place between the enzymes examined and on the different substrates. The relationship between synergy and the composition of the substrates (Table 5.3) does not display a clear correlation – the level at which an enzyme is required is not necessarily related to the percentage of that sugar within the overall composition, as the associations of components within the substrate are not taken into account by the analysis.

Synergistic interactions were observed between EngE and XynA on all substrates where this combination was tested. In most cases, the degree of synergy between EngE and XynA on all the substrates tested was quite high relative to the other enzyme combinations. When the composition of the substrates is examined in Table 5.3, it is apparent that all the substrates contained cellulose and some degree of xylan, although the xylan composition was not characterised in the corn stalk, grass and pineapple pulp, but it is known that arabinoxylan is the main hemicellulose in grasses. These results correspond to other reports in literature (Bura et al. 2009; Garcia-Aparicio et al. 2007; Selig et al. 2008) that xylanase supplementation improved cellulose hydrolysis regardless of the xylan content. Addition of higher concentrations of xylanase allowed lower cellulase enzyme loadings while achieving the same level of cellulose hydrolysis. Kumar and Wyman (2009) demonstrated that supplementing xylanases with cellulase not only improved cellulose hydrolysis but displayed a linear relationship

Table 5.2 A summary of synergy studies conducted in our research group

Substrate	Pretreatment used	Enzymes used	Ratio	Degrees of synergy (DS)	Reference
Corn stalk	Untreated	EngE:XynA	75:25	6.5	Olver et al. (2011)
		EngE:ManA	50:50	3.5	Olver et al. (2011)
		ManA:XynA	75:25	2.6	Olver et al. (2011)
Grass	Untreated	EngE:XynA	75:25	3.3	Olver et al. (2011)
		EngE:ManA	75:25	2.0	Olver et al. (2011)
		ManA:XynA	75:25	1.2	Olver et al. (2011)
Pineapple pulp	Untreated	EngE:XynA	50:50	2.4	Olver et al. (2011)
		EngE:ManA	50:50	2.8	Olver et al. (2011)
		ManA:XynA	75:25	2.5	Olver et al. (2011)
Sugar beet pulp	Lime pretreatment	ArfA:ManA	75:25	1.36	Dredge et al. (2011)
		ArfA:XynA		No synergy	Dredge et al. (2011)
Sugarcane bagasse	Untreated	EngE:ManA	50:50	2.28	Beukes et al. (2008)
		EngE:XynA	25:75	4.65	Beukes et al. (2008)
		XynA:ManA	75:25	3.95	Beukes et al. (2008)
		XynA:EngE:ManA	25:25:50	2.73	Beukes et al. (2008)
	Untreated	ArfA:XynA	12.5:87.5	2.4	Beukes and Pletschke (2010)
		ArfA:ManA:XynA	Various	±1.6	Beukes and Pletschke (2010)
	Lime (CaOH)	ManA:XynA	87.5: 12.5	1.4	Beukes and Pletschke (2010)
		ArfA:XynA	50:50	1.3	Beukes and Pletschke (2010)
		ArfA:ManA	Various	1.8–2.0	Beukes and Pletschke (2010)
		ArfA:ManA:XynA	Various	1.6–2.2	Beukes and Pletschke (2010)
	NH_4OH pretreated	XynA:ManA	75:25	2.85	Beukes and Pletschke (2011)
		ArfA:XynA	ArfA at high levels	3	Beukes and Pletschke (2011)
		ArfA:ManA:XynA	Various	3	Beukes and Pletschke (2011)
	NaOH	ArfA:XynA	37.5:62.5	1.6	Beukes and Pletschke (2011)
		ManA:ArfA	75:25	1.8	Beukes and Pletschke (2011)
		ManA:XynA		No synergy	Beukes and Pletschke (2011)
		ArfA:ManA:XynA	Various	1.4–1.6	Beukes and Pletschke (2011)

between glucose and xylose release. Murashima et al. (2003) observed that synergistic interactions between cellulases and xylanases were only present in simultaneous, not sequential, reactions. The mechanism behind this type of synergy is generally considered to be the fact that the xylan masks the cellulose and prevents access of cellulases to their substrate. As the xylanases hydrolyse the xylan components in the substrate, the cellulose is exposed, allowing the cellulases to access the cellulose. Based on this model, one could predict that the synergy should be sequential, rather than simultaneous. Selig et al. (2008), however, came to the conclusion that the cellulose also masked the xylan in a similar manner as the xylan masked the cellulose. The simultaneous interactions between the two types of enzymes allow for the synchronised hydrolysis of both cellulose and xylan.

The role of XynA on untreated substrates becomes particularly important as XynA contains an acetyl xylan esterase module. Acetate

Table 5.3 Composition of complex substrates studied in our research group

Substrate	Pretreatment	Composition	Reference
Corn stover	Drying and grinding	18 % lignin, 41.7 % cellulose, 20.5 % hemicellulose	Merino and Cherry (2007)
Grass	Drying and grinding	10–30 % lignin, 25–40 % cellulose, 25–50 % hemicellulose (average values)	Howard et al. (2003)
Pineapple pulp	Drying and grinding Soluble sugars removed	19 % cellulose, 22 % hemicellulose, 5 % lignin and 53 % cell soluble matters	Ban-Koffi and Han (1990)
Sugar beet pulp	Lime pretreated	Rhamnose 2.4 %, fucose 0.2 %, arabinose 20.9 %, xylose 1.7 %, mannose 1.1 %, galactose 5.1 %, glucose 21.1 %, galacturonic acid 21.1 %	Micard et al. (1996)
Bagasse	Untreated	42.4 % lignin, 1.1 % arabinose, 28.8 % glucose, 14.2 % xylose, 7.5 % acetate	Beukes and Pletschke (2010)
	Lime pretreated	32.8 % lignin, 1.0 % arabinose, 34 % glucose, 15 % xylose, 5.6 % acetate	
	NH_4OH	8.8 % lignin, 1.24 % arabinose, 54.64 % glucose, 15.46 % xylose	Beukes and Pletschke (2011)
	NaOH	15.65 % lignin, 0 % arabinose, 42.88 % glucose, 17.5 % xylose	

groups on acetylated xylan have been demonstrated to hinder the hydrolysis of xylan, contributing to substrate recalcitrance (Selig et al. 2009). Once alkaline pretreatments have been carried out, acetate groups are generally removed, as can be observed from Table 5.3 for the NH_4OH and NaOH treatments of bagasse, although only a limited reduction in acetate was observed with lime pretreatment.

Combinations of EngE and ManA also displayed synergistic interactions in all cases where these enzymes were used together (Olver et al. 2011; Beukes et al. 2008). This indicates that mannan and cellulose interact within the three-dimensional structure of the lignocellulose substrate and probably hydrogen bonds with the cellulose in a similar manner to xylan. This results in the mannan masking the cellulose and preventing access of the cellulases to their substrate. Combining these enzymes in assays on lignocellulose substrates produces a synergistic interaction, even when negligible amounts of mannan appeared to be present within the substrate. The degrees of synergy found were generally lower than the interactions between EngE and XynA, except in pineapple pulp which probably contains high levels of mannan. This is in agreement with Tenkanen et al. (1999) who indicated that the effect of mannanase on cellulose hydrolysis is much smaller than the effect of xylanase. Varnai et al. (2011) also found a linear relationship between the hydrolysis of mannan and cellulose, indicating that the mannan and xylan formed a "network of polysaccharides around the cellulosic fibres". They also found that the addition of a mannanase had an impact on hydrolysis of the substrate even when the mannan content was only 0.23 % of the dry weight. This may indicate a requirement for the addition of an endo-mannanase to an optimised enzyme cocktail for certain mannan-containing substrates such as softwoods.

Further synergistic relationships that were observed in studies in our laboratory occurred between different depolymerising hemicellulases, such as XynA, ManA and ArfA. ArfA acts as both a depolymerising hemicellulase (on arabinan) and a debranching enzyme (on arabinoxylan).

Limited studies have been carried out on interactions between depolymerising hemicellulases. Our studies demonstrate that XynA and ManA combinations all displayed synergistic interactions, indicating that some association exists between mannan and xylan within lignocellulose substrates. In one case, on NaOH pretreated bagasse, no synergy was observed between XynA and ManA, although high levels of synergy were present for this enzyme combination on untreated bagasse which were lower in the lime pretreated bagasse. The absence of synergy on NaOH pretreated bagasse indicates that the pretreatment disrupted the association between the xylan and mannan within the substrate. It is interesting that the three different alkaline pretreatments displayed different results with respect to the xylan and mannan association. Varnai et al. (2011) also found a synergistic relationship between mannanase and xylanase and concluded that hydrolysis of mannan improved the hydrolysis of xylan. They concluded that this indicated that the xylan and mannan were also closely associated within the substrate and "partially cover each other". The exact nature of the association is not clear.

Synergy between ArfA and ManA on lime pretreated sugar beet pulp was found by Dredge et al. (2011) although the degree of synergy was low (1.36). In the case of sugar beet pulp, the main hemicellulose in this substrate is arabinan (see Table 5.3), and thus, the arabinose residues are not present as substituents on xylan, but as polymers of arabinose. As the ArfA has prominent activity on arabinase, it was able to depolymerise the arabinan. Based on the synergy found between ArfA and ManA, it is therefore hypothesised that the arabinan and mannan are also closely associated within the sugar beet pulp substrate.

Further hemicellulase interactions found in our studies included combinations of ArfA, ManA and XynA, as well as ArfA and XynA. ArfA, acting as an α-L-arabinofuranosidase, cleaves arabinose substituents from arabinoxylan. There are several reports in literature confirming the synergistic interaction between these two enzymes (Shi et al. 2010). The removal of arabinose substituents by the ArfA assists the action of the XynA on the xylan backbone.

Conclusions and Gaps in the Field

Enzyme synergy studies are an important tool to demonstrate the dependence of enzymes on mutual cooperation with other enzymes to degrade the same substrate, such as arabinoxylan, as well as substrates that are closely associated within a larger network, such as lignocellulose. Our studies have demonstrated the importance of hemicellulases, in combination with cellulases or other hemicellulases, cooperating in order to degrade complex substrates. The role of hemicellulases, such as mannanases, in combination with cellulases, is still an important area that has received limited attention. Similarly, the cooperation required between different depolymerising hemicellulases has received very limited attention and can provide an understanding of the mutual association between different hemicellulose substrates within the framework of lignocellulose. This is an important area requiring further investigation.

Acknowledgments JS van Dyk was supported by the Claude Leon Foundation. This material is based upon work supported by the National Research Foundation (NRF) of South Africa and Rhodes University Joint Research Committee (JRC). Any opinion, findings and conclusions or recommendations expressed in this material are those of the author(s) and therefore the NRF does not accept any liability in regard thereto.

References

Albersheim P, Darvill A, Roberts K, Sederoff R, Staehelin A (2011) Plant cell walls: from chemistry to biology. Garland Science, Taylor & Francis Group, New York

Andersen N, Johansen KS, Michelsen M, Stenby EH, Krogh KB, Olsson L (2008) Hydrolysis of cellulose using mono-component enzymes shows synergy during hydrolysis of phosphoric acid swollen cellulose (PASC), but competition on Avicel. Enzyme Microb Technol 42:362–370

Banerjee G, Scott-Craig JS, Walton JD (2010) Improving enzymes for biomass conversion: a basic research perspective. Bioenerg Res 3:82–92

Ban-Koffi L, Han YW (1990) Alcohol production from pineapple waste. World J Microbiol Biotechnol 6(3): 281–284

Beukes N, Chan H, Doi RH, Pletschke BI (2008) Synergistic associations between *Clostridium cellulovorans* enzymes XynA, ManA and EngE against sugarcane bagasse. Enzyme Microb Technol 42:492–498

Beukes N, Pletschke BI (2010) Effect of lime pre-treatment on the synergistic hydrolysis of sugarcane bagasse by hemicellulases. Bioresour Technol 101:4472–4478

Beukes N, Pletschke BI (2011) Effect of alkaline pre-treatment on enzyme synergy for efficient hemicellulose hydrolysis in sugarcane bagasse. Bioresour Technol 102:5207–5213

Bura R, Chandra R, Saddler J (2009) Influence of xylan on the enzymatic hydrolysis of steam-pretreated corn stover and hybrid poplar. Biotechnol Prog 25(2):315–322

Dredge R, Radloff SE, Van Dyk JS, Pletschke BI (2011) Lime pretreatment of sugar beet pulp and evaluation of synergy between ArfA, ManA and XynA from *Clostridium cellulovorans* on the pretreated substrate. 3 Biotech 1:151–159

Garcia-Aparicio MP, Ballesteros M, Manzanares P, Ballesteros I, Gonzalez A, Negro MJ (2007) Xylanase contribution to the efficiency of cellulose enzymatic hydrolysis of barley straw. Appl Biochem Biotechnol 136–140:353–365

Gomez LD, Steele-King CG, McQueen-Mason SJ (2008) Sustainable liquid biofuels from biomass: the writing's on the wall. New Phytol 178:473–485

Hendriks AW, Zeeman G (2009) Pretreatments to enhance the digestibility of lignocellulosic biomass. Bioresour Technol 100:10–18

Howard RL, Abotsi E, Jansen van Rensburg EL, Howard S (2003) Lignocellulose biotechnology: issues of bioconversion and enzyme production. Afr J Biotechnol 2(12):602–619

Kolenova K, Vrsanska M, Biely P (2006) Mode of action of endo-β-1,4-xylanases of families 10 and 11 on acidic xylooligosaccharides. J Biotechnol 121:338–345

Kosugi A, Murashima K, Doi RH (2002a) Characterization of two noncellulosomal subunits, ArfA and BgaA, from *Clostridium cellulovorans* that cooperate with the cellulosome in plant cell wall degradation. J Bacteriol 184(24):6859–6865

Kosugi A, Murashima K, Doi RH (2002b) Xylanase and acetyl xylan esterase activities of XynA, a key subunit of the *Clostridium cellulovorans* cellulosome for xylan degradation. Appl Environ Microbiol 68(12):6399–6402

Kumar R, Wyman CE (2009) Effects of cellulase and xylanase enzymes on the deconstruction of solids from pretreatment of poplar by leading technologies. Biotechnol Prog 25:302–314

Merino ST, Cherry J (2007) Progress and challenges in enzyme development for biomass utilization. Adv Biochem Eng/Biotechnol 108:95–120

Micard V, Renard CMGC, Thibault JF (1996) Enzymatic saccharification of sugar-beet pulp. Enzyme Microb Technol 19:162–170

Murashima K, Kosugi A, Doi RH (2003) Synergistic effects of cellulosomal xylanase and cellulases from *Clostridium cellulovorans* on plant cell wall degradation. J Bacteriol 185(5):1518–1524

Olver B, Van Dyk JS, Beukes N, Pletschke BI (2011) Synergy between EngE, XynA and ManA from *Clostridium cellulovorans* on corn stalk, grass and pineapple pulp substrates. 3 Biotech 1:187–192

Raweesri P, Riangrungrojana P, Pinphanichakam P (2008) α-L-arabinofuranosidase from *Streptomyces* sp. PC22: Purification, characterization and its synergistic action with xylanolytic enzymes in the degradation of xylan and agricultural residues. Bioresour Technol 99:8981–8986

Selig MJ, Knoshaug EP, Adney WS, Himmel ME, Decker SR (2008) Synergistic enhancement of cellobiohydrolase performance on pretreated corn stover by addition of xylanase and esterase activities. Bioresour Technol 99:4997–5005

Selig MJ, Adney WS, Himmel ME, Decker SR (2009) The impact of cell wall acetylation on corn stover hydrolysis by cellulolytic and xylanolytic enzymes. Cellulose 16:711–722

Shi P, Li N, Yang P, Wang Y, Luo H, Bai Y, Yao B (2010) Gene cloning, expression, and characterization of a Family 51 α-L-arabinofuranosidase from *Streptomyces* sp. S9. Appl Biochem Biotechnol 162:707–718

Tamaru Y, Doi RH (1999) Three surface layer homology domains at the N terminus of the *Clostridium cellulovorans* major cellulosomal subunit EngE. J Bacteriol 181(10):3270–3276

Tamaru Y, Karita S, Ibrahim A, Chan H, Doi RH (2000) A large gene cluster for the *Clostridium cellulovorans* cellulosome. J Bacteriol 182(20):5906–5910

Tenkanen M, Tamminen T, Hortling B (1999) Investigation of lignin-carbohydrate complexes in kraft pulps by selective enzymatic treatments. Appl Microbiol Biotechnol 51:241–248

Varnai A, Siika-aho M, Viikari L (2010) Restriction of the enzymatic hydrolysis of steam-pretreated spruce by lignin and hemicellulose. Enzyme Microb Technol 46:185–193

Várnai A, Huikko L, Pere J, Siika-aho M, Viikari L (2011) Synergistic action of xylanase and mannanase improves the total hydrolysis of softwood. Bioresour Technol 102:9096–9104

Manganese Peroxidases: Molecular Diversity, Heterologous Expression, and Applications

6

Samta Saroj, Pragati Agarwal, Swati Dubey, and R.P. Singh

Abstract

Manganese peroxidases (MnPs) are a fascinating group of biocatalysts with various ecological and biotechnological implications. They are involved in the biodegradation of lignocellulose and lignin and participate in the bioconversion of other diverse recalcitrant compounds, like polycyclic aromatic hydrocarbons, chlorophenols, industrial effluents (mostly from the paper and pulp), and textile and petrochemical industries, and bioremediation of contaminated soils. This chapter presents an overview of the structural basis of the catalytic properties of MnPs and the enumeration of the molecular and protein homology characteristics of this enzyme. Multiple developments mainly pertaining to enzyme engineering for improved substrate specificity and stability of MnPs have also been highlighted. Inevitably, the progress in enzyme engineering research and the expression of MnPs have explored the vast genetic diversity of these enzymes with great interest being placed on exploiting these enzymes for a variety of industrial and scientific applications.

Keywords

Manganese peroxidases • Properties • Heterologous production • Molecular cloning • Enzyme engineering research • Crystal structure • Industrial applications

Introduction

Manganese peroxidase (MnP) [EC 1.11.1.13, Mn^{II}:hydrogen-peroxide oxidoreductase] is an extracellular heme enzyme that utilizes hydrogen peroxide (H_2O_2) as an electron-accepting cosubstrate, for catalyzing the peroxide-dependent oxidation of Mn^{II} to Mn^{III}. These enzymes have mainly been isolated from white-rot fungal species like *Phanerochaete chrysosporium, Trametes versicolor, Heterobasidion annosum, and Irpex lacteus* with the ability to degrade lignin (Eriksson et al. 1990; Cai and Tien 1993). A great majority of these enzymes contain a protoporphyrin IX (heme) prosthetic group. Lignin is a het-

S. Saroj • P. Agarwal • S. Dubey • R.P. Singh (✉)
Department of Biotechnology, Indian Institute of Technology Roorkee, Roorkee 247667, India
e-mail: rpsbsfbs@iitr.ernet.in

P. Shukla and B.I. Pletschke (eds.), *Advances in Enzyme Biotechnology*, DOI 10.1007/978-81-322-1094-8_6, © Springer India 2013

erogeneous, optically inactive polymer, consisting of phenylpropanoid subunits. These phenylpropanoid subunits cannot be cleaved by hydrolytic enzymes unlike most other natural polymers, e.g., cellulose, starch, and proteins. Interestingly, some white-rot basidiomycetes that produce peroxidases have ability to degrade lignin (Sarkanen and Ludwig 1971). Peroxidases were first discovered in *Phanerochaete chrysosporium* (Kuwahara et al. 1984; Glenn and Gold 1985; Paszcynski et al. 1985, 1986). During the last decade, work has been done on the heterologous expression of peroxidases using X-ray crystallographic studies and active-site engineering to enhance substrate specificity and the thermal stability of MnP genes (Mino et al. 1988; Benner and Gerloff 1990; Petersen et al. 1993; Li et al. 2001; Sundaramoorthy et al. 2010). Attempts have also been made to determine the regulation of MnP gene at transcriptional and translational levels (Brown et al. 1991; Gettemy et al. 1998; Johansson et al. 2002). The nonspecific and non-stereoselective nature of MnP allows it to degrade a wide range of pollutants, such as polycyclic aromatic hydrocarbons (PAHs), chlorinated phenols, polychlorinated biphenyls, dioxins, pesticides, explosives, and dyes (Levin et al. 2004).

The main objective of this chapter has been to highlight the diversity of MnPs among basidiomycetes, their heterologous production, phylogenetic analysis, structural characteristics, and molecular features. This chapter also describes attempts made to engineer this enzyme for improved substrate specificity and stability and to quantify the utility of this enzyme.

Occurrence and Phylogenetic Analysis of MnPs

Production of extracellular MnP has been mainly observed from certain basidiomycetes, and thus far, no bacterium, yeast/mold, or mycorrhiza-forming basidiomycete have been reported to produce this enzyme (Cairney and Burke 1998; Hatakka 2001). Many ecophysiological groups of basidiomycetes have been found to secrete isoforms of MnP into their microenvironments (Hatakka 1994, 2001; Heinzkill et al. 1998; Steffen et al. 2000). *C. subvermispora* have been found to produce up to 11 different isoforms of MnP (Lobos et al. 1994; Urzua et al. 1995). Various white-rot fungi, which are well characterized for their ligninolytic ability, belong to phylogenetically older families such as *Meruliaceae* (*P. radiata, P. sordida, P. chrysosporium, Merulius* sp. M15), *Coriolaceae* (*B. adusta, C. subvermispora, C. pruinosum, P. tephropora, A. biennis*), and *Polyporaceae* (*T. versicolor, T. gibbosa, T. trogii, T. hirsuta*) as well as litter decomposers of euagaric families such as *Strophariaceae* and *Tricholomataceae* have been found to have notable expression of MnP (Table 6.1). In addition, some marine-derived fungal strains and strains dwelling on decaying sea grass (Raghukumar et al. 1999), cooling-tower wood (Schmidt et al. 1997), and brown coal (Willmann and Fakoussa 1997a, b) have MnP production ability. Sequences of different MnPs were retrieved from GenBank (available at: http://www.ncbi.nlm.nih.gov). For phylogenetic analysis of all MnPs, an alignment was created with ClustalW2 multiple sequence alignment tool (available at http://www.ebi.ac.uk/Tools/msa/clustalw2/). The evolutionary history was determined using the neighbor-joining method (Saitou and Nei 1987). The percentage of replicate trees in which the associated taxa clustered together in the bootstrap test (1,000 replicates) is shown next to the branches (Felsenstein 1985). The evolutionary distances were computed using the Poisson correction method (Zuckerkand and Pauling 1965) and are in the units of the number of amino acid substitutions per site. The analysis involved 53 amino acid sequences. Evolutionary analyzes were conducted in MEGA5 (Tamura et al. 2011). It can be inferred from the phylogenetic tree that MnPs are divergent among different taxonomic groups. Among the 54 amino acid sequences analyzed, 44 basidiomycetes were grouped together at the same phenetic unit, whereas apart from main basidiomycete grouping, *Coprinopsis cinerea*, *Arthromyces ramosus*, *Coprinellus disseminatus*, and *Ganoderma* sp. arose as distinct genetic groups (Fig. 6.1). The MnP variant from *Inonotus hispidus* was the single gene variation ungrouped

Table 6.1 Amino acid sequences retrieved from GenBank and used for alignment and phylogenetic analyzes

S. No.	Species	Phylum	GenBank acc. no.
1.	*Ganoderma applanatum*	Basidiomycota	BAA88392.1
2.	*Ganoderma australe*	Basidiomycota	ABB77244.1
3.	*Ganoderma formosanum*	Basidiomycota	ABB77243.1
4.	*Ganoderma lucidum*	Basidiomycota	ACA48488.1
5.	*Trametes versicolor*	Basidiomycota	CAA83148.1
6.	*Phanerochaete chrysosporium*	Basidiomycota	AAA33743.1
7.	*Pleurotus* sp. *"Florida"*	Basidiomycota	CAB51617.1
8.	*Phylloporia ribis*	Basidiomycota	ADK60897.1
9.	*Pleurotus pulmonarius*	Basidiomycota	AAY42945.1
10.	*Coprinopsis cinerea*	Basidiomycota	CAA49216.1
11.	*Lenzites gibbosa*	Basidiomycota	AEX01147.1
12.	*Bjerkandera adusta*	Basidiomycota	AAY89586.1
13.	*Ceriporiopsis rivulosa*	Basidiomycota	ABB83813.1
14.	*Trametes gibbosa*	Basidiomycota	ADW83732.1
15.	*Agrocybe praecox*	Basidiomycota	ADW41627.1
16.	*Spongipellis* sp. *FERM P-18171*	Basidiomycota	BAE79812.1
18.	*Phlebia radiata*	Basidiomycota	CAC84573.1
19.	*Hericium erinaceus*	Basidiomycota	ADK26471.3
20.	*Agaricus bisporus*	Basidiomycota	CAG27835.1
21.	*Phlebia* sp. *MG60*	Basidiomycota	BAG12560.1
22.	*Arthromyces ramosus*	Basidiomycota	BAA09861.1
23.	*Laccaria bicolor S238N-H82*	Basidiomycota	XP_001888065.1
24.	*Coprinellus disseminatus*	Basidiomycota	AAZ14938.1
25.	*Lentinula edodes*	Basidiomycota	BAG72080.1
26.	*Phlebia* sp. *b19*	Basidiomycota	ABR66918.1
27.	*Phanerochaete flavidoalba*	Basidiomycota	AAM46826.1
28.	*Phanerochaete sordida*	Basidiomycota	BAC06187.1
29.	*Dichomitus squalens*	Basidiomycota	AAF31330.1
30.	*Phellinidium ferrugineofuscum*	Basidiomycota	ACX51165.1
31.	*Lactarius rufus*	Basidiomycota	ACX51162.1
32.	*Lactarius fulvissimus*	Basidiomycota	ACX51161.1
33.	*Hypholoma fasciculare*	Basidiomycota	ACX51160.1
34.	*Hygrophorus agathosmus*	Basidiomycota	ACX51158.1
35.	*Gomphus clavatus*	Basidiomycota	ACX51157.1
36.	*Cortinarius traganus*	Basidiomycota	ACX51156.1
37.	*Cortinarius malachius*	Basidiomycota	ACX51154.1
38.	*Cortinarius infractus*	Basidiomycota	ACX51153.1
39.	*Cortinarius hinnuleus*	Basidiomycota	ACX51152.1
40.	*Cortinarius armillatus*	Basidiomycota	ACX51151.1
41.	*Inonotus hispidus*	Basidiomycota	ADK60893.1
42.	*Fomitiporia mediterranea*	Basidiomycota	ADK60890.1
43.	*Coriolopsis gallica*	Basidiomycota	AAZ16493.1
44.	*Phlebia albomellea*	Basidiomycota	ABT17238.1
45.	*Hymenochaete corrugata*	Basidiomycota	ADK60895.1
46.	*Polyporus brumalis*	Basidiomycota	AEJ38000.1
47.	*Cytidia salicina*	Basidiomycota	ABT17236.1
48.	*Phlebia chrysocreas*	Basidiomycota	ABT17228.1
49.	*Phlebiopsis gigantea*	Basidiomycota	ABT17220.1

(continued)

Table 6.1 (continued)

S. No.	Species	Phylum	GenBank acc. no.
50.	*Pulcherricium caeruleum*	Basidiomycota	ABT17217.1
51.	*Cryptoporus volvatus*	Basidiomycota	ABT17214.1
52.	*Hapalopilus rutilans*	Basidiomycota	ABT17210.1
53.	*Trametes cinnabarina*	Basidiomycota	ADK60908.1
54.	*Pleurotus ostreatus*	Basidiomycota	AAA84397.1

from the main cluster as aforementioned. *Lactarius rufus* along with *Lactarius fulvissimus* formed the third additional genetic group.

The maximum parsimony (MP) method was used to analyze the evolutionary history among different basidiomycete MnPs. The bootstrap consensus tree obtained from 1,000 replicates is taken to represent the evolutionary history of the taxa analyzed (Felsenstein 1985). Branches corresponding to partitions reproduced in less than 50 % of bootstrap replicates are collapsed. The percentage of replicate trees in which the associated taxa clustered together in the bootstrap test (1,000 replicates) is shown next to the branches (Felsenstein 1985). The MP tree was obtained using the close-neighbor-interchange algorithm (Nei and Kumar 2000) with search level 1 in which the initial trees were obtained with the random addition of sequences (ten replicates). Parsimony analysis revealed four distinct main evolutionary lineages among all basidiomycetes (Fig. 6.2). The MnP from *Pleurotus ostreatus* and *Pleurotus* sp. *Florida* evidenced two additional lineages as they were grouped separately from the main cluster. The MnPs from *Pleurotus pulmonarius*, *Hygrophorus agathosmus*, *Laccaria bicolor S238N-H82*, *Agrocybe praecox*, *Phellinidium ferrugineofuscum*, and *Hymenochaete corrugata* are evolutionarily closely related, and they arose as the sister group to the other remaining groups (Fig. 6.2).

Mechanism of Catalysis

Manganese peroxidase (MnP) [EC 1.11.1.13, Mn^{II}:hydrogen-peroxide oxidoreductase, MnP] is an extracellular heme enzyme that utilizes hydrogen peroxide (H_2O_2) as electron-accepting cosubstrate for catalyzing the peroxide-dependent oxidation of Mn^{II} to Mn^{III}. The Mn^{III} formed is highly active, which in turn is stabilized by fungal chelators like oxalate and malonate. These chelators act as physiological regulators to the enzyme as they enhance the enzyme activity due to their ability to facilitate the dissociation of Mn^{III} from the enzyme. The role of oxalate as an extracellular buffering agent has also been reported. It facilitates the ability of the fungus to control the pH of its environment (Timofeevski and Aust 1997; Zapanta and Tien 1997). Calcium sequestration by these chelators acts to increase the pore size of the plant cell wall and assist in the penetration of enzyme molecules. Oxidation of oxalic acid by Mn^{III} produces a formate radical (HCO_2^-) that reacts with dioxygen to form superoxide (O_2^-) and subsequently H_2O_2 (Khindaria et al. 1994; Urzua et al. 1998). Chelated Mn^{III} in turn acts as low molecular weight, diffusible redox mediator that attacks phenolic lignin structures and monomeric phenols, e.g., azo dyes, resulting in the formation of unstable free radicals that tend to disintegrate spontaneously. Characteristic features of the catalytic cycle of MnP resemble to those of other heme-containing peroxidases, such as horseradish peroxidase (HRP), lignin peroxidase, versatile peroxidase, and chloroperoxidases (Cai and Tien 1993; Magliozzo and Marcinkeviciene 1997; Longoria et al. 2008). However, MnP is unique in its ability to utilize Mn^{II} as a reducing substrate to oxidize it to Mn^{III} (Kishi et al. 1994; Sundaramoorthy et al. 1997; Youngs et al. 2000; Deguchi et al. 2002). Spectroscopic studies have revealed that the heme iron of the native enzyme is in the ferric, high-spin, pentacoordinate state and is ligated to the proximal histidine (Mino et al. 1988; Wariishi et al. 1988; Gelpke et al. 2000). The reactions

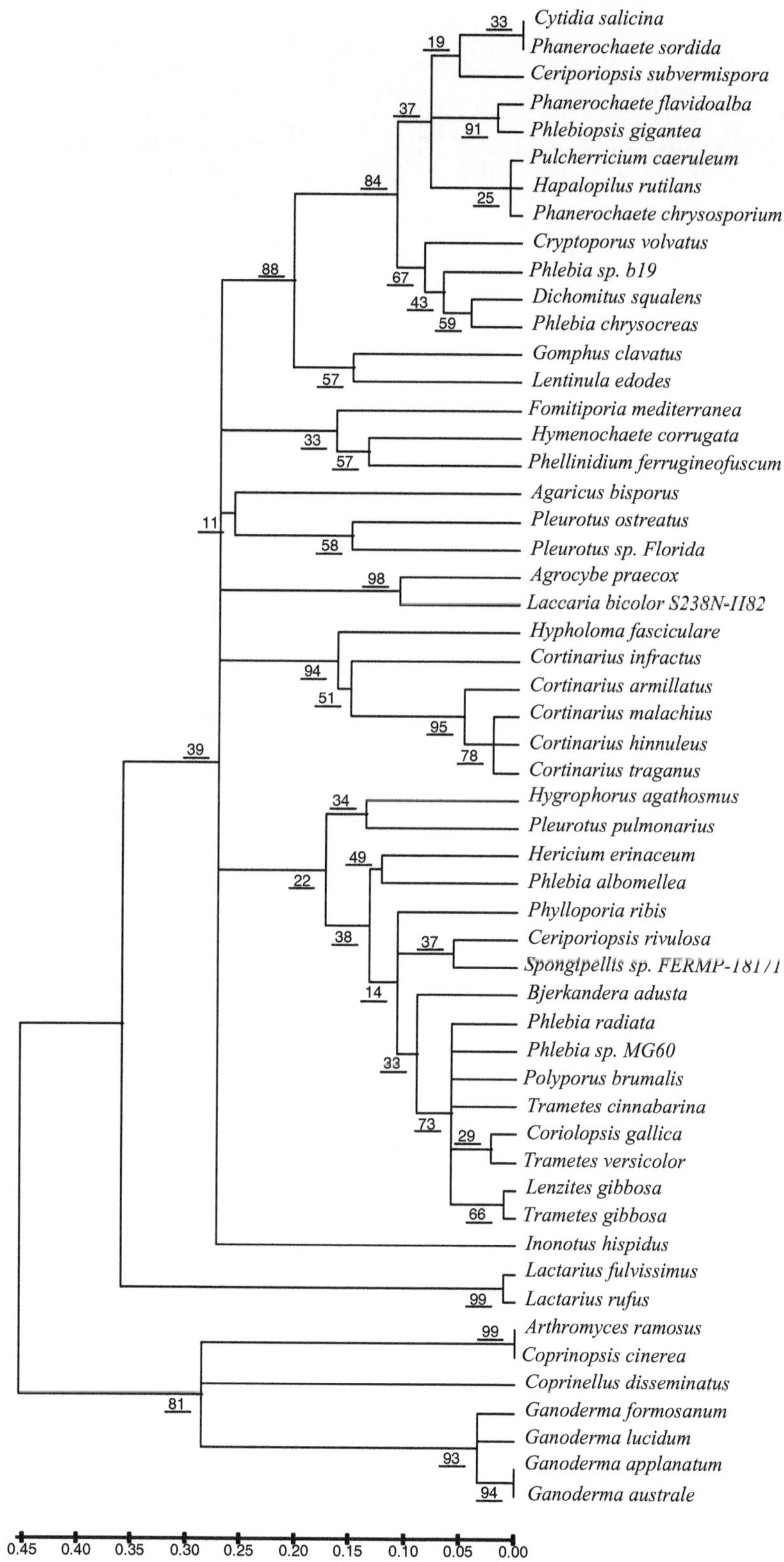

Fig. 6.1 Neighbor-joining tree of manganese peroxidases amino acid sequences. Numbers on the branches are bootstrap values (jackknife values in parentheses) obtained for 1,000 pseudoreplications

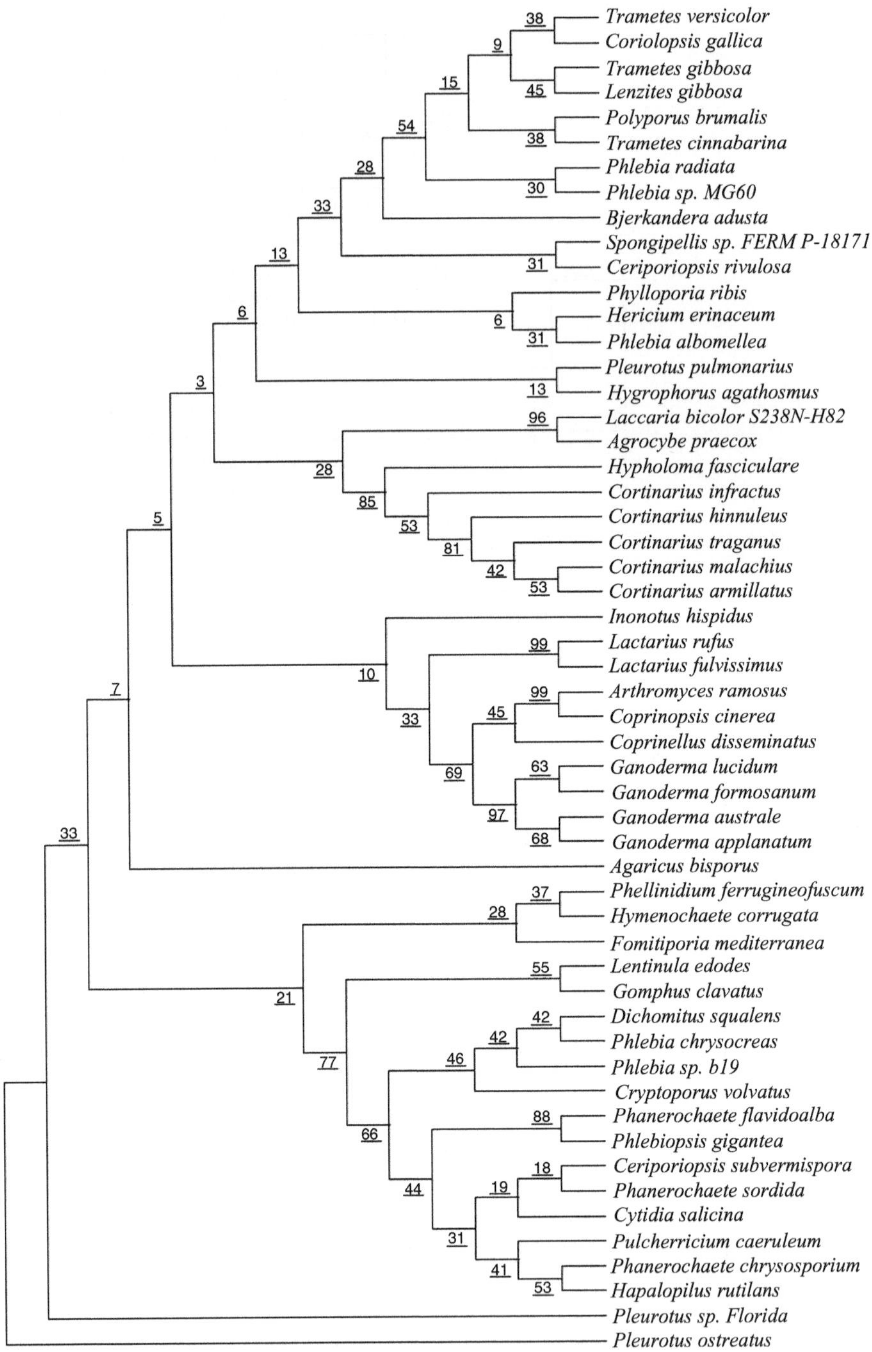

Fig. 6.2 Maximum parsimony analysis of taxa from MnP amino acid sequences. Numbers on the branches are bootstrap values (jackknife values in parentheses) obtained for 1,000 pseudoreplications

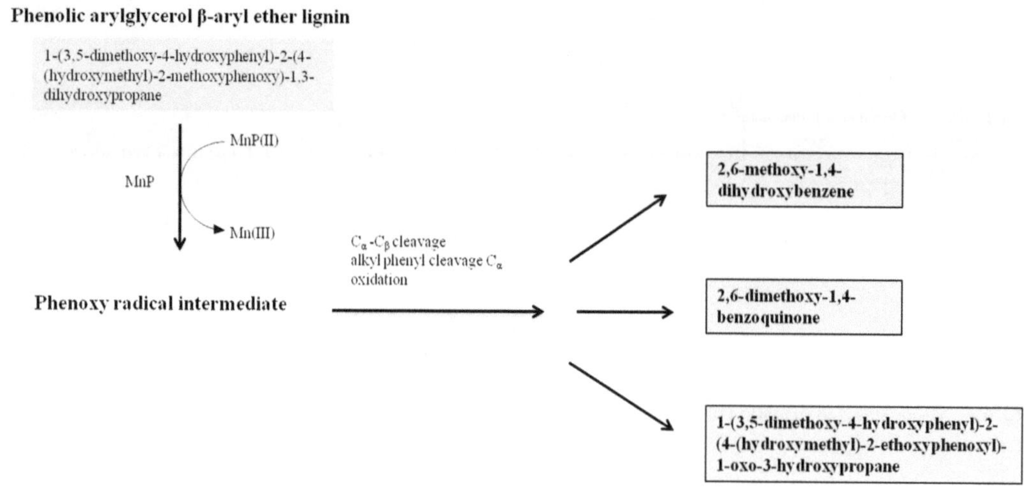

Fig. 6.3 MnP-catalyzed oxidation of phenolic arylglycerol β-aryl ether lignin model compound (Modified according to Tuor et al. 1992)

involved in the MnP catalytic cycle are (Wariishi et al. 1992; Gelpke et al. 1999)

$$MnP + H_2O_2 \rightarrow MnPI + H_2O$$

$$MnPI + Mn^{II} \rightarrow MnPII + Mn^{III}$$

$$MnPII + Mn^{II} \rightarrow MnP + Mn^{III} + H_2O$$

Catalysis of Phenolic Substrates

During the oxidation of phenolic compounds, phenoxy radical intermediates are formed which undergo rearrangements, bond cleavages, and nonenzymatic degradation to yield various breakdown products (Fig. 6.3) (Tuor et al. 1992). Mn^{III} generated by MnP is known to catalyze the oxidation of phenolic substrates, including simple phenols, amines, dyes, and also phenolic lignin substructure and dimers (Wariishi et al. 1989a; Urzua et al. 1995).

Catalysis of Non-Phenolic Substrates

In contrast to LiP-catalyzed reactions, which involve electron abstraction from the aromatic ring, forming a radical cation, Mn^{III} forms reactive radicals in the presence of a second mediator during the oxidation of non-phenolic substrates (Reddy et al. 2003). The presence of thiols, such as glutathione, mediates the oxidation of substituted benzyl alcohols and diarylpropane structures to their respective aldehydes by Mn^{III} (Reddy et al. 2003). In these reactions, thiols are oxidized to thiyl radicals by Mn^{III}, which subsequently removes hydrogen from the substrate to form a benzylic radical. The latter undergoes successive nonenzymatic reactions like addition of O_2 at C1 position of benzylic radical followed by loss of $^{\bullet}OOH$, and homolytic C–O fission at C2 of benzylic radical expels a phenoxy radical that results in the formation of final products (Fig. 6.4) (Wariishi et al. 1989b, c).

Manganese peroxidase generated Mn^{III} has also been coupled with peroxidation of lipids to catalyze C_α–C_β cleavage and to β-aryl ether cleavage of non-phenolic diarylpropane and β-O-4 lignin structures, respectively (Fig. 6.4) (Bao et al. 1994; Daina et al. 2002; Reddy et al. 2003; Kapich et al. 2005). The steps involved in the mechanism are the following: firstly, hydrogen abstraction from the benzylic carbon via lipid peroxy radicals, and secondly, peroxy radicals are formed by addition of O_2 and subsequent oxidative cleavage and nonenzymatic degradation. Absence of exogenous H_2O_2 directs the enzyme to oxidize

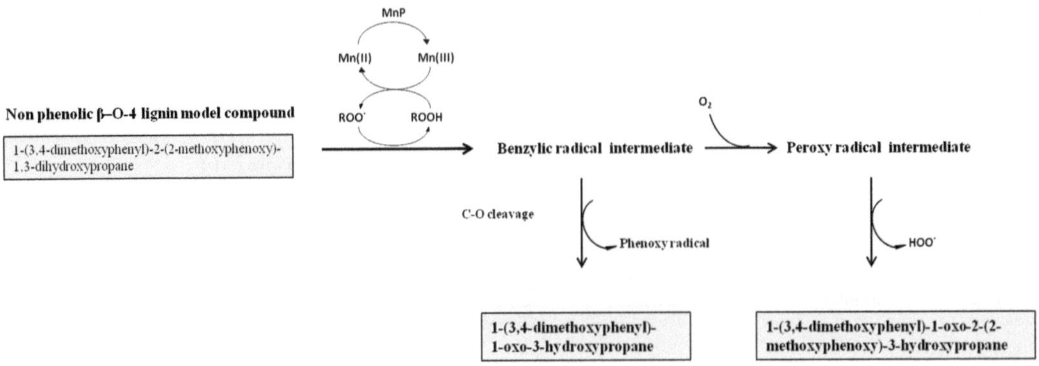

Fig. 6.4 MnP-catalyzed oxidation of non-phenolic β-O-4 lignin model compound (Modified according to Wong 2009)

nicotinamide adenine dinucleotide phosphate (NADPH) (reduced form), glutathione, dithiothreitol, and dihydroxymaleic acid to generate H_2O_2. One can infer (by observing the oxidase activity of MnP) that the H_2O_2 produced may become available for the enzyme to start the peroxidase cycle, thereby, assisting in the degradation of lignin by this fungal species (Paszczynski et al. 1986).

Molecular Cloning and Expression of MnP

In most fungi, MnP appears to be produced as a family of isoenzymes, which may be encoded by structurally related genes (Larrondo et al. 2001; Sakamoto et al. 2009). Commercial applications of MnP require significantly higher levels of extracellular enzyme production by fungal strains; however, the yield of MnP in its native hosts is too low (1.5–5 mgl^{-1}) (Stewart et al. 1996; Li et al. 2001). Therefore, improvement in the yield and reduction in the production cost are the major goals to be focused for commercial exploitation of MnP.

The first extracellular fungal MnP that was characterized and homologously expressed was obtained from *Phanerochaete chrysosporium* (Glenn and Gold 1985). It was found that high concentrations of carbon and nitrogen in the medium significantly affected the transcription of the *mnp*1 gene in *P. chrysosporium* (Mayfield et al. 1994). Sakamoto et al. (2009) studied the transcriptional and translational levels of different isoforms of MnP, i.e., *lemnp*1 and *lemnp*2, in *Lentinula edodes* using sawdust medium and reported *lemnp2* as the major extracellular enzyme.

Heterologous expression of MnP from *Phanerochaete chrysosporium* has been reported in *Aspergillus niger, Pichia pastoris, P. chrysosporium adel, Aspergillus oryzae, and Aspergillus nidulans* (Pribnow et al. 1989; Mayfield et al. 1994; Stewart et al. 1996; Janse et al. 1998; Larrondo et al. 2001; Gu et al. 2003). MnPs from *Trametes versicolor, P. eryngii, P. ostreatus, Dichomitus squalens*, and *Ceriporiopsis subvermispora* have also been cloned and heterologously expressed in *T. versicolor* 9522-1, *Coprinus cinereus, P. chrysosporium*, and *Aspergillus nidulans*, respectively (Ogawa et al. 1998; Camarero et al. 2000; Li et al. 2001; Larrondo et al. 2001; Kim et al. 2005; Yeo et al. 2007). To date there have been no reports of successful expression of fungal MnPs in bacterial expression hosts.

Although the MnP production levels have often been improved significantly by expression in heterologous hosts, the reported levels are still rather low (100–400 mgl^{-1}) for use in industrial applications (Conesa et al. 2000; Punt et al. 2002; Espinosa et al. 2012). The capability of the *P. pastoris* strain to perform various posttranslational modifications, such as heme insertion, glycosylation, folding, and protein secretion, had been reported for successful production of active MnP to a maximum yield of 120 UL^{-1} (Gu et al. 2003). As described by Yeo et al. (2007), increase in MnP activity of transformant TF6 was up to 45 % as compared to the recipient strain. A thermostable recombinant MnP from *D. squalens* has also been heterologously expressed in *P. chrysosporium* and purified. The recombinant protein appeared similar in kinetic and spectral characteristics to the wild-type MnP from *D. squalens* (Li et al. 2001), whereas the homologous expression of *P. chrysosporium* recombinant MnP resulted in 30 % of the level of MnP activity expressed under by wild-type strain (Mayfield et al. 1994). Data indicated that the addition of exogenous Mn^{II}, Cd^{II}, and Zn^{II} conferred additional thermal stability to MnP from *D. squalens and P. chrysosporium* (Li et al. 2001). *Aspergillus* species have proven to be excellent hosts for the expression of heterologous proteins such as those of *A. oryzae.* The same has been shown to be effective in the expression of *P. chrysosporium mnp1* with notable yields (Stewart et al. 1996).

Isozyme multiplicity of MnPs has been observed in various strains of *C. subvermispora* and *P. chrysosporium.* As an expression host, *Aspergillus nidulans* proved to be convenient system for MnPs. It has also been demonstrated that Mn^{II} is the key component that regulates the transcription of different recombinant MnP isoforms in carbon-limited cultures (Banci et al. 1992; Alic et al. 1997; Larrondo et al. 2001). Nutrient limitation such as Mn concentration, culture agitation, heat shock, H_2O_2 concentration, and other chemical stresses have also been reported to significantly regulate transcription of different isozymes of MnP (*mnp1, mnp2, and mnp3*) in *P. chrysosporium* (Janse et al. 1998). In contrast to manganese regulation, Tello et al. (2000) have proposed that the putative MREs (metal response elements) found in the upstream region of MnP genes in *P. chrysosporium* might have a role in the regulation of transcription of genes coding for MnP in filamentous fungi. Metallothionein genes are known to be regulated by MREs through various metals in animal cells, although these sites do not respond to manganese. Therefore, further work is required to decipher both the role of MREs in the upstream region of these genes and the mechanism of transcriptional regulation by manganese in basidiomycetes.

Characteristic Features of MnP Gene

MnP genes among different basidiomycetes (starting from the ATG codon) encompass a genomic region of 1.4–1.9 kbps. Lobos et al. (1998) had shown that MnP genes contain seven short intervening sequences with sizes ranging between 52 and 60 bp. The last intron restrained by *mnp1* and *mnp2* genes of *P. chrysosporium* segregates a codon for proline to give different isozymes (Fig. 6.5) (Godfrey et al. 1990; Mayfield et al. 1994). Sequences at the intron splicing junctions adhere to the GT–AG rule. In turn, three putative internal lariat formation sites match the consensus sequence CTRAY (Padgett et al. 1989). After structural comparison of *mnp* genes from the *Cs-mnp1* gene of *C. subvermispora* and five *mnp* genes from different basidiomycetes, Lobos et al. (1994) revealed an almost perfect alignment between Cs-*mnp1* and *mnp2* of *P. chrysosporium* (Fig. 6.5), and both genes have an additional intron splitting exon 3 of *mnp1* and *mnp3* of *P. chrysosporium* at the codon for the distal histidine, H46 (Mayfield et al. 1994). The pattern of differential distribution of introns observed in the *P. chrysosporium mnp* genes might be because the *mnp2* represents the ancestral gene structure and the *mnp1*and *mnp3* genes arose with the loss of one intron each (Alic et al. 1997). In contrast, only one of the 15 introns (intron 12) in the *P. ostreatus mnp* gene aligns exactly with a *C. subvermispora mnp1* intron (intron 6), and none of the 15 introns of the *P. ostreatus mnp* gene align precisely with introns of the *P. chrysosporium mnp* genes.

The regulatory sequence of *Cs-mnp2B* contains a TATA box and an inverted CAAT (ATTG) element located 92 and 191 bp upstream of the ATG codon (Tello et al. 2000), respectively. Examination of the promoter regions of the *mnp1* and *mnp2* genes (Godfrey et al. 1990; Gold and Alic 1993; Mayfield et al. 1994) revealed the presence of putative MREs within 800 bp of the translation initiation codon. These sequences are identical

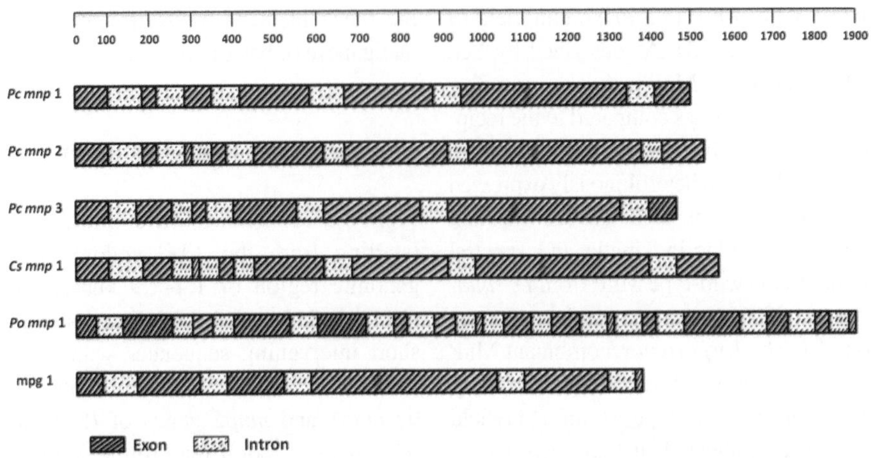

Fig. 6.5 Intron/exon structure of MnP genes *mnp1*, *mnp2*, and *mnp3* from *P. chrysosporium*, *Cs-mnp1* from *C. subvermispora*, *mnp* from *P. ostreatus*, and *mpg1* from *T. versicolor*. The exons are indicated by *open boxes*, whereas the *solid black boxes* correspond to the introns (Modified after Lobos et al. 1998)

to cis-acting MRE sequences responsible for heavy-metal induction of animal cell metallothionein genes (Gettemy et al. 1998). Interestingly, closer examination of the *mnp1* promoter region also revealed the presence of putative HSEs within 400 bp upstream of the *mnp1* translation initiation codon (Lobos et al. 1998). Mn^{2+} regulation of MnPs has been previously highlighted by different research groups (Bonnarme and Jeffries 1990). Godfrey et al. (1990) stated that 1,500 bp of sequence immediately upstream of the MnP translation start site is sufficient to regulate the *ural* reporter in a manner analogous to the regulation of the endogenous MnP genes with respect to Mn, nutrient nitrogen levels, and metabolic phase of growth.

The translocation of 48-bp fragment of promoter region of MnP isozyme 1 to a site 120 bp downstream of its original location has been shown to regulate Mn^{2+}-dependent expression of downstream genes; this suggests the possibility of the presence of at least one Mn^{2+}-responsive *cis* element in the fragment. However, deletion of a 48-bp fragment, located at 521 bp upstream of the translation start codon in the *mnp1* promoter, or replacement of this fragment with an unrelated sequence resulted in *egfp* expression under nitrogen limitation, both in the absence and presence of exogenous Mn^{2+} (Ma et al. 2004).

Orth and coworkers (1994) have analyzed the organization of the *MnP* gene family of *P. chrysosporium* BKM1767 and concluded that the *λMP-1* and *λMP-2* genes hybridized to 3.6 and 3.8 Mb of DNA fragments located on separate chromosomes and in contrast to five LiP genes that are localized to a dimorphic chromosome of about 3.7 and 3.5 Mb (Gaskell et al. 1991).

Crystallographic Analysis of MnP

MnP from different basidiomycetes has been crystallized and subsequently analyzed (Poulos et al. 1993; Sundaramoorthy et al. 1994a, b, 1995; Duenas et al. 1999). MnP is a glycoprotein with the molecular weight of 46 kDa and contains one heme group (Sundaramoorthy et al. 1994a, b). The structural features of manganese peroxidase (pdb 1mnp) from *P. chrysosporium* are displayed/shown in Fig. 6.6, with a resolution of 2.06 Å (retrieved from http://www.rcsb.org/pdb). The substrate-bound MnP (Mn–MnP) consists of 357 amino acids, three sugar residues, a heme prosthetic group, two structural calcium ions, substrate Mn^{II} ion, and 478 solvent molecules, including two glycerol molecules (Sundaramoorthy et al. 2010). The sequence of the heme distal helix Glu35–Ala48 is highly conserved in the key residues (Selvaggini

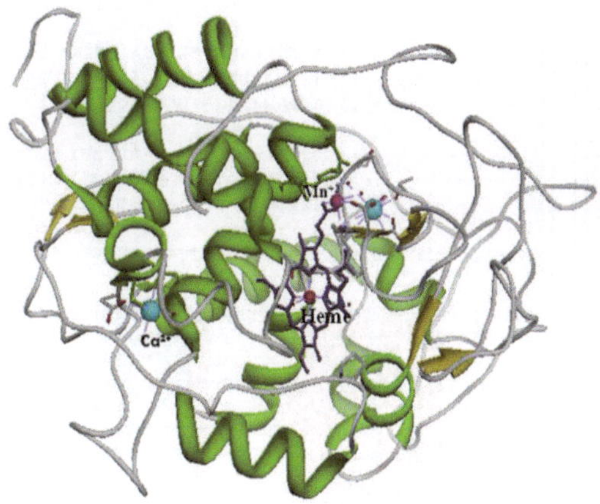

Fig. 6.6 The overall structure of *P. chrysosporium* MnP (pdb 1mnp) as analyzed by UCSF chimera 1.4.1. The cyan spheres are structural Ca^{II} ions conserved in extracellular heme peroxidases. The location of the substrate, Mn^{+2}, near the heme, is indicated. The *purple color* shows active-site structure of MnP. This architecture is highly conserved in heme peroxidase

et al. 1995). The active site consists of a proximal His ligand H-bonded to an Asp residue, and a distal side peroxide-binding pocket consisting of a catalytic His and Arg is the same among all peroxidases (Fig. 6.7). MnP differs with respect to having five rather than four disulfide bonds. The additional disulfide bond is located near the C-terminus of the polypeptide chain. The ligands constituting the Mn^{2+} binding site include Asp179, Glu35, Glu39, a heme propionate, and two water molecules. The overall structure is similar to that of two other fungal peroxidases, i.e., lignin peroxidase from *Phanerochaete chrysosporium* and *Arthromyces ramosus* peroxidase. Like the other fungal peroxidases, MnP also has two structural calcium ions and N-acetylglucosamine residues N-linked to Asn131 (Sundaramoorthy et al. 1994a, b).

Proximal pocket of MnP active site contains His173 that coordinates to heme atom and the side chain of Asp242 interacts with His173 through H bond. The Phe190 residue is positioned just below the heme ring, near the His173 residue. There is a second opening between the heme propionate residues; this region is polar and characterized by residues Arg177, Asp179, Glu35, and Glu39 (Selvaggini et al. 1995). Its ability to oxidize Mn^{2+} with high substrate affinity is related to the presence of a Mn binding site (involving Glu36, Glu40, and Asp179) which enables the oxidation of this cation by the internal heme propionate (Sundaramoorthy et al. 1997).

Multiple sequence alignment of amino acid sequences (retrieved from www.ncbi.nlm.nih.gov/protein/) conducted on several characteristic peroxidases (*P. chrysosporium* MnP (378 amino acids) and LiP (371 amino acids), *T. versicolor* MnP (365 amino acids) and LiP (372 amino acids), *P. ostreatus* MnP (361 amino acids), *P. brasiliensis* Pb01 cytochrome c peroxidase (374 amino acids), and *P. eryngii* versatile peroxidase (370 amino acids), *Brassica rapa* TP7 (296 amino acids), and *Arthromyces ramosus* peroxidase (364 amino acids)) indicated significant conserved structural residues (Fig. 6.7). All the enzymes have the conserved proximal histidine, the distal histidine and the distal arginine. The distal arginine residue has been proposed to participate in the formation of the peroxide-binding pocket together with distal histidine, while Asn131 has been revealed to be the only carbohydrate binding site in MnP1 (Sundaramoorthy et al. 1994a, b, 2005). The iron coordination and the residues involved in the active site are the most conserved region among all the aforementioned fungal peroxidases, as well as cytochrome c peroxidase. The two glutamic acid and the aspartic acid residues present in the manganese binding site are also conserved. It could be inferred from Fig. 6.7 that heme binding sites are conserved among all the characteristic peroxidases with some variations found in the case of the turnip (*Brassica rapa*) and cytochrome c peroxidases. The distal Ca^{2+}-binding residues are conserved in all fungal peroxidases (coordination being completed by two water molecules), whereas some differences exist in the residues binding Ca^{2+} at the proximal side.

Comparative Analysis of MnP, LiP, and VP

Comparative studies were performed between *P. chrysosporium* MnP (pdb 1mnp), *T. cervina* LiP (pdb 3q3u), and *P. eryngii* VP (pdb 3fmu) on the basis of tertiary structure alignment using UCSF

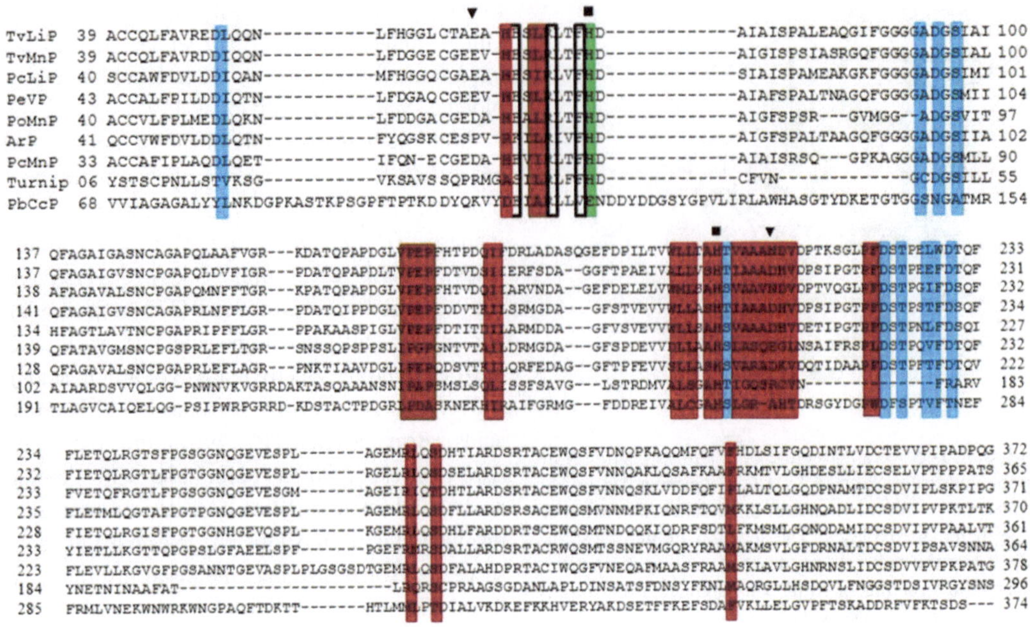

Fig. 6.7 Multiple alignments of amino acid sequences of fungal peroxidases. MnP, LiP, VP, CcP TP, and other peroxidases from *P. chrysosporium* (Pc) and *T. versicolor* (Tv), *P. ostreatus* (Po), *P. brasiliensis* Pb01 (Pb), *P. eryngii* (Pe), *Brassica rapa* turnip (turnip), and *A. ramosus* (ArP) are compared. The alignment was generated and conserved residues were identified by using Clustal X 2.0.11. The sequences highlighted in *red* denote heme binding sites; calcium binding sites are highlighted in *blue*. *Green* shows conserved residues in the substrate binding site. *Blank boxes* encompass the common residues in heme and substrate binding sites. The following marks are used in the consensus line: Mn^{2+} binding site (▼); distal and proximal histidine (■). Accession numbers of the amino acid sequences are as given below: TvLiP, AAA34049.1; TvMnP, CAA83148.1; PcLiP, AAA03748.1; PeVP, AAD54310.1; PoMnP, AAA84397.1; ArP, P28313.3; PcMnP, AAA33743.1; Turnip, P00434.3; PbCcP, EEH41729.1

chimera 1.4.1 software (Fig. 6.8). Structural alignment depicted approximately 50 % similarity of the residues among these peroxidases. Several functionally relevant structurally important features were also observed:

(a) the very close structural similarity between MnP, LiP, and VP active sites, suggesting a similar mode of hydrogen-peroxide activation. The heme prosthetic group is found embedded between the N-terminal and the C-terminal domains, along with the surrounding conserved residues H46 (MnP)/H47 (LiP)/H47(VP), H173 (MnP)/H175 (LiP)/ H169(VP), and R42 (MnP)/R43 (LiP)/ R43(VP);

(b) the substitution of polar residues for the hydrophobic amino acids exposed at the edge of the channel involved in substrate recognition in lignin peroxidase, suggesting that manganese peroxidase does not directly bind aromatic substrates;

(c) the location of residues potentially able to bind Mn^{2+}, spatially positioned on the side of the 3-CH_3 heme edge. The close sequence similarity of all the peroxidases strongly suggests a very similar three-dimensional fold. Lignin peroxidase was characterized by two calcium binding sites involving the side-chain residues, i.e., Asp194, Ser177, Thr196 (proximal site) and Asp65, Asp86 (distal site). The corresponding residues of aligned manganese peroxidase were Asp198, Thr199, Thr219 and Asp64, Asp85 and of versatile peroxidase were Ser195, Asp194, Lys215 and Asp65, Asp86;

(d) the high degree of conservation of the aspartic acid residue at N + 1 site position after the distal histidine suggests that the calcium distal site is involved in maintaining the integrity of

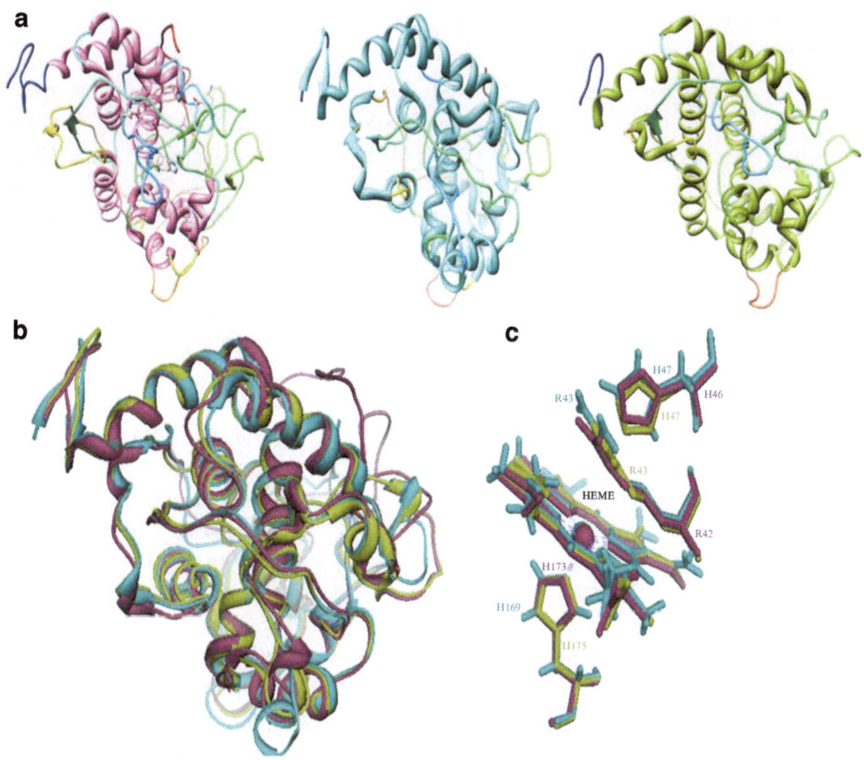

Fig. 6.8 Tertiary structural comparison of MnP, LiP, and VP using UCSF chimera 1.4.1 software. (**a**) Whole structures of *P. chrysosporium* MnP (pdb 1mnp), *T. cervina* LiP (pdb 3q3u), and *P. eryngii* VP (pdb 3fmu) are shown respectively from *left* to *right*. (**b**) Structural alignments of MnP, LiP, and VP are shown in *pink*, *cyan*, and *green*, respectively. (**c**) Structural alignment of the region surrounding heme and the MnP, LiP, and VP residue numbers are shown

the active site is in accordance with Henrisatt et al. (1990) and Welinder (1992);

(e) the sequence corresponding to the solvent-exposed a helix Ala12–Gln25, connected with other portions of the protein by disulfide bridges involving Cys3–Cysl5 and Cysl4–Cys289, is closely similar to the Ala12–Gln25 segment of LiP and VP with some closely related mutations. The coil region located at residues Ala50–Gly65 in the MnP sequence was not completely conserved in the lignin peroxidase and versatile peroxidase. The presence of glycine and proline in this region strongly suggests a coil arrangement (Benner 1989; Benner and Gerloff 1990; Branden and Tooze 1991). The region Pro163–Val175 (heme proximal helix) is highly similar to the aligned segments of lignin peroxidase and versatile peroxidase sequences except for some minor changes in this region.

Superimposition of the MnP, LiP, and VP structures clearly depicts the presence of five disulfide bonds in MnP in contrast to four disulfide bonds present in the LiP and VP. The initial four disulfide bonds in MnP, viz., Cys3–Cys15, Cys33–Cys117, Cys14–Cys289, and Cys253–Cys319, are the same as observed in Lip, VP, and *A. ramosus* peroxidase and also had perfect alignment as portrayed by superimposed structures in Fig. 6.8b. The additional disulfide bond found in MnP (Cys341–Cys348 located near the C-terminus of the polypeptide chain) aids in the formation of the Mn^{II} binding site and is responsible for pushing the C-terminus segment away from the main body of the protein (Sundaramoorthy et al. 1994a, b). LiP, VP, and MnP all have a Phe position 190, but the orientation differs. In LiP the Phe ring is nearly parallel to the proximal His imidazole ring, whereas in MnP the Phe ring is almost perpendicular to the plane of the proximal His.

Table 6.2 Helices distribution among MnP, LiP, and VP

	Helices		
S.No.	MnP	LiP	VP
1.	Ala16-Ile28	Asn11-Leu28	Asn11-Leu28
2.	Gly34-Ile49	Glu35-Val50	Gly35-Ile50
3.	Gly65-Phe70	Ser53-Ala59	Gly63-Ala79
4.	Ile83-His97	Gly69-Thr79	Ile81-His95
5.	Ser101-Ser115	Gly86-Gly102	Ser98-Asn113
6.	Ser147-Gly159	Ser104-Asn119	Ser144-Ala155
7.	Thr162-Leu170	Ser150-Asp160	Ser158-Ile171
8.	Thr199-Val205	Ser164-Ile177	Ser195-Thr201
9.	Leu239-His247	Asp184-Asp188	Gln229-Asp237
10.	Thr251-Gly257	Thr201-Leu208	Thr240-Met247
11.	Glu261-Ala277	Gln234-Asp242	Asn250-Leu267
12.	–	Thr245-Val253	Asp271-Leu275
13.	–	Asn255-Ala271	Ser279-Ile282
14.	–	Ile276-Leu280	Ser299-Val303
15.	–	Ser284-Ile287	–
16.	–	Ser306-Val310	–

Owing to the differences in orientation of the outer propionate in MnP, the distal Arg42 cannot form a hydrogen bond with the propionate, whereas in LiP, the distal arginine directly interacts with the propionate. The second (inner) propionate interacts with Mn^{2+} ion and water molecules, and the peptide NH group of a propionate together with a main chain peptide nitrogen is also found in LiP.

The residues forming the helices found to be structurally aligned in MnP, LiP, and VP models are summarized in Table 6.2.

MnP and Active-Site Engineering

Major challenges remain in understanding the role of functional domains and their structural/functional relationships. The desired features required for the commercial exploitation of these enzymes are stability, notable yields, and enhanced or superannuated activities. Site-directed mutagenesis (SDM) or active-site engineering is an invaluable tool to alter bases at precise positions in the gene. Engineered enzymes are then subjected to analysis for the above-mentioned favorable characteristics using various molecular tools. Site-directed mutagenesis is a major approach that provides opportunities to study unique structural/functional relationships in MnPs and allows the detailed characterization of the Mn^{II} binding site.

Kishi et al. (1994) developed a series of mutants (E35Q, E39Q, and E35Q-D179N) from the gene encoding manganese peroxidase isozyme 1 (*mnp*1) from *Phanerochaete chrysosporium*, using site-directed mutagenesis. The mutations demonstrated that changing any of the acidic amino acid Mn^{II} ligands, Asp179, Glu35, or Glu39, significantly affects the oxidation of Mn^{II}, most probably by decreasing the affinity of the enzyme for Mn^{II}. Asp179, Glu35, and Glu39 residues at the catalytic site are essentially required for Mn^{II} oxidation, since the double mutation, i.e., E35Q–D179N, had almost completely resulted into the loss of Mn^{II} oxidation (Sundaramoorthy et al. 1994a, b). The coordination of Mn^{II} at this site is octahedral, which is typical of Mn^{II} coordination complexes (Demmer et al. 1980). It has been postulated that Glu39 is a Mn ligand and its precise geometry within the Mn binding site of MnP is essential for the efficient binding, oxidation, and release of Mn by this enzyme and that mutation of this ligand decreases both the Mn binding and the rate of Mn oxidation (Martınez 2002; Li et al. 2001).

Miyazaki and Takahashi (2001) found that the site-directed mutagenesis of oxidizable Met273 located near the H_2O_2-binding pocket to a non-oxidizable Leu had resulted into improved stability of IMnP, as it retained more than 60 % of its initial activity in presence of 1 mM H_2O_2 and more than 30 % at a concentration of 3 mM H_2O_2 as compared to wild type that was completely inactivated by 1 mM H_2O_2. Stability in the presence of hydrogen peroxide may be attributed to the above-mentioned mutation that makes it resistant to oxidation by the conformational stabilization around the H_2O_2-binding pocket. Manganese peroxidase (MnP) is susceptible to thermal inactivation due to the loss of calcium. Engineering of a disulfide bond near the distal calcium binding site of MnP by double mutation A48C and A63C showed the improvement in thermal stability as well as pH stability in comparison to native MnP. The disulfide bond adjacent to the distal calcium ligands Asp47 and Asp64 stabilizes the recombinantly expressed MnP against the loss of calcium (Reading and Aust 2000).

Timofeevski et al. (1999) have described that a single mutation (S168W) in rMnP added veratryl alcohol oxidase activity to the enzyme without significantly affecting Mn^{2+} oxidase activity. This surface tryptophan residue, present in various LiP isoenzymes but absent in MnP, may be the site of VA binding and oxidation by LiP. Other research conducted observed how the hydrophobicity of the heme pocket could affect the reactivity of compound I (formed during MnP catalysis). Leu169 and Ser172 were mutated and converted Phe and Ala, respectively. Steady-state kinetics characterization indicated that the Leu169Phe mutation had little effect on activity, whereas the Ser172Ala mutation decreased *k*cat to **45 s^{-1}** as compared to wild type (449 s^{-1}) and also the specificity constant (*k*cat/*K*m) of Ser172Ala mutant decreased from $1.1 \times 10^7\ M^{-1}s^{-1}$ to $5.3 \times 10^5\ M^{-1}s^{-1}$ for Mn^{2+}, but not H_2O_2. It has been shown/demonstrated that compound II is the most sensitive to changes in the heme environment when compared to compound I (Balay et al. 2000).

The role of the axial ligand hydrogen-bonding network on heme reactivity was analyzed by Whitwam et al. (1999); D242 is hydrogen bonded to the proximal His of MnP, in other peroxidases, and this conserved Asp, in turn, is hydrogen bonded to a Trp. In MnP and other fungal peroxidases, the Trp is replaced by a Phe (F190). Both residues are thought to have a direct influence on the catalytic center of the enzyme. Mutagenesis of D242 and F190 has shown that these residues affect the reactivity of the heme active site. The changes in the axial ligand H-bonding network largely influence the reactivity of compound II (Fe^{4+}) and have little influence on the reactivity of compound I (porphyrin cation radical).

Zhang et al. (2009) described the role of Arg42 and Asn131 in the oxidation of 2,6-DMP after performing site-directed mutagenesis with in vitro synthesis. As previously described, R177A and R177K mutants of *P. chrysosporium* had specifically altered binding of Mn, whereas the rate of electron transfer from Mn^{2+} to the oxidized heme was apparently not affected (Gelpke et al. 1999). Whitwam et al. (1997) had suggested that Arg177 may anchor to the carboxylate of Glu35 in the Mn^{2+}-occupied closed configuration of the protein. Shortening the side chain of this residue by one methylene in the E35D mutant probably does not affect the salt bridge to Arg177, but it may restrain the carboxylate of this ligand from making a strong bond with the Mn^{II} atom. This results in a disruption of the ligation for Mn^{II} and hence in the electron-transfer rate as observed by Gelpke et al. (1999).

MnPs and Industrial/Commercial Applications

Industrial applications for manganese peroxidases that have been proposed include bleaching of unbleached kraft pulp in pulp and paper industries, treatment of textile industry effluents, in generating natural aromatic flavors in foods, degradation of environmental pollutants like polyaromatic hydrocarbons (PAHs), azo dyes, TNT, and DTT. Manganese peroxidase is versatile and energy saving, and its ability to bioremediate displays the capacity of this enzyme to be a significant tool for a number of eco-friendly commercial applications.

Ecotoxic organic chemicals generated from textile, pulp, and paper industry effluents are major contributors for the environmental pollution and are the major health hazards for a number of vertebrates and to human population. Detoxification of these compounds poses an immense technical challenge (Evans et al. 2004; Brar et al. 2006). Conventional methods for treatment of various industrial effluents include physical (adsorption, membrane filtration, ion exchange, irradiation, etc.) and chemical (oxidation, coagulation, electrochemical, etc.) processes; these methods have earlier been reviewed extensively (Hao et al. 2000; Forgacs et al. 2004; Joshi et al. 2004; Kuhad et al. 2004). The major drawbacks of physicochemical approaches are that these are prohibitively expensive, less efficient, not versatile, and have interference by other wastewater constituents. Biological methods consisting of biosorption, biodegradation, and enzymatic processes are eco-friendly, simpler, and cost-effective and are receiving greater attention for treatment of industrial effluents (Kuhad et al. 2004; Kaushik and Malik 2009).

Manganese peroxidases are the part of the extracellular oxidative system which evolved in white-rot fungi for lignin degradation (Kirk and Cullen 1991; Hatakka 1994; Vares and Hatakka 1996). Lignin is a heterogeneous, optically inactive polymer consisting of phenylpropanoid interunits, which are linked by several covalent bonds (e.g., aryl-ether, aryl–aryl, carbon–carbon bonds) (Hofrichter 2002). The structure of the lignin polymer implies that lignolytic enzymes possess the ability to oxidize substrates of high redox potential in a nonspecific manner. Paper and pulp industries employ a combination of chlorine-based chemicals and alkaline extraction multistage procedures for bleaching of the kraft pulp. However, chemical bleaching procedures end up with chlorinated organic substances as byproducts which contain toxic, mutagenic, and carcinogenic polychlorinated dioxins, dibenzofurans, and phenols. Discharge of these organic compounds into the effluent generates serious environmental concern. Partially purified manganese peroxidase in the presence of oxalate preparations is known to be effective in decolorizing kraft effluents and oxidizing a broad range of xenobiotic compounds (Harazono et al. 1996; Sasaki et al. 2001). In vitro depolymerization studies using LiP and MnP showed that the enzymes were able to degrade to a variety of aromatic substrates (Conesa et al. 2002).

MnP also demonstrated the ability to decolorize a range of azo and anthraquinone dyes, as well as textile industry effluents in aqueous cultures and in packed bed bioreactors (Robinson et al. 2001; Mielgo et al. 2001; Kasinath et al. 2003; Shin 2004; Yang et al. 2004; Snajdar and Baldrian 2007; Asgher et al. 2008; Sedighi et al. 2009). Susla et al. (2008) have evaluated the contribution of MnP from *D. squalens* and laccase in degradation of azo, anthraquinone, phthalocyanine, and oligocyclic aromatic dyes. MnP has been observed to be capable of decolorizing the mixture of azo dyes at a concentration range of 10–200 mgl^{-1} each (Singh and Pakshirajan 2010), and also MnP from *P. chrysosporium* sp. HSD has been reported to rapidly decolorize a higher concentration (up to 600 mgl^{-1}) of azo dyes (Hailei et al. 2009).

MnP has been shown to have the mineralization ability for many environmental contaminants. Besides having ability to degrade azo, heterocyclic, reactive, and polymeric dyes (Champagne and Ramsay 2005), it can degrade 1.1.1-trichloro-2.2-bis-(4-chlorophenyl) ethane (DDT), 2.4.6-trinitrotoluene (TNT), and polycyclic aromatic hydrocarbons (PAHs) too (Maciel et al. 2010).

Manganese peroxidase (MnP) from two metabolically distinct fungi *Phanerochaete chrysosporium* BKM-F-1767 (ATCC 24725) and *Bjerkandera* sp. BOS55 (ATCC90940) has the ability to degrade (98 %) anthracene to generate anthraquinone in organic solvent mixtures after 6 h of operation under optimal conditions (Eibes et al. 2005). Utilization of MnP (from the basidiomycete *Bjerkandera adusta)* for acrylamide polymerization has also been reported (Iwahara et al. 2000). MnPs from *Phanerochaete chrysosporium* have also been employed in styrene degradation, an important industrial polymer used as a raw material for wrapping and transporting goods. Its disposal poses serious environmental concerns (Soto et al. 1991; Lee et al. 2006).

MnP a redox enzyme has the potential of directly transferring the electrons to the electrodes.

This enables the use of this enzyme for various applications in the development of biosensors, designing effective biofuel cells, and for selective bioorganic synthesis (Maciel et al. 2010).

Future Perspectives

This review highlights the various developments related to the molecular features, cloning, heterologous production, crystal structure refinements of MnPs, and its possible industrial and biotechnological applications. MnPs are promising enzymes and an eco-friendly alternative to the conventional physicochemical processes as presently employed for various such as the pulp and paper, textile, pharmaceutical, and for food industries. However, a major challenge in the commercialization of the MnPs is due to its lower thermal stability. Despite recent progress, our understanding of the process is still limited due to its substrate complexity and because of multiplicity of the peroxidases. Manganese peroxidases currently generated are so far not promising enough for their commercial scale exploitation. Although efforts have been made for improving the thermal stability of MnPs, further efforts are required for the development of designer enzymes with desired levels of thermal stability for its industrial applications. Therefore, the tailor-made enzymes can be designed using a combination of molecular approaches like site-directed mutagenesis, saturation mutagenesis, and directed evolution enabling the industrial exploitation of the unique catalytic abilities of these biocatalysts. The various developments in enzyme engineering research open a wide spectrum of possible applications in the near future.

References

Alic M, Akileswaran L, Gold MH (1997) Characterization of the gene encoding manganese peroxidase isozyme 3 from *Phanerochaete chrysosporium*. Biochim Biophys Acta 1338:1–7

Asgher M, Bhatti HN, Ashraf M, Legge RL (2008) Recent developments in biodegradation of industrial pollutants by white rot fungi and their enzyme system. Biodegradation 19:771–783

Balay KA, Dougherty M, Tien M (2000) Reactivity of manganese peroxidase: site-directed mutagenesis of residues in proximity to the porphyrin ring. Arch Biochem Biophys 382:89–94

Banci L, Bertini I, Pease EA, Tien M, Turano P (1992) 1H NMR investigation of manganese peroxidase from *Phanerochaete chrysosporium*. A comparison with other peroxidases. Biochemistry 31(41):10009–10017

Bao W, Fukushima Y, Jensen KA, Moen MA, Hammel KE (1994) Oxidative degradation of non-phenolic lignin during lipid peroxidation by fungal manganese peroxidase. FEBS Lett 354:297–300

Benner SA (1989) Patterns of divergence in homologous proteins as indicators of tertiary and quaternary structure. Adv Enzyme Regul 28:219–236

Benner SA, Gerloff D (1990) Patterns of divergence in homologous proteins as indicators of secondary and tertiary structure: a prediction of the structure of the catalytic domain of protein kinases. Adv Enzyme Regul 31:121–181

Bonnarme P, Jeffries TW (1990) Mn^{II} regulation of lignin peroxidases and manganese dependent peroxidases from lignin-degrading white rot fungi. Appl Environ Microbiol 56:210–217

Branden C, Tooze J (1991) Introduction to protein structure. Garland Publishing, New York/London

Brar SK, Verma M, Surampalli RY, Misra K, Tyagi RD, Meunier N, Blais JF (2006) Bioremediation of hazardous wastes: a review. Pract Period Hazard, Toxic Radioact Waste Manag 10:59–72

Brown JA, Alic M, Gold MH (1991) Manganese peroxidase gene transcription in *Phanerochaete chrysosporium*: activation by manganese. J Bacteriol 173:4101–4106

Cai D, Tien M (1993) Lignin-degrading peroxidases of *Phanerochaete chrysosporium*. J Biotechnol 30:79–90

Cairney JWG, Burke RM (1998) Do ecto- and ericoid mycorrhizal fungi produce peroxidase activity. Mycorrhiza 8:61–65

Camarero S, Ruiz-Duen FJ, Sarkar S, Martinez MJ, Martinez AT (2000) The cloning of a new peroxidase found in lignocellulose cultures of *Pleurotus eryngii* and sequence comparison with other fungal peroxidases. FEMS Microbiol Lett 191:37–43

Champagne PP, Ramsay JA (2005) Contribution of manganese peroxidase and laccase to dye decoloration by *Trametes versicolor*. Appl Microbiol Biotechnol 69:276–285

Conesa A, van Den Hondel CA, Punt PJ (2000) Studies on the production of fungal peroxidases in *Aspergillus niger*. Appl Environ Microbiol 66:3016–3023

Conesa A, Punt PJ, vanden Hondel CAMJJ (2002) Fungal peroxidases: molecular aspects and applications. J Biotechnol 93:143–158

Daina S, Orlandi M, Bestetti G, Wiik C, Elegir G (2002) Degradation of β-5 lignin model dimers by *Ceriporiopsis subvermispora*. Enzyme Microb Technol 30:499–505

Deguchi T, Matsubara M, Nishida T (2002) NADH oxidation by manganese peroxidase with or without alpha-hydroxy acid. Biosci Biotechnol Biochem 66(4):717–721

Demmer H, Hinz I, Keller-Rudex H, Koeber K, Kottelwesch H, Schneider D (1980) In: Schleitzer-Rust E (ed) Coordination compounds of manganese, vol 56, 8th edn. Springer, New York, pp 1–185

Duenas FJR, Martnez MJ, Martnez A (1999) Molecular characterization of a novel peroxidase from the ligninolytic fungus *Pleurotus eryngii*. Mol Microbiol 31:223–235

Eibes G, Lú-Chau T, Feijoo G, Moreira MT, Lema JM (2005) Complete degradation of anthracene by Manganese Peroxidase in organic solvent mixtures. Enzyme Microb Technol 37:365–372

Eriksson KEL, Blanchette RA, Ander P (1990) Microbial and enzymatic degradation of wood and wood components. Springer, Berlin/Heidelberg/New York

Espinosa DVC, Absalón AE, Sanchez N, Loera O, Vazquez RR, Fernández FJ (2012) Heterologous expression of manganese peroxidase in *Aspergillus niger* and its effect on phenanthrene removal from soil. J Mol Microbiol Biotechnol 21:120–129

Evans FF, Rosado S, Sebastian GV, Casella R, Machado PLOA, Holmstrom C, Kjelleberg S, Van Elsas JD, Seldin L (2004) Impact of oil contamination and biostimulation on the diversity of indigenous bacterial communities in soil microcosms. FEMS Microbiol Ecol 49:295–305

Felsenstein J (1985) Confidence limits on phylogenies: an approach using the bootstrap. Evolution 39:783–791

Forgacs E, Cserhati T, Oros G (2004) Removal of synthetic dyes from wastewaters: a review. Environ Int 30:953–971

Gaskell J, Dieperink E, Cullen D (1991) Genomic organization of lignin peroxides genes of *Phanerochaete chrysosporium*. Nucleic Acids Res 19:599–603

Gelpke SMD, Loccoz PM, Gold MH (1999) Arginine 177 is involved in Mn^{II} binding by manganese peroxidase. Biochemistry 38:11482–11489

Gelpke MD, Youngs HL, Gold MH (2000) Role of arginine 177 in the MnII binding site of manganese peroxidase. Studies with R177D, R177E, R177N, and R177Q mutants. Eur J Biochem 267(24):7038–7045

Gettemy JM, Ma B, Alic M, Gold MH (1998) Reverse transcription-PCR analysis of the regulation of the manganese peroxidase gene family. Appl Environ Microbiol 64:569–574

Glenn JK, Gold MH (1985) Purification and characterization of an extracellular Mn^{II}-dependent peroxidase from the lignin-degrading basidiomycete, *Phanerochaete chrysosporium*. Arch Biochem Biophys 242:329–341

Godfrey BJ, Mayfield MB, Brown JA, Gold MH (1990) Characterization of a gene encoding a manganese peroxidase from *Phanerochaete chrysosporium*. Gene 93:119–124

Gold MH, Alic M (1993) Molecular biology of the lignin degrading basidiomycetes *Phanerochaete chrysosporium*. Microbiol Rev 57:605–622

Gu L, Lajoie C, Kelly C (2003) Expression of a *Phanerochaete chrysosporium* manganese peroxidase gene in the yeast *Pichia pastoris*. Biotechnol Prog 19:1403–1409

Hailei W, Ping L, Min P, Zhijun Z, Guangli Y, Guosheng L, Jianming Y (2009) Rapid decolourization of azo dyes by a new isolated higher manganese peroxidase producer: Phanerochaete sp. Hsd Biochem Eng J 46:327–333

Hao OJ, Kim H, Chiang PC (2000) Decolourization of wastewater. Crit Rev Environ Sci Technol 30:449–505

Harazono K, Kondo R, Sakai K (1996) Bleaching of hardwood kraft pulp with manganese peroxidase from *Phanerochaete sordida* YK-624 without addition of MnSO(inf$_4$). Appl Environ Microbiol 62(3):913–917

Hatakka A (1994) Lignin-modifying enzymes from selected white-rot fungi: production and role in lignin degradation. FEMS Microbiol Rev 13:125–135

Hatakka A (2001) Biodegradation of lignin. In: Hofrichter M, Steinbüchel A (eds) Biopolymers. Lignin, humic substances and coal. Wiley-VCH, Weinheim, pp 129–180

Heinzkill M, Bech L, Halkier T, Schneider P, Anke T (1998) Characterization of laccases and peroxidases from wood-rotting fungi (family *Coprinaceae*). Appl Environ Microbiol 64:1601–1606

Henrisatt B, Saloheimo M, Lavaitte S, Knowles JKC (1990) Structural homology among the peroxidase enzyme family revealed by hydrophobic cluster analysis. Protein Struct Funct Genet 8:251–257

Hofrichter M (2002) Review: lignin conversion by manganese peroxidase (MnP). Enzyme Microb Technol 30:454–466

Iwahara K, Hirata M, Honda Y, Watanabe T, Kuwahara M (2000) Free-radical polymerization of acrylamide by manganese peroxidase produced by the white-rot basidiomycete *Bjerkandera adusta*. Biotechnol Lett 22:1355–1361

Janse BJH, Gaskell J, Akhtar M, Cullen D (1998) Expression of *Phanerochaete chrysosporium* genes encoding lignin peroxidases, manganese peroxidases, and glyoxal oxidase in wood. Appl Environ Microbiol 64:3536–3538

Johansson T, Nyman PO, Cullen D (2002) Differential regulation of *mnp2*, a new manganese peroxidase-encoding gene from the ligninolytic fungus *Trametes versicolor* PRL 572. Appl Environ Microbiol 68: 2077–2080

Joshi M, Bansal R, Purwar R (2004) Colour removal from textile effluents. Indian J Fibre Text Res 29:239–259

Kapich AN, Steffen KT, Hofrichter M, Hatakka A (2005) Involvement of lipid peroxidation in the degradation of a non-phenolic lignin model compound by manganese peroxidase of the litter-decomposing fungus *Stropharia coronilla*. Biochem Biophys Res Commun 330:371–377

Kasinath A, Novotny C, Svobodova K, Patel KC, Saseck V (2003) Decolorization of synthetic dyes by *Irpex lacteus* in liquid cultures and packed-bed bioreactor. Enzyme Microb Technol 32:167–173

Kaushik P, Malik A (2009) Fungal dye decolourization: recent advances and future potential. Environ Int 35:127–141

Khindaria A, Grover TA, Aust SD (1994) Oxalate-dependent reductive activity of manganese peroxidase from *Phanerochaete chrysosporium*. Arch Biochem Biophys 314:301–306

Kim Y, Yeo S, Kum J, Song HG, Choi HT (2005) Cloning of a manganese peroxidase cDNA gene repressed by manganese in *Trametes versicolor*. J Microbiol 43:569–571

Kirk TK, Cullen D (1991) Enzymology and molecular genetics of wood degradation by white-rot fungi. Wiley, New York

Kishi K, Wariishi H, Marquez L, Dunford BH, Gold MH (1994) Mechanism of manganese peroxidase II reduction. Effect of organic acid and chelators and pH. Biochemistry 34:8694–8701

Kuhad RC, Sood N, Tripathi KK, Singh A, Ward OP (2004) Developments in microbial Methods for the treatment of dye effluents. Adv Appl Microbiol 56:185–213

Kuwahara M, Glenn JK, Morgan MA, Gold MH (1984) Separation and characterization of two extracellular H_2O_2-dependent oxidases from ligninolytic cultures of *Phanerochaete chrysosporium*. FEBS Lett 169:247–250

Larrondo LF, Lobos S, Stewart P, Cullen D, Vicuña R (2001) Isoenzyme multiplicity and characterization of recombinant manganese peroxidases from *Ceriporiopsis subvermispora* and *Phanerochaete chrysosporium*. Appl Environ Microbiol 67:2070–2075

Lee JW, Lee SM, Hong EJ, Jeung EB, Kang HY, Kim MK, Choi IG (2006) Estrogenic reduction of styrene monomer degraded by *Phanerochaete chrysosporium* KFRI 20742. J Microbiol 44:177–184

Levin L, Papinutti L, Forchiassin F (2004) Evaluation of Argentinean white rot fungi for their ability to produce lignin-modifying enzymes and decolorize industrial dyes. Bioresour Technol 94:169–176

Li D, Youngs HLA, Gold MH (2001) Heterologous expression of a thermostable manganese peroxidases from *Dichomitus squalens* in *Phanerochaete chrysosporium*. Arch Biochem Biophys 385:348–356

Lobos S, Larram J, Salas L, Cullen D, Vicuña R (1994) Isoenzymes of manganese dependent peroxidase and laccase produced by the lignin degrading basidiomycete Ceriporiopsis subvermispora. Microbiology 14:2691–2698

Lobos S, Larrondo L, Salas L, Karahanian E, Vicun R (1998) Cloning and molecular analysis of a cDNA and theCs-mnp1 gene encoding a manganese peroxidase isoenzyme from the lignin-degrading basidiomycete *Ceriporiopsis subvermispora*. Gene 206:185–193

Longoria A, Tinoco R, Duhalt VR (2008) Chloroperoxidase-mediated transformation of highly halogenated monoaromatic compounds. Chemosphere 72:485–490

Ma B, Mayfield MB, Godfrey BJ, Gold MH (2004) Novel promoter sequence required for manganese regulation of manganese peroxidase isozyme 1 gene expression in *Phanerochaete chrysosporium*. Eukaryot Cell 3:579–588

Maciel MJ, Silva AC, Ribeiro HCT (2010) Industrial and biotechnological applications of ligninolytic enzymes of the basidiomycota: a review. Elect J Biotechnol 13(6). (ISSN: 0717-3458) doi:org/10.2225/, full-text2

Magliozzo RS, Marcinkeviciene JA (1997) The role of Mn(II)-peroxidase activity of mycobacterial catalase-peroxidase in activation of the antibiotic isoniazid. J Biol Chem 272(14):8867–8870

Martınez AT (2002) Molecular biology and structure-function of lignin-degrading heme peroxidases. Enzyme Microb Technol 30:425–444

Mayfield MB, Godfrey BJ, Gold MH (1994) Characterization of the *mnp2* gene encoding manganese peroxidase isozyme 2 from the basidiomycete *Phanerochaete chrysosporium*. Gene 142:231–235

Mielgo I, Moreira MT, Feijoo G, Lema JM (2001) A packed-bed fungal bioreactor for the continuous decolourisation of azo-dyes (Orange II). J Biotechnol 89:99–106

Mino Y, Wariishi H, Blackburn NJ, Loehrldx TM, Gold MH (1988) Spectral characterization of manganese peroxidase, an extracellular heme enzyme from the lignin-degrading basidiomycete, *Phanerochaete chrysosporium*. J Biol Chem 263:7029–7036

Miyazaki C, Takahashi H (2001) Engineering of the H_2O_2-binding pocket region of a recombinant manganese peroxidase to be resistant to H_2O_2. FEBS Lett 509(1):111–114

Nei M, Kumar S (2000) Molecular evolution and phylogenetics. Oxford University Press, New York

Ogawa K, Yamazaki T, Hasebe T, Kajiwara S, Watanabe A, Asada Y, Shishido K (1998) Molecular breeding of the basidiomycete *Coprinus cinereus* strains with high lignindecolorization and β-degradation activities using novel heterologous protein expression vectors. Appl Microbiol Biotechnol 49:285–289

Orth AB, Rzhetskaya M, Cullen D, Tien M (1994) Characterization of a cDNA encoding a manganese peroxidase from *Phanerochaete chrysosporium*: genomic organization of lignin and manganese peroxidase-encoding genes. Gene 148:161–165

Padgett R, Konarska M, Grabowski P, Hardy S, Sharp P (1989) Lariat RNAs as intermediates and products in the splicing of messenger RNA products. Science 225:898–903

Paszcynski A, Huynh V-B, Crawford R (1985) Enzymatic activities of an extracellular, manganese-dependent peroxidase from *Phanerochaete chrysosporium*. FEMS Microbiol Lett 29:37–41

Paszcynski A, Huynh V-B, Crawford R (1986) Comparison of ligninase-1 and peroxidase-M2 from the white-rot fungus *Phanerochaete chrysosporium*. Arch Biochem Biophys 244:750–765

Petersen JF, Tams JW, Vind J, Svensson A, Dalbøge H, Welinder KG, Larsen S (1993) Crystallization and X-ray diffraction analysis of recombinant *Coprinus cinereus* peroxidase. J Mol Biol 232:989–991

Poulos TL, Edwards SL, Wariishi H, Gold MH (1993) Crystallographic refinement of lignin peroxidase at 2 Å. J Biol Chem 268:4429–4440

Pribnow D, Mayfield MB, Nipper VJ, Brown JA, Gold MH (1989) Characterization of a cDNA encoding a manganese peroxidase, from the lignin-degrading basidiomycete *Phanerochaete chrysosporium*. J Biol Chem 264:5036–5040

Punt PJ, Biezen NV, Conesa A et al (2002) Filamentous fungi as cell factories for heterologous protein production. Trends Biotechnol 20:200–206

Raghukumar C, D'Souza TM, Thorn RG, Reddy CA (1999) Lignin-modifying enzymes of Flavodon flavus, a basidiomycete isolated from a coastal marine environment. Appl Environ Microbiol 65:2103–2111

Reading NS, Aust SD (2000) Engineering a disulfide bond in recombinant manganese peroxidase results in increased thermostability. Biotechnol Prog 16:326–333

Reddy GVB, Sridhar M, Gold MH (2003) Cleavage of nonphenolic β-1 diarylpropane lignin model dimers by manganese peroxidase from Phanerochaete chrysosporium. Eur J Biochem 270:284–292

Robinson T, Chandran B, Nigam P (2001) Studies on the production of enzymes by white-rot fungi for the decolourisation of textile dyes. Enzyme Microb Technol 29:575–579

Saitou N, Nei M (1987) The neighbor-joining method: a new method for reconstructing phylogenetic trees. Mol Biol Evol 4:406–425

Sakamoto Y, Nakade K, Nagai M, Uchimiya H, Sato T (2009) Cloning of Lentinula edodes lemnp2, a manganese peroxidase that is secreted abundantly in sawdust medium. Mycoscience 50:116–122

Sarkanen KV, Ludwig CH (1971) Lignins: occurrence, formation and structure. Wiley–Interscience, New York

Sasaki T, Kajino T, Li B, Sugiyama H, Takahashi H (2001) New pulp biobleaching system involving manganese peroxidase immobilized in a silica support with controlled pore sizes. Appl Environ Microbiol 67(5):2208–2212

Schmidt O, Schmitt U, Moreth U, Potsch T (1997) Wood decay by the white-rotting basidiomycete *Physisporinus vitreus* from a cooling tower. Holzforschung 51:193–200

Sedighi M, Karimi A, Vahabzadeh F (2009) Involvement of ligninolytic enzymes of *Phanerochaete chrysosporium* in treating the textile effluent containing Astrazon Red FBL in a packed-bed bioreactor. J Hazard Mater 169(1–3):88–93

Selvaggini C, Salmona M, Gioia LD (1995) Manganese peroxidase from *Phanerochaete chrysosporium*. A homology-based molecular model. Eur J Biochem 228:955–961

Shin KS (2004) The role of enzymes produced by white-rot fungus *Irpex lacteus* in the decolorization of the textile industry effluent. J Microbiol 42:37–41

Singh S, Pakshirajan K (2010) Enzyme activities and decolourization of single and mixed azo dyes by the white-rot fungus *Phanerochaete chrysosporium*. Int biodeterior Biodegrad 64:1–5

Snajdar J, Baldrain P (2007) Temperature effect the production, activity and stability of ligninolytic enzymes in *Pleurotus ostreatus* and *Trametes versicolor*. Folia Microbiol 52(5):498–502

Soto AM, Justicia H, Wray JW, Sonnens Chein C (1991) *p*-nonyl-phenol: an estrogenic xenobiotic released from "modified" polystyrene. Environ Health Perspect 92:167–173

Steffen K, Hofrichter M, Hatakka A (2000) Mineralization of 14C-labelled lignin and production of ligninolytic enzymes by selected litter decomposing basidiomycetes. Appl Microbiol Biotechnol 54:819–825

Stewart P, Whitwam RE, Kersten PJ, Cullen D, Tien M (1996) Efficient expression of a *Phanerochaete chrysosporium* manganese peroxidase gene in *Aspergillus oryzae*. Appl Microbiol Biotechnol 62:860–864

Sundaramoorthy M, Kishi K, Gold MH, Poulos TL (1994a) Preliminary crystallographic analysis of manganese peroxidase from Phanerochaete chrysosporium. J Mol Biol 238:845–848

Sundaramoorthy M, Kishi K, Gold MH, Poulos TL (1994b) The crystal structure of manganese peroxidase from *Phanerochaete chrysosporium* at 2.06-A resolution. J Biol Chem 269:32759–32767

Sundaramoorthy M, Terner J, Poulos TL (1995) The crystal structure of chloroperoxidase: a heme peroxidase-cytochrome P450 functional hybrid. Structure 3:1367–1377

Sundaramoorthy M, Kishi K, Gold MH, Poulos TL (1997) Crystal structures of substrate binding site mutants of manganese peroxidase. J Biol Chem 272:17574–17580

Sundaramoorthy M, Youngs HL, Gold MH, Poulos TL (2005) High-resolution crystal structure of manganese peroxidase: substrate and inhibitor complexes. Biochemistry 44:6463–6470

Sundaramoorthy M, Gold MH, Poulos TL (2010) Ultrahigh (0.93 Å) resolution structure of manganese peroxidase from *Phanerochaete chrysosporium*: implications for the catalytic mechanism. J Inorg Biochem 104:683–690

Susla M, Novotny C, Erbanova P, Svobodova K (2008) Implication of *Dichomitus squalens* manganese-dependent peroxidase in dye decolorization and cooperation of the enzyme with laccase. Folia Microbiol 53:479–485

Tamura K, Peterson D, Peterson N, Stecher G, Nei M, Kumar S (2011) MEGA5: molecular evolutionary genetics analysis using maximum likelihood, evolutionary distance and maximum parsimony methods. Mol Biol Evol 28(10):2731–2739

Tello M, Corsini G, Larrondo L, Salas L, Lobos S, Vicun R (2000) Characterization of three new manganese peroxidase genes from the ligninolytic basidiomycete *Ceriporiopsis subvermispora*. Biochim Biophys Acta 1490:137–144

Timofeevski SL, Aust SD (1997) Effects of Mn^{2+} and oxalate on the catalytic activity of manganese peroxidase. Biochem Biophys Res Commun 239:645–649

Timofeevski SL, Nie G, Reading NS, Aust SD (1999) Addition of veratryl alcohol oxidase activity to manganese peroxidase by site-directed mutagenesis. Biochem Biophys Res Commun 256:500–504

Tuor U, Wariishi H, Schoemaker HE, Gold MH (1992) Oxidation of phenolic arylglycerol β- aryl ether lignin model compounds by manganese peroxidase from Phanerochaete chrysosporium: oxidative cleavage of an α-carbonyl model compound. Biochemistry 31:4986–4995

Urzua U, Larrondo LF, Lobos S, Larraın J, Vicuna R (1995) Oxidation reactions catalyzed by manganese peroxidase isoenzymes from *Ceriporiopsis subvermispora*. FEBS Lett 371:132–136

Urzua U, Kersten PJ, Vicuna R (1998) Manganese peroxidase-dependent oxidation of glyoxylic and oxalic acids synthesized by *Ceriporiopsis subvermispora* produces extracellular hydrogen peroxide. Appl Environ Microbiol 64:68–73

Vares T, Hatakka A (1996) Lignin-degrading activity and ligninolytic enzymes of different white-rot fungi: effects of manganese and malonate. Can J Bot 75:61–71

Wariishi H, Akaleswaran L, Gold MH (1988) Manganese peroxidase from the basidiomycete *Phanerochaete chrysosporium*: spectral characterization of oxidized states and the catalytic cycle. Biochemistry 27: 5365–5370

Wariishi H, Dunford HB, MacDonald ID, Gold MH (1989a) Manganese peroxidase from the lignin-degrading basidiomycete *Phanerochaete chrysosporium*. Transient state kinetics and reaction mechanism. J Biol Chem 264:3335–3340

Wariishi H, Valli K, Renganathan V, Gold MH (1989b) Oxidative cleavage of a phenolic diarylpropane lignin model dimer by manganese peroxidase from *Phanerochaete chrysosporium*. Biochemistry 28:6017–6023

Wariishi H, Valli K, Renganathan V, Gold MH (1989c) Thiol-mediated oxidation of nonphenolic lignin model compounds by manganese peroxidase of *Phanerochaete chrysosporium*. J Biol Chem 264: 14185–14191

Wariishi H, Valli K, Gold MH (1992) Manganese(II) oxidation by manganese peroxidase from the basidiomycete *Phanerochaete chrysosporium*. J Biol Chem 267:23688–23695

Welinder KG (1992) Superfamily of plant, fungal and bacterial peroxidases. Curr Opin Struct Biol 2:388–393

Whitwam RE, Brown KR, Musick M, Natan MJ, Tien M (1997) Mutagenesis of the Mn^{2+} binding site of manganese peroxidase affects oxidation of Mn^{2+} by both compound I and compound II. Biochemistry 36: 9766–9773

Whitwam RE, Koduri RS, Natan M, Tien M (1999) Role of axial ligands in the reactivity of Mn peroxidase from *Phanerochaete chrysosporium*. Biochemistry 38:9608–9616

Willmann G, Fakoussa RM (1997a) Biological bleaching of water-soluble coal macromolecules by a basidiomycete strain. Appl Microbiol Biotechnol 47:95–101

Willmann G, Fakoussa RM (1997b) Extracellular oxidative enzymes of coal attacking fungi. Fuel Process Technol 52:27–41

Wong DWS (2009) Structure and action mechanism of ligninolytic enzymes. Appl Biochem Biotechnol 157:174–209

Yang HS, Yoon JS, Kim MN (2004) Effects of storage of a mature compost on its potential for biodegradation of plastics. Polym Degrad Stab 84:411–417

Yeo PSN, Song HG, Choi HT (2007) Generation of a transformant showing higher manganese peroxidase (Mnp) activity by overexpression of Mnp Gene in *Trametes versicolor*. J Microbiol 45:213–218

Youngs HL, Gelpke SMD, Li D, Sundaramoorthy M, Gold MH (2000) The role of Glu39 in MnII binding and oxidation by manganese peroxidase from *Phanerochaete chrysoporium*. Biochemistry 40:2243–2250

Zapanta LS, Tien M (1997) The roles of veratryl alcohol and oxalate in fungal lignin degradation. J Biotechnol 53:93–102

Zhang X, Wang Y, Wang L, Chen G, Liu W, Gao P (2009) Site-directed mutagenesis of manganese peroxidase from *Phanerochaete chrysosporium* in an in vitro expression system. J Biotechnol 139:176–178

Zuckerkand E, Pauling L (1965) Evolutionary divergence and convergence in proteins. In: Evolving genes and proteins. Academic, New York

Advance Techniques in Enzyme Research

7

Debamitra Chakravorty and Sanjukta Patra

Abstract

Enzyme research encompasses the purification, characterization, identification, and applications of enzymes. Enzyme technology has entered a modern era in which bioinformatics, genetic engineering, and high-throughput screening by miniaturization of instruments and robotic handling have altogether made existing processes economical and time saving. These advances resulted in enzymes to be industrial workhorses. This chapter brings forward advances in existing techniques along with novel and innovative methodologies and instrumentations applied in high-throughput enzyme research. Highlights are the applications of advanced recombinant DNA technology like the Gateway® cloning and CODEHOP PCR for the fast and efficient cloning of enzyme genes; computational biology for in silico enzyme characterization like monitoring behavior of enzymes in real time through molecular dynamics simulations; modern analytical techniques like microfluidics, affinity ultrafiltration, automated electrophoresis, and ATP-affinity chromatography; and the use of nanoparticles, nanofibers, and mesoporous silica in enzyme stabilization.

Keywords

Enzyme • Research • Cloning • Screening • Purification • Gateway cloning • CODEHOP

Introduction

Enzymes are ubiquitous natural biocatalysts with well-recognized potential applications (Kim et al. 2006a). The global market for industrial enzymes is expanding exponentially, and a number of biotechnology companies are continuously prospecting for new and adapting existing enzymes. Enzyme technology is an interdisciplinary science and is recognized by the Organization for Economic Cooperation and Development (OECD) as an important component of sustainable industrial development (OECD 1998) as they find practical applications in large number of fields; recent developments are in the areas of organosynthesis,

D. Chakravorty • S. Patra (✉)
Department of Biotechnology, Indian Institute of Technology, Guwahati, India
e-mail: sanjukta@iitg.ernet.in

P. Shukla and B.I. Pletschke (eds.), *Advances in Enzyme Biotechnology*,
DOI 10.1007/978-81-322-1094-8_7, © Springer India 2013

pharmaceuticals, biosensors, bioremediation, and biofuel cells (Kim et al. 2006b). Enzyme research evolved in the twentieth century; since then the scope of enzymology has expanded, from an early focus on the chemical and catalytic mechanisms of individual enzymes to more recent efforts to understand enzyme behavior in the context of functional biological systems in the milieu of many interacting enzymes and proteins (Zalatan and Herschlag 2009). This was possible due to a plethora of new enzymes to study and the curiosity of the researchers to analyze how enzymes could achieve their enormous rate enhancements and exquisite specificities. Enzyme technology thus entered a new era and has become much more economical which is shaping the discovery, development, purification, and application of biocatalysts to a much greater extent (Bailen et al. 2002). Modern researchers focus on complex subjects of enzyme research in the context of the complex biological system where they are naturally found. Recent trend shows development in strategies for enzyme production, such as modernization of enzyme fermentation processes and their recombinant expression in hosts like *Escherichia coli*. It has also been realized that many enzymes are expensive and must be recovered and reused; thus, techniques like enzyme stabilization through immobilization and enzyme engineering are being considered (Bailen et al. 2002). In this chapter we shall discuss the modern technologies developed for enzyme research.

Enzyme Discovery and Screening

High-throughput screening methods for enzymes have been recently developed to perform a key role in the search for new enzymes and biocatalysts (Wahler and Reymond 2001). For such processes the modern trend being set is towards robotic handling and miniaturization of instruments as it improves reproducibility and authenticity of the resulting data. Basically there can be two main approaches to enzyme screening: one can be the forward approach and the other is the reverse approach. The forward approach starts with the native enzyme of interest followed by its purification and characterization, whereas the reverse approach starts with the gene of interest followed by its amplification, recombinant expression, and purification.

Forward Enzyme Screening Approach

In the forward enzyme screening approach, environmental samples are assayed for enzyme of interest, based on their physicochemical properties such as temperature, pH, solvent stability, and their substrate specificity or substrate enantioselectivity or stereospecificity. For this purpose, recent advances in enzymological techniques have seen the development of high-throughput analytical instruments like parallel capillary electrophoresis and thermistor arrays to study the physicochemical properties of particular enzyme of interest. Thermistors are small thermometers that monitor temperature changes in wells of microtiter plates and provide an advantage over the traditional infrared (IR) thermography, which monitors temperature from IR radiation (Wahler and Reymond 2001). On the simple principle that enzyme reactions can be assayed by recording changes in their physicochemical properties, Copeland and Miller attached the fluorescent pH indicator to a solid support to explore the screening of peptide-based catalysts prepared by combinatorial chemistry techniques. The indicator was either directly attached to the bead or incorporated into a polymeric gel. Klein and Reymond in 2001 performed enantioselective assay for alcohol dehydrogenases using fluorogenic substrates to measure the rate of oxidation of each enantiomer to ketone in the presence of bovine serum albumin, which catalyzes β-elimination to form the strongly fluorescent product umbelliferone (Klein and Reymond 2001). In similar lines Olsen et al. in 2000 reported an elegant imaging experiment that relies on the fluorogenic phospholipid substrates for phospholipase A2 (PLA2) to visualize the activity and localization of this enzyme in zebrafish larvae, based on the principle of fluorescence resonance energy transfer (FRET) (Olsen et al. 2000). A similar FRET-based assay has been reported to detect enzyme activities in cells using fluorescence-activated cell sorting (FACS) (Olsen et al. 2000). Another interesting example is in the use of fluorogenic and chromogenic substrates for enzyme assays. For example, Wahler et al. employed fluorogenic substrates to identify caspase activity in bacteria (Wahler and Reymond 2001). Further progress in enzyme research in this area was made by Aw et al. who described continuous assessment of enzymatic activity using nanodroplet microarrays (Aw et al. 2010). By uniformly coating fluorogenic substrates on slides, they successfully generated surfaces capable of detecting enzymatic activity offering an unprecedented ability for performing solution-phase enzymatic assays in nanoliter volumes on microarrays, in contrast to microliter volumes typically required in microplate-based assays, thereby reducing the amounts of reagents required anywhere from a hundred- to a thousandfold. This new approach thus provides a potentially more cost-effective and label-free enzyme screening technique. A single slide is able to accommodate several thousand assays, facilitating the assessment of both dose- and time-dependent inhibition parameters in a single run.

High-throughput screening for enzymes is often aimed not only at activity but also at substrate specificity for the application of enzymes in fine chemical synthesis. Recent progress in analysis of substrate specificity and enantioselectivity of enzymes was by Hwang and Kim (2004), who developed a new high-throughput screening method using staining solution of $CuSO_4$/MeOH for screening ω-transaminase for multiple substrate specificities and their enantioselectivities based on UV-visible spectrometry (Hwang and Kim 2004). Staining solution of $CuSO_4$/MeOH forms a blue complex with the α-amino acid produced during ω-transaminase reaction. This complex can be easily quantified using UV–vis spectrophotometer at 595 nm (Hwang and Kim 2004). In similar lines, the idea of computer-aided substrate screening of enzymes is new and innovative combination of computational skills and enzymology which was recently developed by Xu et al. based on the enzyme structure, and they successfully screened substrates of *Candida antarctica* lipase B (CALB). In this method restricted molecular docking was employed to predict the energetically favorable poses of substrate–enzyme complexes (Xu et al. 2009). Thus, in the future this approach combined with homology modeling, wherein structure of an enzyme can be modeled from its primary amino acid sequence based on a template structure, can be applied to screen substrates for novel and newly characterized enzymes whose X-ray structure is still undetermined.

Some of the other latest innovations in detection and quantification of enzymes include the use of microfluidics, which involves the manipulation of liquid samples in miniaturized systems, often on a chip (Smith 2005). Moreover in 2010, a research team at the Georgia Institute of Technology has developed a technique called gelatin zymography to detect and quantify the nature of mature cathepsin K in femtomoles (Li et al. 2010). Thus, we can use the aforementioned techniques with other enzymes too with modifications done on the enzyme of interest.

Reverse Enzyme Screening Approach

The recent progress in sequencing of genomes of various organisms and the availability of this multitude of data conveniently structured in freely accessible databases like the National Center for Biotechnology Information (www.ncbi.nlm.nih.gov/) and the UniProt (www.uniprot.org/) in the World Wide Web has made it possible to selectively screen for genes expressing enzyme of interest followed by their primer base polymerase chain reaction (PCR) amplification and expression in recombinant host. The PCR-based amplification of gene can follow three strategies. First is amplification of genes whose sequence is already known by conventional primer design and by traditional PCR and, second, amplification of homologous enzyme genes from uncharacterized genomes though the design of degenerate primers followed by PCR amplification by a process known as gene walking. For example, Hallin and Lindgren (1999) designed two sets of PCR primers using consensus regions in gene sequences encoding the two forms of nitrite reductase (Nir), a key enzyme in the denitrification pathway, and successfully amplified *nir* genes in nine species within four genera of nitrifying organisms (Hallin and Lindgren 1999).

The third strategy is amplification of genes from distantly related organisms. In this regard, a recent methodology has been developed to isolate genes coding for functionally equivalent enzyme from distantly related species by Rose et al. (2003) through a new primer design strategy for PCR amplification, based on consensus-degenerate hybrid oligonucleotide primers (CODEHOPs). They have also developed an interactive computational program to design CODEHOP PCR primers from conserved blocks of amino acids within multiple aligned protein sequences. Each CODEHOP consists of a pool of related primers containing all possible 3–4 nucleotide sequences encoding highly conserved amino acids within a 3′ degenerate core, and a longer 5′ non-degenerate clamp region contains the most probable nucleotide predicted for each flanking codon. The primer design software and the CODEHOP PCR strategy have been utilized for the identification and characterization of new gene orthologs and paralogs in different plant, animal, and bacterial species (Rose et al. 2003).

As seen from the above examples, PCR is the most valuable tool to amplify genes of enzyme of interest. Recent developments in conventional PCR techniques are self-sustained sequence replication (3SR) (Guatelli et al. 1990), nucleic acid sequence-based amplification (NSBA) (Compton 1991), Strand Displacement Amplification (SDA) (Walker et al. 1992), and loop-mediated isothermal amplification PCR (LAMP) (Notomi et al. 2000). PCR uses heat denaturation of double-stranded DNA to promote next round of DNA synthesis; NSBA and 3SR eliminate this step of heat denaturation by using a set of transcription and reverse transcription reactions for gene amplification. SDA eliminates the heat denaturation by employing a set of restriction enzyme digestions with modified nucleotides as substrate. Notomi et al. (2000) have developed loop-mediated isothermal amplification PCR (LAMP) which has emerged as the most cost-effective solution for the technique known as real-time PCR. This method can amplify a few copies of DNA to 10^9 in less than an hour under isothermal conditions and with greater specificity (Notomi et al. 2000). This method relies on single temperature incubation of 60–65 °C for displacement DNA synthesis negating the requirement for expensive thermal cyclers. It is performed by Bst DNA polymerase from *Bacillus stearothermophilus*, with high strand displacement activity and a set of two specially designed inner and two outer primers. The inner primers are called the forward inner primer (FIP) and the backward inner primer (BIP), respectively, and each contains two distinct sequences corresponding to the sense and antisense sequences of the target DNA. Detection of amplification products of the targeted DNA can be easily done by photometry for turbidity caused by increasing quantity of magnesium pyrophosphate in solution or by the use of SYBR Green which specifically intercalates only to double-stranded DNA and emits green light ($\lambda_{max} = 520$ nm) (Mori et al. 2001).

Another approach of gene amplification, worth mentioning here, for enzymes, whose sequence is not yet available, is that the sequencing of the N- and C-terminal of proteins, which have not yet been sequenced, can lead to the design of primers and amplification of the gene of the enzyme by degenerate primers coupled with 3′ or 5′ RACE (rapid amplification of cDNA ends) PCR. RACE is a technique which can be used to obtain the full-length sequence of an RNA transcript found within a cell. RACE can provide the sequence of an RNA transcript from a small known sequence within the transcript all the way to the 5′ end (5′ RACE-PCR) or 3′ end (3′ RACE-PCR) of the RNA (Sambrook et al. 2001). The PCR product can then be used for cloning and expression of the enzyme gene. Sakamoto et al. (2011) used this technique and amplified endo-β-1,3-glucanase (GLU1), from the fruiting body of the fungi, *Lentinula edodes*. The *glu1* gene was isolated by rapid amplification of cDNA ends (RACE)-PCR using primers designed from the N-terminal amino acid sequence of GLU1 protein (Sakamoto et al. 2011).

The next step in the reverse enzyme approach, after amplification of the gene of interest for the enzyme, is cloning the amplified gene in a suitable vector followed by its transformation into competent hosts. Traditionally the steps in cloning of an enzyme gene involve DNA isolation followed by restriction digestion of the DNA into desired fragments and then ligating them into the selected vector system. Though these aforementioned processes turn out to be successful, the entire process are fairly time consuming. Thus, sufficient groundwork and experimentation have been carried out to modernize the entire protocol. The recent advances in the cloning techniques have led to the development of more specific, highly reproducible, and guaranteed cloning strategies known as the zero background TA cloning, TOPO cloning, and the Gateway® cloning.

TA cloning is the cloning technique that negates the use of restriction enzymes. The technique is dependent on the ability of adenine and thymine to hybridize in the presence of a DNA ligase. With the property of Taq polymerase terminal transferase activity, deoxyadenosine is added to the 3′ end of double-stranded PCR-amplified DNA duplex leaving a 3′ overhang. Such inserts can be directly cloned to linearized vectors with 5′-deoxythymidine overhang with the aid of a DNA ligase (Chen and Janes 2002). A modified version of TA cloning is known as the zero background TA (ZeBaTA) cloning. The improved system takes advantage of the restriction enzyme XcmI whose restriction site is CCAATACT/TGTATGG to generate a T overhang after digestion and the negative selection marker gene ccdB to eliminate the self-ligation background after transformation which is quite common in the case of simple TA cloning (Chen and Janes 2002). TOPO cloning is also a modified version of the TA cloning strategy and has been invented by Invitrogen. In this method direct insertion DNA takes place in a matter of few minutes with the aid of Taq or Pfu polymerase and vaccinia virus DNA topoisomerase I which specifically recognizes DNA sequence 5′-(C/T) CCTT-3′ into linearized plasmid vectors with 3′ T overhangs, without the requirement for any ligation reaction. Taq polymerase has a nontemplate-dependent terminal transferase activity that adds a single deoxyadenosine (A) to the 3′ ends of PCR products. The topoisomerase I remains covalently bound to the vector and binds to duplex DNA at specific sites and cleaves the phosphodiester backbone after 5′-CCCTT in one strand (Shuman 1991) of a covalent bond between the 3′ phosphate of the cleaved strand and a tyrosyl residue (Tyr-274) of topoisomerase I.

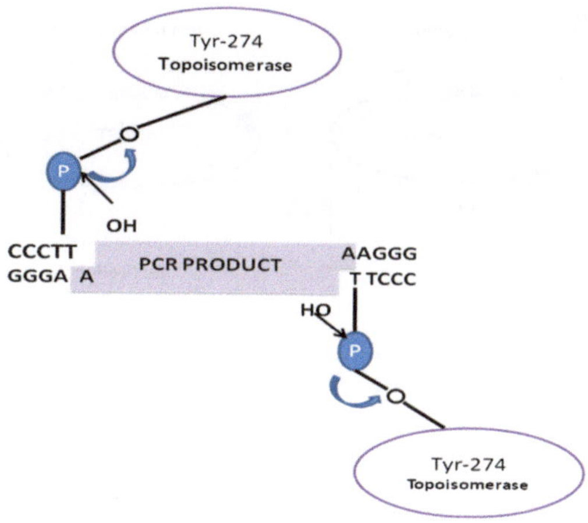

Fig. 7.1 Schematic illustration of the TOPO cloning procedure

The phosphotyrosyl bond between the DNA and enzyme can subsequently be attacked by the 5′ hydroxyl of the original cleaved strand, reversing the reaction and releasing topoisomerase (Shuman 1994). Figure 7.1 illustrates the TOPO cloning procedure.

Gateway Cloning Technology is a powerful new methodology that can greatly facilitate enzyme expression, cloning of PCR products, and analysis of gene function by replacing restriction endonucleases and ligase with site-specific recombination. This technology had been invented and commercialized by Invitrogen since the late 1990s. The overall reaction of Gateway ® cloning has been illustrated in Fig. 7.2. It mainly involves two site-specific recombination reactions known as the LR reaction and the BP reaction. In the LR reaction a recombination reaction between an Entry Clone and a Destination Vector occurs with the action of LR Clonase. An Entry Clone, containing a gene flanked by recombination sites attL1 and attL2, recombines with a Destination Vector (pDESTTM) to yield an Expression Clone and a by-product plasmid with attB1, attB2 and attP1, attP2 sites, respectively. The by-product plasmid contains the ccdB gene and hence gives rise to no colonies when using standard strains of *E. coli*. In BP reaction the gene of interest in the Expression Clone (between attB sites) is transferred into a Donor Vector (containing attP sites) to produce

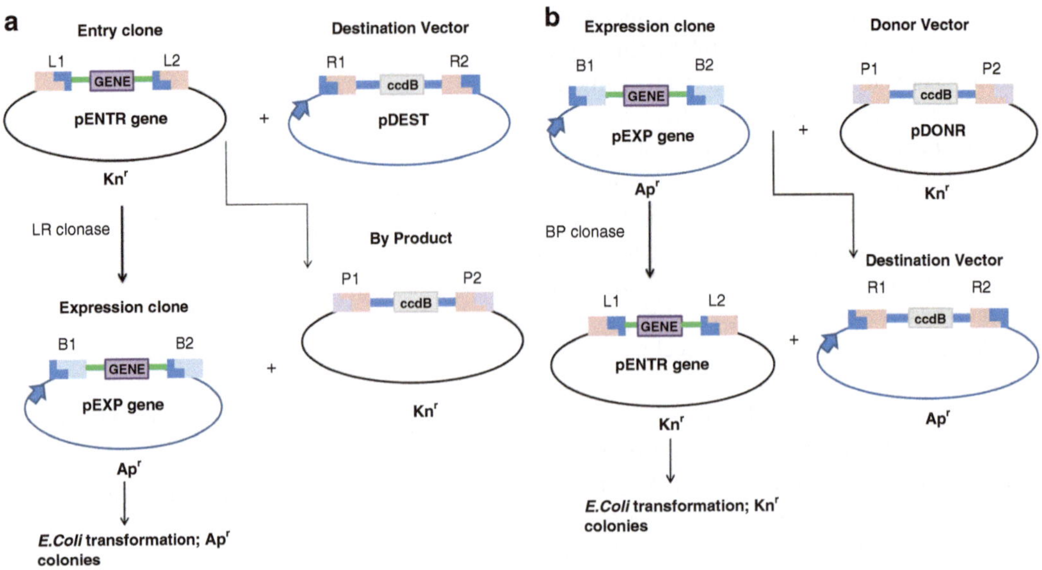

Fig. 7.2 The overall process of Gateway ® cloning. (**a**) The LR reaction. (**b**) The BP reaction. The sites labeled *L*, *R*, *B*, and *P* are, respectively, the *attL*, *attR*, *attB*, and *attP* recombination sites for bacteriophage lambda in *E. coli*. *attL1* reacts only with *attR1*, and *attL2* reacts only with *attR2*. *attB1* reacts only with *attP1* and *attB2* with *attP2* (This figure has been printed with permission from www.lifetech.com/gateway)

a new Entry Clone (attL sites) catalyzed by the BP Clonase enzyme.

Another series of dual expression vectors were constructed by Madzak et al. containing an hp4d promoter which allows for expression of more than one expression cassettes in a single vector and expresses different proteins efficiently and constantly without multiple influences by nutritional and environmental factors in medium, such as carbon/nitrogen sources and pH values (Madzak et al. 2000). This cassette is integrated into the genome of a genetically modified *Yarrowia lipolytica* strain by homologous recombination. This efficient hp4d promoter expression cassette system has been applied to produce various enzymes, such as β-galactosidase, prorennin, cytokinin oxidase, lipase, and laccases successfully (Chuang et al. 2010).

Transformation into competent host cell or expression host is the major vital step for a successful cloning experiment, and enzyme expression level control is still a big challenge in enzyme evolution (Cambon et al. 2010). If this fails we will not be able to recover or express the enzyme of interest. The choice of host affects not only the expression of the enzyme but also the way in which the product can be subsequently purified. Traditional strategies for recombinant enzyme expression involve transfecting cells with a DNA vector that contains the template and then culturing the cells so that they transcribe and translate the desired protein. Typically, the cells are then lysed to extract the expressed protein for subsequent purification. Both prokaryotic and eukaryotic in vivo protein expression systems which consist of a suitable vector with an appropriate promoter and other necessary regulatory sequences along with the gene of interest are widely used. The selection of the system depends on the type of protein, the requirements for functional activity, and the desired yield. *Escherichia coli* have been used extensively till date as expression hosts due to its rapid growth rate, capacity for continuous fermentation, and relatively low cost. Problems associated with such host system are codon bias, correct protein folding, and disulfide bridge formation. Moreover formation of inclusion bodies in such prokaryotic host system is another great obstacle. There are surplus of information available over the Internet on the modern expression host systems developed in the last decade; some of the most relevant developments in this area are discussed here. In this regard the most promising modern expression hosts that were developed are the Rosetta-gami™ B host strains developed by Novagen, as they successfully express the eukaryotic proteins that contain codons rarely used in *E. coli* (Ren 2007). These strains supply tRNAs for AGG, AGA, AUA, CUA, CCC, and GGA codons on a compatible chloramphenicol-resistant plasmid named pRARE. Elorza et al. expressed human alpha galactosidase A, which contains many non-optimized codons disulfide bonds in its native structure in Rosetta-gami™ B host strains (Xiao et al. 2007). Mai et al. have developed the shuttle vector pIKM1 consisting of selection marker which is the thermostable kanamycin cassette from *Streptococcus faecalis* plasmid pKD102for for transformation of *Thermoanaerobacterium* strain JW/SL-YS485 (Mai et al. 2000). Kushnir et al. developed a new inducible protein expression system which was able to overexpress functional proteins, based on the protozoan host *Leishmania tarentolae*. This strain can co-express T7 RNA polymerase and tetracycline repressor and thus can be stably transformed with the heterologous target gene under control of the T7 promoter/TET operator assembly, which can initiate transcription upon addition of tetracycline to the culture medium (Kushnir et al. 2005). Enzymes from extremophiles are hard to express in normal preexisting host systems. Recently Cambon et al. developed a new *Yarrowia lipolytica* JMY1212 expression system which is an efficient tool for rapid and reliable kinetic analysis of improved enzymes (Cambon et al. 2010). This strain allows for targeted integration of the expression cassette into the genome and homogenous transformants (Cambon et al. 2010). Most recently a method designated Plasmid Artificial Modification (PAM) was proposed by Suzuki and Yasui (2011). This PAM method promises higher transformation efficiency as the presence of PAM plasmid encoding the modification enzymes ensures that recombinant plasmids harboring the gene of interest are modified such that it is protected from restriction endonuclease digestion in the target bacterium (Suzuki and Yasui 2011). Another example of an excellent expression system developed by the company TAKARA (http://www.takara-bio.com) is the *Brevibacillus* expression system. *Brevibacillus* is a gram-positive microbe, characterized by its ability to secrete/produce large amount of proteins. This system is recommended for the production of functional proteins that are naturally produced intracellularly but become insolubilized and cannot undergo in vitro refolding when produced by *Escherichia coli*. The target protein can be easily purified using a histidine-tagged protein purification column. The His-Tag can be removed by exposing the purified protein to enterokinase treatment. In similar lines, Biomedal CASCADE™ was developed as a bacterial protein expression system that provides tightly regulated, high-level expression. The system makes use of linked regulatory circuits to amplify gene expression levels when induced, maintaining low basal expression levels under noninducing conditions. Its features are tight control of gene expression, activity at low temperature, low sensitivity to media formulation, and expression of heteromultimeric proteins. Recently a trend has moved towards eukaryotic expression system, and in this regard *Drosophila* expression system (DES®) by Invitrogen and *Leishmania tarentolae* expression system known as LEXSY by Jena Bioscience were developed.

Other than the aforesaid techniques, a new system known as cell-free enzyme expression or in vitro expression has recently emerged as a strong tool in enzyme research as it is simple, inexpensive, and can translate proteins or enzymes in hours from PCR fragments without the need for *Escherichia coli* cloning, and thus it is easily adapted for high-throughput, automation, and/or miniaturization procedures (He 2008). Cell-free protein expression makes use of cellular extracts or purified molecular components to direct protein synthesis from added DNA template in test tubes (He 2008). Finally, cell-free systems are capable of synthesizing large protein populations in a single reaction, offering an exploitable system for developing proteomic tools (He 2008). The cell-free environment is also helpful in the study of protein translation and synthetic biology applications (He 2008). PURExpress™ In Vitro Protein Synthesis Kit from New England Biolab and TNT® Quick Coupled Transcription/Translation Systems are examples of such technology.

Other most promising recently developed and high-fidelity reverse enzyme screening methods are the processes known as metagenomics, diversity generation by error-prone polymerase chain reaction (PCR), and gene shuffling of an existing enzyme's gene or gene family (Wahler and Reymond 2001). Metagenomics is the new rising science used to explore the unculturable microbes directly from the environmental samples. It is the outcome of the strength of genomics, bioinformatics, and system biology and is a multistage process that implies direct cloning of environmental DNA into large clone libraries to facilitate the analysis of the genes and the sequences within the libraries (Handelsman 2004). In the recent past, employing metagenomics, Tirawongsaroj and co-workers isolated novel thermostable lipolytic enzymes from a Thailand hot spring metagenomic library (Tirawongsaroj et al. 2008). Chu et al. also identified two novel esterase genes from a marine metagenomic library through functional screening of lipolytic clones (Chu et al. 2008), and Wang et al. with the aid of bioinformatics and the huge freely accessible databases available on the web designed metagenomic gene-specific primers and isolated 23 truncated lipase gene fragments from 60 metagenomic samples (Wang et al. 2010). They designed primers based on well-known conserved motifs of lipase genes.

From the aforesaid we can say that bioinformatics or in silico biology related to genes of enzymes can play a prominent role in screening and functional analysis of novel enzymes. In 2005 Ishikawa and co-workers discovered a novel restriction endonuclease by genome comparison and application of a wheat germ-based cell-free translation assay from the hyperthermophilic archaeon *Pyrococcus abyssi* (Ishikawa et al. 2005). Their methodology involved identification by comparison of candidate genes from the already sequence and related genomes of the hyperthermophilic archaea *Pyrococcus abyssi* and *Pyrococcus horikoshii* by using certain conserved genes as genomic landmarks (Ishikawa et al. 2005). This method is much more fast and economical than the traditional method used to screen for restriction endonucleases wherein individual strains are cultured and their extracts assayed for the ability to produce specific fragments from small DNA molecules. The method is also beneficial in the case when we are dealing with extremophiles who are fastidious and economically hefty to laboratory culture and maintenance.

Enzyme Purification

An apparent quote from Efraim Racker, an eminent Austrian biochemist, is "Don't waste clean thinking on a dirty enzyme." This quote stresses on enzyme purification as the first criteria before analyzing its properties. The science, art, and practice of protein and enzyme purification are known for more than a century, yet, in many respects, the field is only now evolving (Ward and Swiatek 2009). Each enzyme possesses unique properties, which can be exploited for its purification (Puig et al. 2001), and modern techniques aim for enzyme purification with a minimum of labor and cost to remove all contaminants while retaining as much as possible of the enzyme of interest. Chromatography is called the protein or enzyme purification workhorse for both native and recombinant proteins (Smith 2005). The outline of the general scheme of enzyme purification is illustrated in Fig. 7.4 below. Today numerous companies like Qiagen, Applied Biosystems, BioChrom, and Amersham Biosciences offer improvements to these techniques by introducing new tweaks to improve efficacy of the procedure. Here we discuss the recent strategies for purifying native and recombinant enzymes. The generalized steps for enzyme purification are given in Fig. 7.3.

Among recently developed column chromatographic matrices, BioChrom has developed a polymeric material, Hydrocell. The high level of cross-linking in the polymer beads is intended to give better and faster protein separation compared to conventional silica-based HPLC

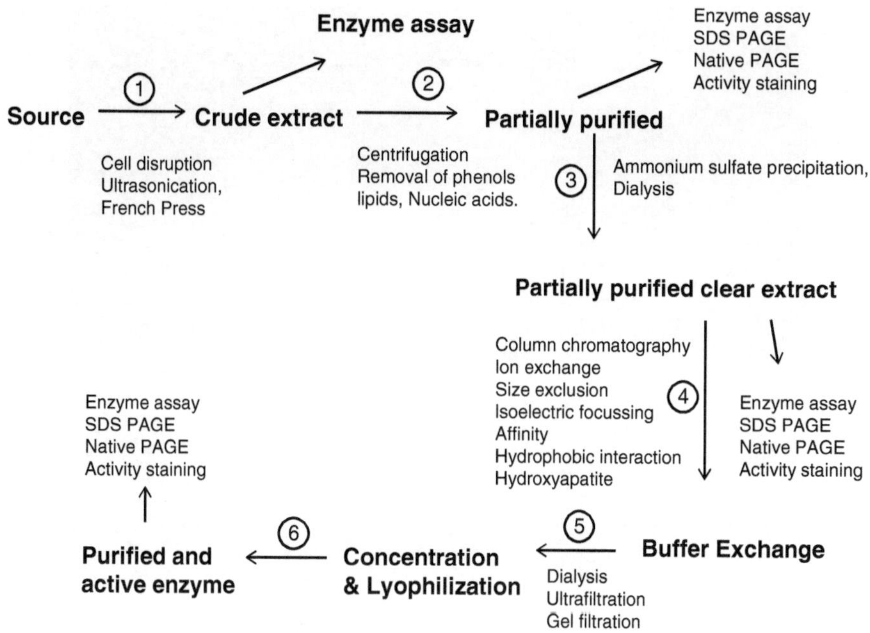

Fig. 7.3 Generalized steps for enzyme purification

columns. GE Healthcare has also recently launched a new chromatography matrix for the purification of glutathione S-transferase (GST) and His-tagged proteins which combines small-bead technology with affinity purification. The relatively small bead size results in distinctly separated protein elution peaks.

Another relatively new method of enzyme purification is affinity ultrafiltration (Galaev 1999). In affinity ultrafiltration, affinity separation is combined with membrane filtration technique. The enzyme to be purified is complexed with a macroligand composed of a soluble polymer along with target protein-specific affinity ligands. The complex is trapped by an ultrafiltration membrane, whereas unwanted proteins pass through the membrane (Galaev 1999). Several enzymes like urokinase have been purified through this method (Male et al. 1990). Two-Step Affinity Purification System by Qiagen combines sequential affinity purification steps and yields ultrapure (>98 %) protein. This system can be useful with enzymes in the presence of high concentrations of chelating compounds or from eukaryotic cell lysates. In similar lines, magnetic beads have become a popular choice for protein and enzyme purification due to their speed, ease of use, and affordability. They possess an iron core surrounded by an inert polymer material (Fig. 7.4). When subjected to magnetic field, the beads behave like magnets simplifying purification procedures negating the use of centrifugation. Such beads also come along with specific molecular tags suitable for affinity purification. Some companies such as Dynal Biotech, Polysciences, and BioScience Beads have capitalized on magnetic beads as one of their leading products (Smith 2005).

For purification of small quantity of proteins and enzymes, there is an increasing trend in the use of mass spectroscopy in protein and enzyme analysis. For this purpose, Vivascience has developed Vivapure MALDI-Prep Micro spin columns, which contain the C-18 (the number of carbons in the alkyl chains attached to silica particles in conventional column chromatography membrane), for desalting and concentrating protein or enzyme digests for mass spectroscopy. Companies like Millipore are providing solutions for purification of bulk amount of enzymes. They have developed

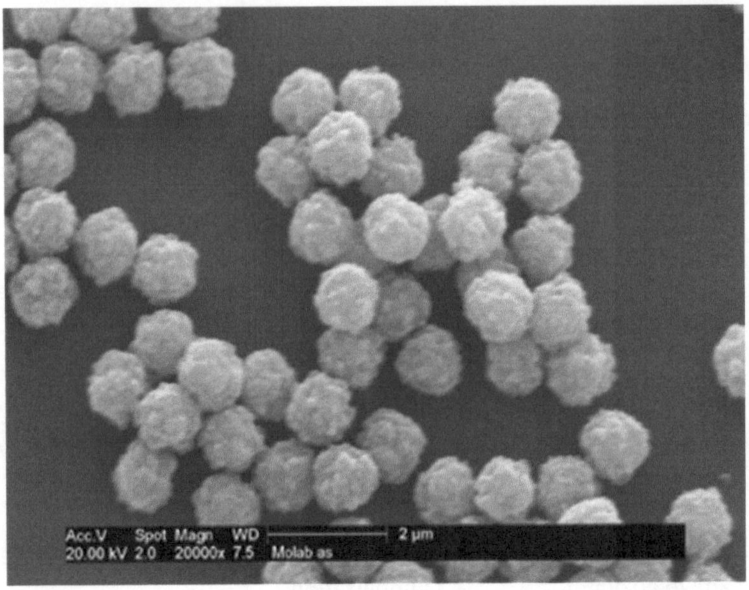

Fig. 7.4 Magnetic Dynabeads from Dynal Biotech (Courtesy of Dynal Biotech) (This figure has been reprinted with permission from Smith 2005; Striving for purity: advances in protein purification. Nature Methods 2, 71–77 (2005))

the Amicon concentrators which can be used to concentrate and desalt from 4 to 15 ml of solution (Smith 2005). A recent breakthrough in mass spectrometry was made by scientists at ETH Zurich, along with AB SCIEX, a global leader in life science analytical technologies; they have developed MS/MS^{ALL} with SWATH™ Acquisition, a groundbreaking that successfully quantifies nearly peptides and proteins in a sample from a single analysis. These techniques employ data-independent methods to systematically generate complete, high-specificity fragment ion maps that can be queried for the presence and quantity of any protein of interest using a targeted data analysis strategy (Framingham, Mass, April 17, 2012). This technique will thus aid in enzyme sequencing and identification in a much faster rate in the near future.

Recombinant protein expression and purification calls for production of large amounts of an affinity-tagged protein so that a single purification step using affinity chromatography is sufficient to achieve the desired level of purity. In affinity chromatography, the protein or enzyme of interest is purified by its ability to bind a specific ligand that is immobilized on a chromatographic bead material (matrix). This matrix is usually packed into a column. Crude cell lysates are loaded onto the column under conditions that ensure specific binding of the protein to the immobilized ligand. Other enzymes that do not bind the immobilized ligand are washed through. Elution of the bound enzyme of interest can be achieved by changing the experimental conditions to favor desorption.

There is a variety of new commercially available fusion systems where enzyme genes are tagged. This is designed to ease enzyme purification followed by cleavage of enzyme of interest. A variety of tags are available for such purpose and are listed in Table 7.1. The tagged enzyme is purified through washing in an affinity chromatography medium and the tagged enzyme is eluted. Modern columns developed in this regard are His GraviTrap for histidine-tagged proteins. HiTrap columns are suitable for use with a syringe or peristaltic pump for histidine-, GST-, MBP-, and Strep-tagged proteins (HisTrap, GSTrap™, MBPTrap™, and StrepTrap™ columns, respectively).

After affinity purification of the recombinant enzyme, cleavage of the protein from the tag system calls for proteases like factor Xa, thrombin, or enterokinase. The limitation is that they may

Table 7.1 Fusion vector systems with tags for recombinant protein purification

Company	Vector system	Tags (N/C terminus)	Purification
Novagen, Inc. (Merck)	T7 pET	T7-tag (monoclonal antibody)	1. Ni affinity column purification followed by Exopeptidase cleavage of N-terminal His tag
		His Tag (Histidine$_6$ tag for metal chelation chromatography)	2. Antibody affinity systems
		S-tag (RNAse S-protein)	
		HSV-tag (monoclonal antibody)	
		pelB/ompT (potential periplasmic localization)	
Pharmacia Biotech (GE Healthcare)	pGEX vectors	Glutathione S-transferase (GST)	Glutathione Sepharose 4B affinity chromatography
Eastman Kodak	Flag system	Flag marker octapeptide	Affinity chromatography and an amino-terminal Flag peptide can be removed by the protease, enterokinase
Promega	PinPoint™ Xa Protein Purification System	Biotinylated	Affinity-purified using the SoftLink™ Soft Release Avidin Resin from promega
Stratagene	VariFlex™	Solubility enhancement tags (SETs)	Enhances solubility of the fusion peptide by providing a net negative charge preventing protein aggregation. Proteins can be purified through affinity chromatography through Streptavidin Resin and tags can be removed by Thrombin digestion
Thermo Scientific	Myc-tag	Derived from the c-myc gene product	Immunoprecipitation purification
Invitrogen	Champion™ pET SUMO	Small ubiquitin like modifier (SUMO) fusion	Produces soluble protein in *Escherichia coli*, for purification the SUMO moiety can be cleaved by the highly specific and active SUMO (ULP-1) protease at the carboxyl terminal, producing a native protein
Takara	pCold TF DNA Vector	Cold shock expression vector that expresses Trigger Factor (TF) chaperone as a soluble tag along with His-Tag sequence	The chaperone ensures co-translational protein folding. Affinity purification followed by Thrombin or Factor Xa cleavage
Takara	Single Protein Production system	MazF tag	Utilizes *E. coli* protein MazF which is an endoribonuclease and cleaves mRNA at ACA sequence. Thus transcript of interest should lack ACA sequence so that it cleaves ones derived from the host proteins or others at ACA sequences
BioRad	Profinity eXact fusion-tag system	Prodomain of the subtilisin protease	Utilizes a modified form of the subtilisin protease, which is immobilized onto a chromatographic support and used to generate pure, tag-free target protein in a single step

(continued)

Table 7.1 (continued)

Company	Vector system	Tags (N/C terminus)	Purification
European Molecular Biology Laboratory	TAP tag	Calmodulin binding peptide (CBP) from the N-terminal, followed by tobacco etch virus protease (TEV protease) cleavage site and Protein A, which binds tightly to IgG	Tagged protein binds to beads coated with IgG, the TAP tag is then broken apart by an enzyme, and finally a different part of the TAP tag binds reversibly to beads of a different type
Schaffer et al. 2010	SnAvi-tag	Fluorescent tag	Enhanced green fluorescent protein (EGFP) to enable in vivo localization studies, the TEV-protease recognition motif to allow elution under non-denaturing conditions

cleave the protein of interest also. Thus, new proteases were discovered that have made their way into commercial expression vector systems. PreScission protease by GE Healthcare Life Sciences (Amersham Pharmacia) is one such product based on the rhinovirus protease. New England Biolabs markets a system named IMPACT where the cleavage is affected by the activity of a self-splicing protein called an intein (Chong et al. 1998). For tag-free expression, Wacker Biotech ESETEC® technology provides an innovative and highly efficient *Escherichia coli* expression system, which enables secretion of native or recombinant enzyme products into the fermentation broth. This simplifies primary recovery and purification processes. The system is based on a two-step export mechanism: first, the target product is transported across the cytoplasmatic membrane into the periplasmatic space. During this step the signal peptides are cleaved off, releasing the native product. The second step is mediated by a unique feature of the proprietary WACKER Secretion Strain. The correctly folded product is secreted from the periplasmatic space across the outer membrane into the culture broth for easy recovery. Although the tagged system allows for single-step enzyme purification through affinity chromatography, it suffers from the disadvantage that affinity tags may sometimes interfere with the post-purification use of the protein. The solution to such a problem is known as Capture, Intermediate Purification, and Polishing (CIPP) technique. In the capture phase, enzymes are isolated, concentrated, and stabilized. During the intermediate purification phase, bulk impurities such as other proteins and nucleic acids, endotoxins, and viruses are removed. In the polishing phase, trace amounts or closely related substances are removed (GE Healthcare 2009). Another worth-mentioning innovation is in the use of adenosine triphosphate (ATP)-affinity chromatography. This has been widely used to purify various ATP-binding enzymes such as kinases and β- and γ-glutamate decarboxylase using affinity purification by ATP-matrix attachment (Jeansonne et al. 2006; Bendz et al. 2007; Dhillon et al. 2009).

Recent trend in enzyme detection after purification has seen the development of Caliper Life Sciences' LabChip90, an automated electrophoresis system and an alternative to the conventional slab gel electrophoresis system such as SDS-PAGE. It is less expensive than traditional gel electrophoresis and increases sample throughput by two- to threefold. This system integrates and automates all of the processing steps right from sampling to separation to detection and to quantification via the software that comes with the instrument (Smith 2005).

Enzyme Engineering

Enzyme engineering leads to tailor-made physical and catalytic properties of enzymes. This approach can be used either to study the function of an enzyme or to produce an enzyme of desired physicochemical properties. Two popular approaches are directed evolution and rational design. An effective directed evolution strategy combines different mutagenesis methods with efficient high-throughput screening or selection assays. Rational design is pillared on the three-dimensional enzyme structure and identification of amino acids involved in enzyme functionality or catalysis and constructing mutants by single-site or multisite, site-directed mutagenesis (Cambon et al. 2010). Song et al. (2008) used a modified method of site-directed mutagenesis using RecA-mediated homologous recombination and temperature-sensitive replications, site-directed mutagenesis of the catalytic triad amino acid cysteines in tandem ketosynthase domains to test the function they play in OZM biosynthesis (Song et al. 2008). A modified and better method for multisite-directed mutagenesis was developed by Tian et al. (2010) which is based on polymerase chain reaction (PCR), DpnI digestion, and overlap extension. The method does not require 5′-phosphorylated primers and ligation and, thus, significantly simplifies the routine work and reduces the experimental cost for multisite-directed mutagenesis (Tian et al. 2010). Today, rational engineering is also benefitted through in silico approaches involving multiple web-based tools and software packages which aid in saving time, effort, and money. For example, protein mutant data is being compiled in the protein mutant database (PMD). This database provides information on the results of functional and/or structural influences that can be brought about by amino acid mutations. It covers natural as well as artificial mutants, including random and site-directed ones. The advantage of PMD is that it is based on literature, not on proteins (Kawabata et al. 1999). A gist of the various other web-based tools has been listed in Table 7.2.

Apart from the aforesaid, a methodology in enzyme evolution known as synthetic gene technology has recently evolved which results in de novo synthesis of entire protein-coding sequences of the enzyme from preannealed oligonucleotides (Gustafsson et al. 2004). Through this strategy, genes of over 1 Kb can now be efficiently and economically synthesized in a matter of weeks. In creating the synthetic gene, degeneracy of the genetic code can be used to generate nucleotide sequences that have useful properties such as large numbers of endonuclease restriction sites, optimized primer sites for the polymerase chain reaction (PCR) and sequencing, and desired levels of GC content and codon bias (Withers-Martinez et al. 1999).

Enzyme Structure and Function Determination

NMR spectroscopy and X-ray crystallography are the routine approaches that provide an atomic-level, 3D structural model of a protein or an enzyme. Crystallization has been automated which allows the use of small protein sample volumes, and development of crystallization robots has contributed to increasing the efficiency of screening crystallization conditions (Liu and Hsu 2005). The applications of these high-resolution approaches, however, are limited by enzyme size, conformational flexibility, and aggregation propensity. Mass spectrometry (MS) has become the recent method of choice for studying protein structure, dynamics, interactions, and function. Hamuro et al. (2003) published the use of rapid analysis of protein or enzyme structure by hydrogen/deuterium exchange mass spectrometry (Hamuro et al. 2003). Recent developments in mass spectrometry as the method of choice for the high-throughput identification of proteins and their modifications led to the concept of protein cross-linking to determine structural information about protein like enzymes in combination with mass spectrometry (Leitner et al. 2010). The overall process in protein structure determination has been illustrated in Fig. 7.5.

The knowledge of three-dimensional structure and space by these techniques is still limited. Thus, computational methods such as comparative and de novo approaches and molecular dynamic simulations are intensively used as alternative tools to predict the three-dimensional structures and dynamic behavior of enzymes (Liu and Hsu 2005).

Table 7.2 Web tools and software packages for predicting protein and enzyme stability on point mutations

Sl No	Web tool/software	Applications	Reference
1	HotSpot Wizard	HotSpot Wizard is a web server for automatic identification of 'hot spots' for engineering of substrate specificity, activity or enantioselectivity of enzymes and for annotation of protein structures	Pavelka et al. (2009)
2	iPTREE-STAB	It is a web based tool which discriminates the stability of proteins and predicts their stability changes (ΔΔG) upon single amino acid substitutions from amino acid sequence	Huang et al. (2007)
3	Eris server	It is a protein stability prediction server. Eris server calculates the change of the protein stability induced by mutations (ΔΔG)	Yin et al. (2007)
4	I-Mutant	Support Vector Machines based Predictor of Protein stability changes upon Single Point Mutation from the Protein Sequence and Structure. The tool correctly predicts whether the protein mutation stabilises or destabilises the protein in 80 % of the cases when the three-dimensional structure is known and 77 % of the cases when only the protein sequence is available	Capriotti et al. (2005)
5	CUPSAT	A web tool to analyse and predict protein stability changes upon point mutations	Parthiban et al. (2006)
6	MUpro	It is a set of machine learning programs to predict how single-site amino acid mutation affects protein stability	Cheng et al. (2006)
7	Swiss-PdbViewer	This software allows browsing a rotamer library in order to change amino acids sidechains. This can be very useful to quickly evaluate the putative effect of a mutaion before actually doing the lab work	http://spdbv.vital-it.ch/

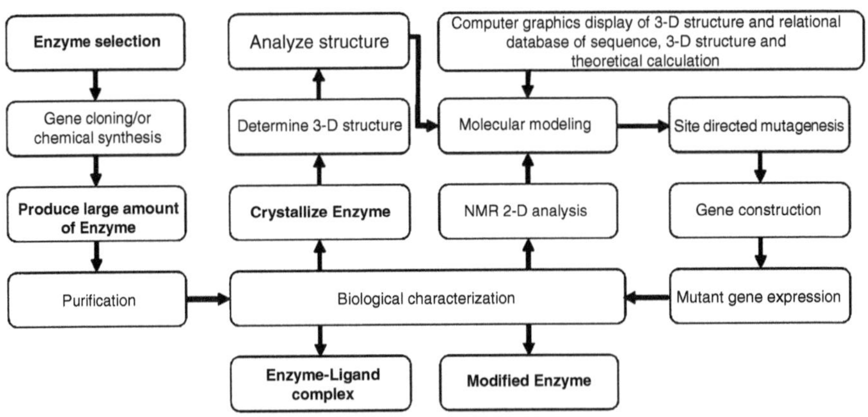

Fig. 7.5 Flowchart showing the overall process of protein or enzyme structure determination (This figure has been adapted from Liu and Hsu (2005))

Currently high-throughput genome sequencing programs coupled have provided researchers with a perplexing array of sequence and biological data to contend with. Thus, application of bioinformatics tools is necessary to process and prioritize the plethora of data. For this purpose freely accessible enzyme databases like BRENDA have been recently developed that have been classified by the International Union of Biochemistry and Molecular Biology (IUBMB). Every classified enzyme is characterized with respect to its catalyzed biochemical reaction. Kinetic properties of the corresponding reactants, i.e., substrates and products, are also described in detail (Scheer et al. 2011). Other existing enzyme databases are Enzyme nomenclature database (http://enzyme.expasy.org/) and IntEnz (http://www.ebi.ac.uk/intenz/). In addition to enzyme databases, software has been developed to analyze and present functional and statistical data for enzyme kinetics; various software tools have been developed. EZ-Fit™ by Perrella Scientific is one such software. This software has features for nonlinear regression of enzyme inhibition data. It makes data entry simple followed by selection of an equation from the menu of commonly used inhibition models. EZ-Fit does the rest automatically, curve-fits the data to one or more models, displays the results, and draws a graph of the data and curve (Perrella 1999). VisualEnzymics by SoftZymics is another such tool which presents a custom visual interface. Through this software we can see curve fits in real time providing the user with interactive control. It possesses models for one substrate rate saturation data, one substrate one inhibitor data, pH rate profiles, exponential data, dose response data, two substrate data, one substrate one activator data, binding data, and tight binding data. Users can access the web link http://www.softzymics.com/index.htm for further details.

Nowadays, a plethora of enzyme sequence data exist in the NCBI (http://www.ncbi.nlm.nih.gov) and UniProt (www.uniprot.org) databases but lack crystallized structure information in the Protein Data Bank (www.rcsb.org). Thus, their functional elucidation becomes difficult. In modern era to speed up work, protein structure models are designed computationally based on the principle of homology with known target protein structures, and the process is known as homology modeling or comparative modeling. Homology modeling relies on the production of an alignment that maps residues in the query sequence to residues in the template sequence. The quality of the model depends on the quality of the sequence alignment and template structure (Venselaar et al. 2010). Today various software packages exist which involve rigorous mathematical models to build the protein model structures. Some of them are, namely, Modeller (Fiser and Sali 2003), Swiss Model (Arnold et al. 2009), and I-Tasser (Roy et al. 2010).

Enzyme–substrate or inhibitor binding is of critical importance for biomedical research; in vitro analysis of the same is time consuming and requires the assistance from patient unbiased researchers. To speed up work and reduce the workload due to trial and error methodologies, computational programs have been developed known as molecular docking to analyze enzyme–ligand binding. Molecular docking is the computational modeling of the quaternary structure of complexes formed by two or more interacting biological macromolecules enabling us to predict how small molecules, such as substrates or drug candidates, bind to a receptor of known 3D protein or enzyme structures. Popular program packages are AutoDock (Goodsell and Olson 1990), Gold (Jones et al. 1995), and FlexX (Rarey et al. 1996) and are available to carry out the docking studies.

Apart from enzyme structure prediction, we can enjoy the benefits of computational programming and simulate the behavior of enzymes in real time up to femto (10^{-15}) second scale. This process is known as molecular dynamics simulation. Enzyme–substrate binding, enzyme–inhibitor interaction, enzyme folding, and thermodynamics can be investigated through MD simulation. Software packages regularly utilized for MD simulation are NAMD (Nelson et al. 1996), AMBER (Ponder and Case 2003), and GROMACS (Van der Spoel et al. 2005). This computational method calculates the time-dependent behavior of a molecular system, and it is now routinely used to examine the structure, dynamics, and thermodynamics of biological molecules and their complexes. Numerous enzymes like protein tyrosine phosphatase 1B (Peters et al. 2000) and metaloenzyme thiocyanate hydrolase (Peplowski and Nowak 2008) have been investigated through this approach. Buch et al. (2011) successfully reconstructed the complete binding process of the enzyme–inhibitor complex trypsin–benzamidine by performing 495 molecular dynamics simulations (Buch et al. 2011).

Enzyme Stabilization

It is a fact that enzymes function in industrial solvents with enhanced technological utility (Klibanov 2001). Enzyme-catalyzed reactions in organic solvents and supercritical fluids have found numerous potential applications, some of which have been successfully commercialized (Klibanov 2001). Research over the past few decades has shown that in such solvents, enzymes can catalyze reactions which are seemingly impossible in water (Klibanov 2001). A priori, it can be of concern that when an enzyme is exposed to an organic solvent, it can denature, but enzymes were instead discovered to be much stable and to retain "molecular memory" in such solvents. The reason behind is that reactions such as deamidation of thermolabile residues and hydrolysis of peptide bonds do not ensue in solvents in the absence of water. The stability of enzymes in organic solvents is also due to their conformational rigidity in the dehydrated state (Klibanov 2001). Furthermore, proteolytic degradation does not occur in the absence of water as they are insoluble in organic solvents (Klibanov 2001). Irrespective of the aforesaid, large-scale usage of organic solvents provided with certain limitations, such as increased polarity of organic solvents, typically results in reduced catalytic activity (Lee and Dordick 2002). Recent advances in nanotechnology have promised solution in this booming area of enzyme stabilization, suggesting the usage of inorganic nonaqueous solvents or the room temperature ionic liquids (RTILs) for reaction medium engineering or by means of nanostructures with large surface area for the immobilization of enzyme molecules (Kim et al. 2006a). Ionic liquids have proved to be a better reaction media for enzymes since they have essentially zero vapor pressure, are highly stable to the range of operating conditions in bioreactors, are stable to air and water, and enable the choice of a wide range of substrates that is insoluble in most organic solvents (Kim et al. 2006b). For example, Erbeldinger and colleagues used 1-butyl-3-methylimidazolium hexafluorophosphate as the solvent for thermolysin-catalyzed synthesis of Z-aspartame (Erbeldinger et al. 2000). Along similar lines, lipase-catalyzed transamination of carboxylic acids with ammonia has been established in 1-butyl-3-methylimidazolium tetrafluoroborate (Kim et al. 2006b).

Other than room temperature ionic liquids, nanostructures like nanoparticles, nanofibers, mesoporous silica, and nanoparticles prepared via sol–gel encapsulation have been instrumental in enzyme stabilization. More recent development in this area is the "single-enzyme nanoparticles (SENs)" which is a new enzyme composite of nanometer scale by Kim and Grate (Kim et al. 2006a). Nanostructures are better carriers for immobilization of enzymes since their reduced size can generally improve the efficiency of immobilized enzymes due to larger surface area per unit mass, pore sizes tailored to protein molecule dimensions, functionalized surfaces, multiple sites for interaction or attachment, and reduced mass-transfer limitations (Kim et al. 2006a). A recent report using magnetic nanoparticles for the enzyme immobilization was demonstrated with covalently attached lipase on the magnetic γ-Fe_2O_3 nanoparticles and showed stability for a month (Dyal et al. 2003). One disadvantage in using nanoparticles is that their dispersion in reaction solutions and the subsequent recovery for reuse are often found to be a daunting task. Thus, a replacement of nano-particles can be the electrospun nanofibers which provide a large surface area for the attachment or entrapment of enzymes and the enzyme reaction. Another recent achievement in nanoenzymology is the nanoparticles called Single-Enzyme Nano-particles (SENs) (Kim et al. 2006a). Each enzyme molecule is surrounded with a porous composite organic/inorganic network of less than a few nanometers thick. Converting free enzymes to SENs can result in significantly more stable catalytic activity, while the nanoscale structure of the

SEN does not impose a serious mass-transfer limitation on substrates (Kim et al. 2006a).

Interestingly another breakthrough in nanoenzymology was reported in November 2006 when it was discovered that silica-spun FMS pores which are hexagons of approximately 30 nm in diameter and mimic the crowding of cells can be used for refolding of enzymes and upon refolding the enzyme gets reactivated and becomes capable of catalyzing thousands of reactions a second (PNNL 2006).

Improved Bioseparation Techniques: Downstream Processing of Enzymes

Downstream processing of enzymes holds an important position in industries as the bulk of the product cost is determined by the steps leading to purification of enzymes. Thus, biotechnology companies are prospecting for new and improved processing methods. New improved methods over the conventional chromatography and membrane filtration techniques are the cloud point extraction (CPE) and field-assisted separation technologies like MAGSEP (magnetic separator) and ELECSEP (electrophoretic separator) for the quantitative separation of enzymes are gaining more importance.

Cloud point extraction technique has been used to separate proteinase 3 (Pr3) from neutrophil azurophilic granules, tyrosinase from mushroom pileus by Garcia-Carmona and co-workers, cholesterol oxidase extracted from *Nocardia rhodochrous* by Kula and co-workers, and hexokinase and lactate dehydrogenase from aqueous solutions using Triton X-114 as the surfactant. The technique in brief can be described as surface-active agents (surfactants, detergents) can aggregate in aqueous solution to form colloidal-sized clusters referred to as micelles (normal micelles). The minimum concentration of surfactant required for this phenomenon to occur is called the critical micelle concentration (CMC). Upon heating, aqueous solutions of many nonionic surfactants become turbid at a temperature known as the cloud point (or the lower consulate temperature), above which there is a separation of the solution into two phases (Minuth et al. 1996).

Affinity chromatography techniques are the most applied tools available for downstream processing. However, porous affinity supports can only work in a later stage of purification in clear solutions and not in early stages when fouling compounds are present in the system. Nonporous support particles can replace such a system if and only if the size of nonporous support particles is in the range of 0.1–1 μm. Magnetic separation seems to be the only feasible method for the recovery of such small particles from the biological debris. For separating enzymes, an appropriate affinity ligand is usually immobilized on a magnetic carrier, such as silanized magnetite, or on polymeric magnetic (chitin) or magnetizable particles.

The electrophoretic separation of proteins without gels has been an innovative goal in separation research. Electrophoretic separations are influenced by factors such as size (or molecular weight), shape, secondary structure, and charge of the macromolecule or cell. Ohmic heating hinders the scale-up of electrophoresis. The heat generated is equal to the product of the current and voltage, and this heat can cause free convection and mixing within the system. Too much heat can denature the labile biomolecules or cells. To overcome such problems, the multistage electrophoretic method was developed which is a combination of free electrophoresis and multistage extraction. The apparatus used is known as advanced separation apparatus (ADSEP). This ADSEP was modified to act as an ELECSEP by replacing the chamber bottoms with metal cover plates. In these systems electrodes are kept over these cover plates with gaskets in between them. Each cavity has a height of few millimeters ensuring the fluid within it to remain isothermal during the application of an electric field, which transfers the molecules from the bottom to the top cavity. As each molecule is transferred to the new cavity, it is either drawn into the upper cavity by the electric field or left in the lower cavity, depending on its electrophoretic mobility (Karumanchi et al. 2002).

Conclusion

Modern enzyme research is engrossed in developing better ways of isolation, purification, characterization, and stabilization of enzymes in an application-oriented approach. Numerous companies like Qiagen, Applied Biosystems, BioChrom, and Amersham Biosciences are involved in such research to improve efficacy of the whole procedure. Improvement of existing methods in enzyme research is being carried out through the interplay of multidisciplinary subjects and multitasking through a single approach or equipment. To mention a few trends, miniaturization of equipments, robotic handling of high-end equipments, and development of label-free techniques by the use of biosensors to detect as low as femto (10^{-15}) molar quantity of enzymes are the modern approaches. Available biological databases and knowledge bases of enzymes have added new tweak to existing methodologies by saving time and effort and thus increasing industrial potential for the applicability of enzymes. Modern PCR like self-sustained sequence replication (3SR), loop-mediated isothermal amplification PCR, RACE, and cloning techniques like Gateway and TOPO cloning strategies is bringing enzyme research to the forefront. Breakthroughs in evolution of enzyme research that have been brought about by approaches like metagenomics are leading to the isolation of enzymes even from uncultured sources and cell-free in vitro enzyme expression along with their affinity tag-free expression and in silico analysis of their kinetics with substrates and inhibitors are possible through the aid of rigorous mathematical algorithm packages known as molecular dynamics simulation and molecular docking. Another worth-mentioning approach is the synthetic gene technology which can be used to design and synthesize enzyme genes with desired properties. Improvements have also occurred in the downstream processing and stabilization of enzymes. Conclusively we can say that enzyme research has really evolved and has entered a new era, and we expect more innovations in this field in the near future.

References

Arnold K, Bordoli L, Kopp J, Schwede T (2009) The SWISS-MODEL workspace: a web-based environment for protein structure homology modelling. Bioinformatics 22(2):195–201

Aw KL, Yao SQ, Uttamchandani M (2010) Nanodroplet microarrays for high-throughput enzyme screening. Methods Mol Biol 669:79–93

van Bailen JB, Li Z (2002) Enzyme technology: an overview. Curr Opin Biotechnol 13(4):338–344

Bendz H, Ruhland SC, Pandya M et al (2007) Human heat shock protein 70 enhances tumor antigen presentation through complex formation and intracellular antigen delivery without innate immune signaling. J Biol Chem 282(43):31688

Buch I, Giorgino T, De Fabritiis G (2011) Complete reconstruction of an enzyme-inhibitor binding process by molecular dynamics simulations. Proc Natl Acad Sci USA 108(25):10184–10189

Cambon E, Piamtongkam R, Bordes F et al (2010) A new *Yarrowia lipolytica* expression system: an efficient tool for rapid and reliable kinetic analysis of improved enzymes. Enzym Microb Technol 47:91–96

Capriotti E, Fariselli P, Casadio R (2005) I-Mutant2.0: predicting stability changes upon mutation from the protein sequence or structure. Nucleic Acids Res 33:W306–W310

Chen B-Y, Janes HW (2002) PCR cloning protocols. Humana Press, New Jersey

Cheng J, Randall A, Baldi P (2006) Prediction of protein stability changes for single-site mutations using support vector machines. Proteins Struct Funct Bioinform 62(4):1125–1132

Chong S, Montello GE, Zhang A et al (1998) Utilizing the C-terminal cleavage activity of a protein splicing element to purify recombinant proteins in a single chromatographic step. Nucleic Acids Res 26:5109–5115

Chu X, He H, Guo C, Sun B (2008) Identification of two novel esterases from a marine metagenomic library derived from South China Sea. Appl Microbiol Biotechnol 80(4):615–625

Chuang L-T, Chen D-C, Nicaud J-M, Madzak C, Chen Y-H, Huang Y-S (2010) Co-expression of heterologous desaturase genes in *Yarrowia lipolytica*. New Biotechnol 27(4):277–282

Compton J (1991) Nucleic acid sequence-based amplification. Nature 350(6313):91–92

Dhillon AS, Yip YY et al (2009) The C-terminus of Raf-1 acts as a 14-3-3-dependent activation switch. Cell Signal 21(11):1645

Dyal A, Loos K, Noto M, Chang SW, Spagnoli C, Shafi KV, Ulman A, Cowman M, Gross RA (2003) Activity of *Candida rugosa* lipase immobilized on gamma-Fe2O3 magnetic nanoparticles. J Am Chem Soc 125(7):1684–1685

Erbeldinger M, Mesiano AJ, Russell AJ (2000) Enzymatic catalysis of formation of Z-aspartame in ionic liquid – an alternative to enzymatic catalysis in organic solvents. Biotechnol Prog 16(6):1129–1131

Fiser A, Sali A (2003) Modeller: generation and refinement of homology-based protein structure models. Methods Enzymol 374:461–491. doi:10.1016/S0076-6879(03)74020-8

Framingham M (April 17, 2012). Business Wire Data-independent SWATH™ acquisition on the TripleTOF™ 5600 System enables comprehensive proteome-wide quantitation. Mol Cell Proteomic. http://www.businesswire.com/news/home/20120417005204/en/Breakthrough-Proteomics-Technique-SWATH%E2%84%A2-Acquisition-Worldwide-Scientific

Galaev IY (1999) New methods of protein purification. Affinity ultrafiltration. Biochemistry (Mosc) 64(8): 849–856

Gateway® Cloning. Life Technologies. USA. www.lifetech.com/gateway

GE Healthcare (2009) Strategies for protein purification handbook. www.gelifesciences.com/protein-purification

Goodsell DS, Olson AJ (1990) Automated docking of substrates to proteins by simulated annealing. Proteins Struct Funct Genet 8:195–202

Guatelli JC, Whitfield KM, Kwoh DY, Barringer KJ, Richman DD, Gingeras TR (1990) Isothermal in vitro amplification of nucleic acids by a multienzyme reaction modeled after retroviral replication. Proc Natl Acad Sci 87:1874–1878

Gustafsson C, Govindarajan S, Minshull J (2004) Codon bias and heterologous protein expression. Trends Biotechnol 22:346–353

Hallin S, Lindgren P-E (1999) PCR detection of genes encoding nitrite reductase in denitrifying bacteria. Appl Environ Microbiol 65(4):1652–1657

Hamuro Y, Coales SJ, Southern MR et al (2003) Rapid analysis of protein structure and dynamics by hydrogen/deuterium exchange mass spectrometry. J Biomol Tech 14(3):171–182

Handelsman J (2004) Metagenomics: application of genomics to uncultured microorganisms. Microbiol Mol Biol Rev 68(4):669–685

He M (2008) Cell-free protein synthesis: applications in proteomics and biotechnology. N Biotechnol 25:126–132

Huang L-T, Gromiha MM, Ho S-Y (2007) iPTREE-STAB: interpretable decision tree based method for predicting protein stability changes upon mutations. Bioinformatics 23(10):1292–1293

Hwang B-Y, Kim B-G (2004) High-throughput screening method for the identification of active and enantioselective ω-transaminases. Enzym Microb Technol 34(5):429–436

Ishikawa K, Watanabe M, Kuroita T et al (2005) Discovery of a novel restriction endonuclease by genome comparison and application of a wheat-germ-based cell-free translation assay: PabI (5′-GTA/C) from the hyperthermophilic archaeon *Pyrococcus abyssi*. Nucleic Acids Res 33(13):e112

Jeansonne DP, Bordes TJ et al (2006) A rapid ATP affinity-based purification for the human non-receptor tyrosine kinase c-Src. Protein Expr Purif 46:240

Jones G, Willett P, Glen RC (1995) Molecular recognition of receptor sites using a genetic algorithm with a description of desolvation. J Mol Biol 245:43–53

Kawabata T, Ota M, Nishikawa K (1999) The protein mutant database. Nucleic Acids Res 27(1):355–357

Kim J, Grater JW, Wang P (2006a) Nanostructures for enzyme stabilization. Chem Eng Sci 61(3):1017–1026

Kim J, Jia H, Wang P (2006b) Challenges in biocatalysis for enzyme-based biofuel cells. Biotechnol Adv 24:296–308

Klein G, Reymond JL (2001) An enzyme assay using pM. Angew Chem Int Ed Engl 40:1771–1773

Klibanov AM (2001) Improving enzymes by using them in organic solvents. Nature 409:241–246

Karumanchi R, Doddamane SN, Sampangi C, Todd PW (2002) Field-assisted extraction of cells, particles and macromolecules. Trends Biotechnol 20(2):72–78

Kushnir S, Gase K, Breitling R, Alexandrov K (2005) Development of an inducible protein expression system based on the protozoan host *Leishmania tarentolae*. Protein Expr Purif 42(1):37–46

Lee M-Y, Dordick J-S (2002) Enzyme activation for nonaqueous media. Curr Opin Biotechnol 13(4): 376–384

Leitner A, Walzthoeni T, Kahraman A (2010) Probing native protein structures by chemical cross linking, mass spectrometry, and bioinformatics. Mol Cell Proteomic 9:1634–1649

Li WA, Barry ZT, Cohen JD, Wilder CL, Deeds RJ, Keegan PM, Platt MO (2010) Detection of femtomole quantities of mature cathepsin K with zymography. Anal Biochem 401(1):91–98

Liu H-L, Hsu J-P (2005) Recent developments in structural proteomics for protein structure determination. Proteomics 5:2056–2068

Madzak C, Tréton B, Blanchin-Roland S (2000) Strong hybrid promoters and integrative expression/secretion vectors for quasi-constitutive expression of heterologous proteins in the yeast *Yarrowia lipolytica*. J Mol Microbiol Biotechnol 2(2):207–216

Mai V, Wiegel J (2000) Advances in development of a genetic system for *Thermoanaerobacterium* spp.: expression of genes encoding hydrolytic enzymes, development of a second shuttle vector, and integration of genes into the chromosome. Appl Environ Microbiol 66(11):4817–4821

Male KB, Nguyen AL, Luong JHT (1990) Isolation of urokinase by affinity ultrafiltration. Biotechnol Bioeng 35(1):87–93

Minuth T, Thommes J, Kula MR (1996) A closed concept for purification of the membrane-bound cholesterol oxidase from *Nocardia rhodochrous* by surfactant-based cloud-point extraction, organic-solvent extraction and anion-exchange chromatography. Biotechnol Appl Biochem 23:107

Mori Y, Nagamine K, Tomita N, Notomi T (2001) Detection of loop-mediated isothermal amplification reaction by turbidity derived from magnesium pyrophosphate formation. Biochem Biophys Res Commun 289:150–154

Nelson M, Humphrey W, Gursoy A et al (1996) NAMD – a parallel, object-oriented molecular dynamics program. Int J Supercomput Appl High Perform Comput 10:251–268

Notomi T, Okayama H, Masubuch H et al (2000) Loop-mediated isothermal amplification of DNA. Nucleic Acids Res 28(12):e63

OECD (1998) Biotechnology for clean industrial products and processes. OECD, Paris

Olsen M, Iverson B, Georgiou G (2000) High-throughput screening of enzyme libraries. Curr Opin Biotechnol 11:331–337

Parrella FW (1999) EZ-Fit™ by Perrella Scientific, enzyme inhibition kinetics software. http://www.jlc.net/~fperrell/index.htm

Parthiban V, Gromiha MM, Schomburg D (2006) CUPSAT: prediction of protein stability upon point mutations. Nucleic Acids Res 34:W239–W242

Pavelka A, Chovancova E, Damborsky J (2009) HotSpot Wizard: a web server for identification of hot spots in protein engineering. Nucleic Acids Res 37(2): W376–W383

Peplowski L, Nowak W (2008) Molecular dynamics simulations of the metaloenzyme thiocyanate hydrolase with non-corrinoid Co(III) in active site. CBSB08 Proceed NIC Workshop 40:353–356

Peters GH, Frimurer TM, Andersen JN, Olsen OH (2000) Molecular dynamics simulations of protein-tyrosine phosphatase 1B. II. substrate-enzyme interactions and dynamics. Biophys J 78(5):2191–2200

PNNL Media Contacts November 28, 2006. Night of the living enzyme 509:375–3776

Ponder JW, Case DA (2003) Force fields for protein simulations. Adv Protein Chem 66:27–85

Puig O, Caspary F, Rigaut G et al (2001) The tandem affinity purification (TAP) method: a general procedure of protein complex purification. Methods 24:218–229

Rarey M, Kramer B, Langauer T et al (1996) A fast flexible docking method using an incremental construction algorithm. J Mol Biol 261:470–489

Ren X (2007) Rosetta-gami ™ B host strains from Novagen. Biocompare. http://www.biocompare.com/Articles/ProductReview/981/Rosetta-gami--B-Host-Strains-From-Novagen.html

Rose TM, Henikoff JG, Henikoff S (2003) CODEHOP (COnsensus-DEgenerate Hybrid Oligonucleotide Primer) PCR primer design. Nucleic Acids Res 31(13):3763–3766

Roy A, Kucukural A, Zhang Y (2010) I-TASSER: a unified platform for automated protein structure and function prediction. Nat Protoc 5:725–738

Sakamoto Y, Nakade K, Konno N (2011) Endo-β-1,3-glucanase GLU1, from the fruiting body of *Lentinula edodes*, belongs to a new glycoside hydrolase family. Appl Environ Microbiol 77(23):8350–8354

Sambrook J, Fritsch EF, Maniatis T (2001) Molecular cloning: a laboratory manual. Cold Spring Harbor Laboratory, Cold Spring Harbor

Schaffer U, Schlosser A, Müller KM, Schäfer A, Katava N, Baumeister R, Schulze E (2010) SnAvi–a new tandem tag for high-affinity protein-complex purification. Nucleic Acids Res 38(6):e91. doi:10.1093/nar/gkp1178

Scheer M, Grote A, Chang A, Schomburg I, Munaretto C, Rother M, Söhngen C, Stelzer M, Thiele J, Schomburg D (2011) BRENDA, the enzyme information system in 2011. Nucleic Acids Res 39:D670–D676. doi:10.1093/nar/gkq1089

Shuman S (1991) Recombination mediated by vaccinia DNA topoisomerase I in *Escherichia coli* is sequence specific. Proc Natl Acad Sci 88:10104–10108

Shuman S (1994) Novel approach to molecular cloning and polynucleotide synthesis using vaccinia DNA topoisomerase. J Biol Chem 269:32678–32684

Smith C (2005) Striving for purity: advances in protein purification. Nat Methods 2:71–77. doi:10.1038/nmeth0105-71

Song D, Coughlin J et al (2008) Alternative method for site-directed mutagenesis of complex polyketide synthase in *Streptomyces albus* JA3453. Acta Biochim Biophys Sin 40:319–326

Suzuki T, Yasui K (2011) Plasmid artificial modification: a novel method for efficient DNA transfer into bacteria. Methods Mol Biol 765:309–326

Tian J, Liu Q, Dong S et al (2010) A new method for multi-site-directed mutagenesis. Anal Biochem 406(1):83–85

Tirawongsaroj P, Sriprang R, Harnpicharnchai P et al (2008) Novel thermophilic and thermostable lipolytic enzymes from a Thailand hot spring metagenomic library. J Biotechnol 133(1):42–49

Van Der Spoel D, Lindahl E, Hess B, Groenhof G, Mark AE, Berendsen HJ (2005) GROMACS: fast, flexible, and free. J Comput Chem 26(16):1701–1718. doi:10.1002/jcc.20291

Venselaar H, Joosten RB, Vroling B et al (2010) Homology modelling and spectroscopy, a never-ending love story. Eur Biophys J 39:551–563

Wahler D, Reymond J-L (2001) High-throughput screening for biocatalysts. Curr Opin Biotechnol 12:535–544

Walker GT, Fraiser MS, Schram JL, Little MC, Nadeau JG, Malinowski DP (1992) Strand displacement amplification–an isothermal, in vitro DNA amplification technique. Nucleic Acids Res 20(7):1691–1696

Ward WW, Swiatek G (2009) Protein purification. Curr Anal Chem 5(2):1–21

Wang Q, Wu H, Wang A, Du P, Pei X, Li H, Yin X, Huang L, Xiong X (2010) Prospecting metagenomic enzyme subfamily genes for DNA family shuffling by a novel PCR-based approach. J Biol Chem 285(53):41509–41516. doi:10.1074/jbc.M110.139659

Withers-Martinez C, Carpenter EP, Hackett F, Ely B, Sajid M, Grainger M, Blackman MJ (1999) PCR-based gene synthesis as an efficient approach for expression of the A+T-rich malaria genome. Protein Eng 12:1113–1120

Xiao H, Tanaka T, Ogawa M, Yada RY (2007) Expression and enzymatic characterization of the soluble recombinant plasmepsin I from *Plasmodium falciparum*. Protein Eng Des Sel 20(12):625–633

Xu T, Zhang L, Wang X, Wei D, Li T (2009) Structure-based substrate screening for an enzyme. BMC Bioinform 10:257. doi:10.1186/1471-2105-10-257

Yin S, Ding F, Dokholyan NV (2007) Eris: an automated estimator of protein stability. Nat Methods 4(6):466–467

Zalatan JG, Herschlag D (2009) The far reaches of enzymology. Nat Chem Biol 5:516–520

Regulatory Motif Identification in Biological Sequences: An Overview of Computational Methodologies

8

Shripal Vijayvargiya and Pratyoosh Shukla

Abstract

The transcription factor binding sites (TFBS), also called as motifs, are short, recurring patterns in DNA sequences that are presumed to have a biological function. Identification of the motifs from the promoter region of the genes is an important and challenging problem, specifically in the eukaryotic genomes. In this chapter, an overview of motif identification methods has been presented. The computational methods for motif identification are classified as enumerative methods, probabilistic methods, phylogeny-based methods, and machine learning methods. The chapter also presents the standard evaluation scheme for accuracy of prediction.

Keywords

TFBS • Motif identification

Introduction

Understanding the regulatory networks of higher organisms is one of the main challenges of functional genomics. Gene regulation is a finely controlled mechanism. The main part of regulation is performed by the specific proteins called transcription factors (TFs) binding to specific transcription factor binding sites (TFBS) in regulatory regions associated with genes. A TFBS is also known as motif. A motif is a pattern of nucleotide bases or amino acids, which captures a biologically meaningful feature common to a group of nucleic acid or protein sequences. Regulatory motifs capture the patterns of DNA bases responsible for controlling when and where a gene is expressed. Typically, regulatory motifs describe TFBSs embedded in the DNA sequences upstream of a gene's transcription start site (TSS). More rarely, regulatory signals may occur downstream of the TSS and even within coding sequences.

Identification of the regulatory regions and binding sites is a prerequisite for understanding gene regulation (Lockhart and Winzeler 2000).

S. Vijayvargiya (✉)
Department of Computer Science and Engineering, Birla Institute of Technology, Mesra Extension Center Jaipur, 27, Malviya Industrial Area, Jaipur 302017, Rajasthan, India

Department of Biotechnology, Birla Institute of Technology (Deemed University), Extension Centre, Jaipur, India
e-mail: shripalvijay@rediffmail.com

P. Shukla
Enzyme Technology and Protein Bioinformatics Laboratory, Department of Microbiology, M.D. University, Rohtak, India
e-mail: pratyoosh.shukla@gmail.com

P. Shukla and B.I. Pletschke (eds.), *Advances in Enzyme Biotechnology*,
DOI 10.1007/978-81-322-1094-8_8, © Springer India 2013

Initially the experimental techniques like DNase footprinting assay and the electrophoretic mobility shift assay (EMSA) have been used to discover and analyze DNA binding sites. However, the development of DNA microarrays and fast sequencing techniques has led to new methods for in vivo identification of binding sites, such as ChIP-chip and ChIP-seq (Elnitski et al. 2006). Experimental identification and verification of such elements are challenging and costly; therefore, much effort has been put into the development of computational approaches. A good computational method can potentially provide high-quality prediction of the binding sites and reduce the time required for experimental verification.

Computational discovery of the regulatory elements is possible because they occur several times in the same genome and they may be evolutionarily conserved (Sandve and Drablos 2006). This means that searching for overrepresented motifs across regulatory regions may discover novel regulatory elements. However, this simple looking problem turns out to be a tough problem, made difficult by a low signal-to-noise ratio. This is because of the poor conservation and short length of the transcription factor binding sites in comparison to the length of promoter sequences. Many computational techniques and tools have been developed for the motif identification, but many of the existing tools for regulatory motif discovery have some limitations like the limited applicability of current nucleotide background models, rapid failure with increasing sequence length, and a tendency to report false positives rather than true transcription factor binding sites (Tompa et al. 2005; Hu et al. 2005). This chapter presents a survey of the experimental as well as the computational methods developed for the motif identification. It mainly focuses on the computational methods.

Transcription Factor Binding Sites (Motifs)

The DNA and genes give only static and general view of the genome. The body of an advanced organism like human is composed of several kinds of different tissues, consisting of cells that are also dynamic and changing over time, although the basic DNA is the same across the body and across time. The dynamics of organisms are handled by the gene regulatory mechanisms. So understanding the process that regulates gene expression and identification of those regulating element is a major challenge of biology. The main idea in gene expression is that every gene contains the information to produce a protein, which performs most of the biological functions of an organism. The main part of regulation is performed by specific proteins called transcription factors (TFs) that regulate the production of RNA and proteins from genes. This regulation is achieved by the TFs binding to DNA near genes, thus influencing the recruitment of RNA polymerase. RNA polymerase is a protein that performs the translation of genes into RNA, the first step in translating genes to proteins. The regions where TFs bind are often called regulatory regions. The region just before the gene, called upstream region, is the most basic regulatory region, but TFs can also bind in regulatory region that are situated after the gene (downstream), within the gene (introns), or further upstream.

Gene regulation is a finely controlled mechanism, and the TFs do not attach randomly to the DNA. As both the TFs and the DNA are molecules containing a structured organization of positive and negative charges, binding of a TF to DNA will depend on whether these charges can be aligned in a complementary way that forms strong physical bonds between the molecules. Because of this, each TF will have its own sequence-specific requirement for binding to DNA.

Determining where in the DNA each TF can bind is important for several reasons. The regulation of genes by TFs is a basic component of a very complex system of interactions between genes. Knowing the exact locations where TFs can bind is an important step toward determining how genes are regulated by a given TF, and it may also explain how slight sequence variations between individuals in the regulatory regions may influence, for instance, the risk for a specific disease (Fig. 8.1).

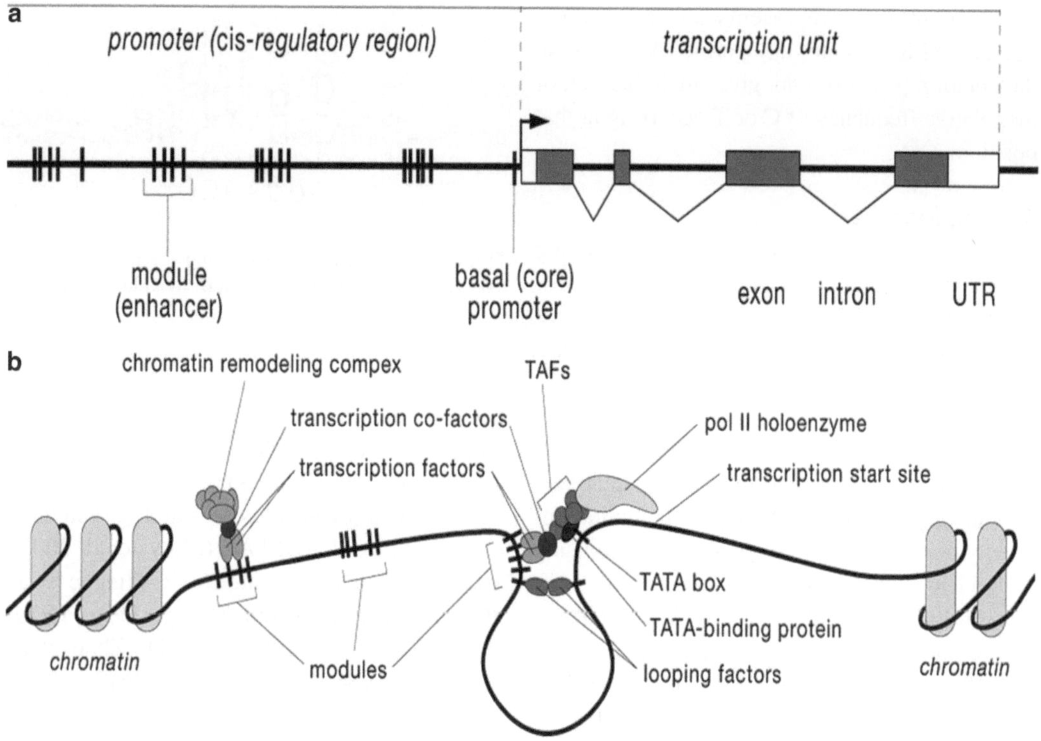

Fig. 8.1 Gene regulation. (**a**) Promoter region showing distribution of motifs in upstream sequence and transcription unit showing exons, introns, and UTR. (**b**) Shows binding of TFs with regulatory factors to initiate transcription (Figure adapted from Wray et al. (2003))

Representations and Scoring Matrices

Motifs are generally represented as consensus IUPAC strings, position frequency matrices (PFMs), position weight matrices (PWMs), or position-specific scoring matrices (PSSMs) in databases. Commonly, motifs in noncoding DNA sequences are conserved but still tend to be degenerate, which can influence the interaction between TFs and motifs. Therefore, after the motif data are collected and aligned from experimental or computational results, relevant consensus IUPAC strings can be constructed by selecting a degeneracy base pair symbol for each position in the alignment (Wasserman and Sandelin 2004). The motif data can also be modeled as PFM by aligning identified sites and counting the frequency of each base pair at each position of the alignment (Vavouri and Elgar 2005). Moreover, by using sequence logos, PWM can be displayed with color and height proportional to the base pair frequency and information content for each position by formulas (Schneider and Stephens 1990).

Consensus Sequence

A consensus sequence refers to the most common nucleotide at a particular position after multiple sequences are aligned. A consensus sequence is a way of representing the results of a multiple sequence alignment, where related sequences are compared to each other and similar functional sequence motifs are found. A consensus sequence shows that which residues are most abundant in the alignment at each position. The consensus sequences are represented using the following notation:

A[CT]N{ A }YR

In this notation, A means that an A is always found in that position; [CT] stands for either C or T; N stands for any base; and {A} means any base except A. Y represents any pyrimidine (C, T), and R indicates any purine (A, G). Here, the notation [CT] does not give any indication of the relative frequency of C or T occurring at that position.

Scoring Matrices

In probabilistic models the collection of binding sites is represented using a profile matrix. A profile is a matrix of numbers containing scores for each residue or nucleotide at each position of a fixed-length motif. There are two types of weight matrices:

1. A position frequency matrix (PFM) records the position-dependent frequency of each residue or nucleotide. PFMs can be computationally discovered by tools such as MEME using hidden Markov models.
2. A position weight matrix (PWM) contains log-odds weights for computing a match score. A cutoff is needed to specify whether an input sequence matches the motif. PWMs are calculated from PFMs.

Sequence Logo

A sequence logo is a graphical representation of the sequence conservation of nucleotides (in a strand of DNA/RNA) or amino acids (in protein sequences). This is a graphical representation of the consensus sequence in which the size of a symbol is related to the frequency that a given nucleotide (or amino acid) occurs at a certain position. In sequence logos the more conserved the residue, the larger the symbol for that residue is drawn, the less frequent, the smaller the symbol (Schneider and Stephens 1990). Sequence logos can be used to represent conserved DNA binding sites, where transcription factors bind (Fig. 8.2).

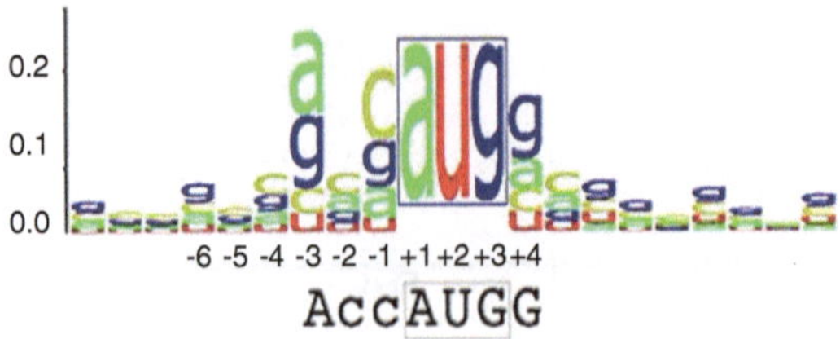

Fig. 8.2 A sequence logo showing the most conserved bases around the initiation codon from all human mRNAs

Regulatory Motif Databases

There are several private and public databases devoted to compilation of experimentally reported, and sometimes computationally predicted, motifs for different transcription factors in different organisms. Motifs are generally represented as consensus IUPAC strings, position frequency matrices (PFMs), position weight matrices (PWMs), or position-specific scoring matrices (PSSMs) in databases. Known regulatory motif profiles are cataloged in databases such as TRANSFAC (Matys et al. 2003), JASPAR (Sandelin et al. 2004; Vlieghe et al. 2006), SCPD (Zhu and Zhang 1999), TRRD (Kolchanov et al. 2000), TRED (Zhao et al. 2005), and ABS (Blanco et al. 2006).

Experimental Methods for Motif Identification

Since identification of regulatory regions and binding sites is a prerequisite for understanding gene regulation (Lockhart and Winzeler 2000; Stormo 2000), various experimental and computational techniques have been employed for this purpose. Earlier, the experimental techniques of choice to discover and analyze DNA binding sites have been the DNase footprinting assay and the electrophoretic mobility shift assay (EMSA). However, the development of DNA microarrays and fast sequencing techniques has led to new, massively parallel methods for in vivo identification of binding sites, such as ChIP-chip and ChIP-seq (Elnitski et al. 2006). Following is the brief description of experimental strategies.

Chromatin immunoprecipitation (ChIP) provides a powerful in vivo strategy to determine its target locations for a known protein. Using formaldehyde, the proteins are cross-linked to the DNA, which is then fragmented into 100–500 bp long pieces. A protein-specific antibody, coupled to a retrievable tag, is used to pull down (precipitate) the DNA-protein complex from the pool of DNA fragments. Finally, the associated DNA is recovered, sequenced, and analyzed – either through amplification or through the use of DNA microarrays.

In a DNA microarray, probe sequences of known DNA molecules are placed on an array of inert substrate, thus forming a collection of microscopic spots. By measuring the hybridization levels of target sequences, one can determine their enrichment under different conditions or locations. Using the DNA purified by ChIP, the precise location of the binding regions on the sequence can be identified. This technique known as ChIP-on-chip provides an efficient and scalable way for the identification of binding sites of DNA-binding proteins. Through recently developed genome-wide analyses, one can determine the binding sites of a protein throughout the genome (Ren et al. 2000; Horak and Snyder 2002). Methods generating higher resolution and coverage (Boyer et al. 2005; Odom et al. 2006) have also been proposed. The method ChIP-chip has become popular due to its ability to identify the motifs in an unbiased manner. However, the dependence on a highly TF-specific antibody is usually a major hurdle in performing ChIP-chip experiments.

Computational Methods for Motif Identification

Experimental identification and verification of motifs are challenging and costly, so much effort has been put into the development of computational approaches. Over the years, many algorithms and computational techniques have been proposed for the motif identification. The computational motif identification schemes can be classified on the basis of various criteria. One of them is based on the type of DNA sequence information employed by the algorithm to discover the motifs. The algorithms available for motif finding can be classified into three major classes (Das and Dai 2007):

1. Those that use promoter sequences from coregulated genes from a single genome
2. Those that use orthologous promoter sequences of a single gene from multiple species (i.e., phylogenetic footprinting)
3. Those that use promoter sequences of coregulated genes as well as phylogenetic footprinting

Motif identification algorithms can also be classified based on the representation of motifs and the combinatorial approach used in the design of the algorithms like:

1. Consensus sequence-based or regular expression-based counting methods that mostly rely on exhaustive enumeration like Weeder (Pavesi et al. 2004), YMF (Sinha and Tompa 2003), and MITRA (Eskin and Pevzner 2002)
2. Matrix-based methods that use probabilistic sequence models where the model parameters are estimated using maximum-likelihood principle, for example, MEME (Bailey et al. 2006) and Gibbs sampler (Lawrence et al. 1993)
3. Feature based (Chin and Leung 2008; Sharon et al. 2008)

Motif recognition is NP-complete and therefore cannot be solved in polynomial time unless P=NP (Evans et al. 2003). Nonetheless, numerous methods and tools have been developed for the motif identification problem, including MEME (Bailey and Elkan 1995a), AlignACE (Roth et al. 1998), REDUCE (Bussemaker et al. 2001), Winnower (Pevzner and Sze 2000), PROJECTION (Buhler and Tompa 2002), MITRA (Eskin and Pevzner 2002), MDScan (Liu et al. 2002), YMF (Sinha and Tompa 2003), pattern-driven approaches (Sze et al. 2004), Weeder (Pavesi et al. 2004), DME (Smith et al. 2005), PSM1 (Rajasekaran et al. 2005), VAS (Chin and Leung 2006), MEME (Bailey et al. 2006), RISOTTO (Pisanti et al. 2006), PMSprune (Davila et al. 2007), the voting algorithm (Chin and Leung 2005), MCLWMR (Boucher et al. 2007), and Trawler (Ettwiller et al. 2007). Despite these available tools, the effective and efficient identification of motifs within datasets of interest remains a challenging problem, particularly when studying datasets derived from mammals, such as those from mice and humans. Tompa et al. (2005) have evaluated 13 different motif discovery tools and showed that many of the tools are inefficient when used on datasets derived from organisms higher than yeast.

Enumerative Methods

The word-based enumerative methods guarantee global optimality, and they are appropriate for short motifs and are therefore useful for motif finding in eukaryotic genomes where motifs are generally shorter than prokaryotes.

Sagot (1998) introduced a word-based approach for motif finding that is based on the representation of a set of sequences with a suffix tree. Representation of upstream sequences as suffix trees gives a large number of possible combinations; therefore, efficiency of the algorithm is a challenging issue. The motif-finding algorithms Weeder (Pavesi et al. 2001) and MITRA (Mismatch Tree Algorithm) (Eskin and Pevzner 2002) are also based on the suffix tree and its variant. The algorithms WINNOWER (Pevzner and Sze 2000) and cWINNOWER (Liang 2003) use word-based approach combined with graph-theoretic methods for motif finding.

Zaslavsky and Singh (2006) used a combinatorial optimization framework for motif finding that couples graph pruning techniques with a novel integer linear programming formulation. They also proposed an approach for determining statistical significance of uncovered motifs. Zhang and Zaki (2006a) proposed the algorithm SMOTIF to solve the structured motif search problem. Given one or more sequences and a structured motif, SMOTIF searches the sequences for all occurrences of the motif. Further, they proposed another algorithm, called EXMOTIF (Zhang and Zaki 2006b). On given some sequences and a structured motif template, EXMOTIF extracts all frequent structured motifs that have quorum q. This algorithm can also extract the composite regulatory binding sites in DNA sequences.

Pisanti et al. (2006) proposed a consensus-based algorithm, called RISOTTO. This algorithm exhaustively enumerates all motifs of a certain size by collecting their occurrences, at a given distance, from a set of coregulated DNA sequences (Carvalho et al. 2006; Marsan and Sagot 2000). RISOTTO, being a method based on the detection of overrepresentation of motifs in coregulated DNA sequences, faces problems in detecting weak motifs. Carvalho et al. (2011) extended RISOTTO by post-processing its output with a greedy procedure that uses prior information. They combined position-specific priors from different sources into a scoring criterion that guides the greedy search procedure. The rationale behind their approach was that the combinatorial algorithm could exploit the full space of possible motifs pointing out good candidates. Afterwards a greedy search is performed over these initial guesses, and good motifs are up weighted by the prior. The reduction of the search space attained in the greedy search by using the output of a combinatorial algorithm improves efficiency of their algorithm, called GRISOTTO.

The word-based methods can also be very fast when implemented with optimized data structures such as suffix trees (Sagot 1998) and are a good choice for finding totally constrained motifs, i.e., all instances are identical. However, for typical transcription factor motifs that often have several weakly constrained positions, word-based methods can be problematic, and the result often needs to be post-processed with some clustering system (Vilo et al. 2000). Word-based methods also suffer from the problem of producing too many spurious motifs.

Probabilistic Methods

Probabilistic or randomized approaches make certain decisions randomly. This concept extends the classical model of deterministic algorithms. The probabilistic approach involves representation of the motif model by a position weight matrix (Bucher 1990). Probabilistic methods have the advantage of requiring few search parameters but rely on probabilistic models of the regulatory regions, which can be very sensitive with respect to small changes in the input data. Many of the algorithms developed from the probabilistic approach are designed to find longer or more general motifs that are required for transcription factor binding sites. Therefore, they are more appropriate for motif finding in prokaryotes, where the motifs are generally longer than eukaryotes. However, these algorithms are not guaranteed to find globally optimal solutions, since they employ some form of local search, such as Gibbs sampling, expectation maximization (EM), or greedy algorithms, that may converge to a locally optimal solution.

The Gibbs sampling (Lawrence et al. 1993) algorithm is one of the simplest Markov chain Monte Carlo algorithms. By Gibbs sampling, the joint distribution of the parameters will converge to the joint probability of the parameters in the given dataset. Gibbs sampling strategies claim to be fast and sensitive. Roth et al. (1998) developed a motif-finding tool AlignACE, which is based on the Gibbs sampling algorithm.

EM for motif finding was introduced by Lawrence and Reilly (1990), and it was an extension of the greedy algorithm for motif finding by Hertz et al. (1990). The EM algorithm is used to estimate the probability density of a given dataset by employing the Gaussian mixture model. The probability density of a dataset is modeled as the weighted sum of a number of Gaussian distributions. No alignment of the sites is required, and the basic model assumption is that each sequence must contain at least one common site. The uncertainty in the location of the sites is handled by employing the missing information principle to develop an EM algorithm. This approach allows for the simultaneous identification of the sites and characterization of the binding motifs. The main advantage of EM is its fast speed, while the disadvantage is that it requires "appropriate" starting values and is difficult to deal with constrained parameters.

The MEME algorithm by Bailey and Elkan (1995b) extended the EM algorithm for identifying motifs in unaligned biopolymer sequences. The aim of MEME is to discover new motifs in a set of biopolymer sequences where little is known in advance about any motifs that may be present. MEME incorporated three novel ideas for discovering motifs. First, subsequences that actually occur in the biopolymer sequences are used as starting points for the EM algorithm to increase the probability of finding globally optimum motifs. Second, the assumption that each sequence contains exactly one occurrence of the shared motif is removed. Third, a method for probabilistically erasing shared motifs after they are found is incorporated so that several distinct motifs can be found in the same set of sequences.

Further, Bailey et al. (2009) developed the MEME suite. In MEME suite the MEME motif discovery algorithm is complemented by the GLAM2 algorithm which allows the discovery of motifs containing gaps. They used three sequence scanning algorithms, MAST, FIMO (Frith et al. 2008), and GLAM2SCAN (Bailey and Gribskov 1998), that allow scanning numerous DNA and protein sequence databases for motifs discovered by MEME and GLAM2.

Li (2009) proposed the method GADEM, which combines spaced dyads and an expectation-maximization (EM) algorithm. In this method candidate words (four to six nucleotides) for constructing spaced dyads are prioritized by their degree of overrepresentation in the input sequence data. Spaced dyads are converted into starting position weight matrices (PWMs). GADEM then employs a genetic algorithm (GA), with an embedded EM algorithm to improve starting PWMs, so as to guide the evolution of a population of spaced dyads toward one whose entropy scores are more statistically significant. Spaced dyads whose entropy scores reach a prespecified significance threshold are declared motifs.

Generally, PWM-based approaches assume independence between the base positions of the sequence motif and suffer from high false-positive rates. However, recent studies have shown that the independent assumption is not true and modeling the dependencies in motifs could lead to better predictions (Bulyk et al. 2002). Examples include feature-based method (Sharon et al. 2008; Chin and Leung 2008), HMM-based method (Marinescu et al. 2005), and Markov chain-based method (Wang et al. 2006).

Phylogeny-Based Methods

The major advantage of phylogenetic footprinting over the coregulated genes approach is that the latter requires a reliable method for identifying coregulated genes. Whereas in using phylogenetic footprinting approach, it is possible to identify motifs specific to even a single gene, as long as they are sufficiently conserved across the many orthologous sequences considered. The rapid accumulation of genomic sequences from a wide variety of organisms makes it possible to use the phylogenetic footprinting approach for motif finding. The standard method used for phylogenetic footprinting is to construct a global multiple alignment of the orthologous promoter sequences and then identify conserved region in the alignment using a tool such as CLUSTALW (Thompson et al. 1994). However, it has been observed (Tompa 2001) that this approach to phylogenetic footprinting does not always work. The reason is that if the species are too closely related, the sequence alignment is obvious but uninformative, since the functional elements are not sufficiently better conserved than the surrounding nonfunctional sequence. On the other hand, if the species are too distantly related, it is either difficult or impossible to find an accurate alignment. To overcome this problem, one of the several existing motif-finding algorithms such as MEME, Consensus, and Gibbs sampler has been used for phylogenetic footprinting. Cliften et al. (2001) used AlignACE for motif finding by comparative DNA sequence analysis of several species of *Saccharomyces* and reported some successes where the global multiple alignment tools failed. McCue et al. (2001) used Gibbs sampler for motif finding using phylogenetic footprinting in proteobacterial genomes. That the use of such general motif discovery algorithms can be problematic in phylogenetic footprinting has been pointed out by Blanchette and Tompa (2002). These motif-finding algorithms do not take into account the phylogenetic relationship of the given sequences since these methods assume the input sequences to be independent. Therefore, the datasets containing a mixture of some closely related species would have an unduly high weight in the choice of motifs reported. Even if these methods were modified to weigh the input sequences unequally, this would still not capture the information in an arbitrary phylogenetic tree.

Carmack et al. (2007) developed a scanning algorithm, PhyloScan, which combines evidence from matching sites found in orthologous data from several related species with evidence from multiple sites within an intergenic region to better detect regulons.

Some algorithms integrate two important aspects of a motif's significance, i.e., overrepresentation of motifs in promoter sequences of coregulated and cross-species conservation, into one probabilistic score.

Based on the Consensus algorithm (Hertz et al. 1990), Wang and Stormo (2003) developed the motif-finding algorithm PhyloCon (Phylogenetic Consensus). Phylogenetic Consensus (PhyloCon) takes into account both conserved orthologous genes and coregulated genes within a species. The key idea of PhyloCon is to compare aligned sequence profiles from orthologous genes or coregulated genes rather than unaligned sequences.

Sinha et al. (2004) developed the algorithm PhyME based on a probabilistic approach that handles data from promoters of coregulated genes and orthologous sequences. PhyME integrates two different axes of information content in evaluating the significance of candidate motifs. One axis is the overrepresentation that depends on the number of occurrences of motifs in each species. The other axis is the level of conservation of each motif instance across species. Siddharthan et al. (2005) developed the algorithm PhyloGibbs that combines the motif-finding strategies of phylogenetic footprinting and Gibbs sampling into one integrated Bayesian framework.

Zhang et al. (2010) recently developed an algorithm named GLECLUBS (Global Ensemble and Clustering of Binding Sites) for genome-wide de novo prediction of motifs in a prokaryotic genome (Zhang et al. 2009). GLECLUBS employs a phylogenetic footprinting technique to first identify all possible motifs, and then clusters similar motifs. In order to harvest as many as possible true motifs by phylogenetic footprinting, GLECLUBS uses multiple complementary motif-finding tools instead of using only a single tool and considers multiple outputs of each tool. Also, GLECLUBS assumes that only a small portion of predicted motifs by phylogenetic footprinting are true motifs and that the vast majority of them are spurious predictions. Therefore, the clustering step of GLECLUBS aims to discriminate true motifs from spurious ones using an iterative filtering procedure.

Machine Learning–Based Methods

Liu et al. (2004) developed the algorithm FMGA based on genetic algorithms (GAs) for finding motifs in the regulatory regions. The crossover is implemented with specially designed gap penalties to produce the optimal child pattern. The mutation in GA is performed by using position weight matrices to reserve the completely conserved positions. This algorithm also uses a rearrangement method based on position weight matrices to avoid the presence of a very stable local minimum, which may make it quite difficult for the other operators to generate the optimal pattern. The authors reported that FMGA performs better in comparison to MEME and Gibbs sampler algorithms.

Liu et al. (2006) developed a self-organizing neural network structure for motif finding in DNA and protein sequences. The network contains several layers with each layer performing classifications at different levels. The authors maintained a low computational complexity through the use of layered structure so that each pattern's classification is performed with respect to a small subspace of the whole input space. The authors also maintain a high reliability of their search algorithm using self-organizing neural network since it will grow as needed to make sure that all input patterns are considered and are given the same amount of attention. From simulation results, the authors reported that their algorithm outperformed the algorithms MEME and Gibbs sampler in certain aspects and their algorithm also works well for long DNA sequences.

Chan et al. (2009) proposed a new generalized model, which tackles the width uncertainty in the motif widths by considering and evaluating a wide range of interests simultaneously. Moreover, they also proposed a meta-convergence framework for genetic algorithms to provide multiple overlapping optimal motifs simultaneously in an effective and flexible way. Incorporating Genetic Algorithm with Local Filtering (GALF) for searching, the new algorithm GALF-G (G for generalized) was proposed based on the generalized model and meta-convergence framework.

The GA-based hybrid schemes have also been proposed. One such method is GARPS that combines GA and Random Projection Strategy (RPS) to identify planted (l, d) motifs. In the method GARPS, RPS is used to find good starting positions by introducing position-weighted function. Then, GA is used to refine the initial population obtained from RPS (Hongwei et al. 2010).

Chengwei Lei and Jianhua Ruan (2010) developed a motif-finding algorithm (PSO+) using the particle swarm optimization (PSO) which is a population-based stochastic optimization technique. They proposed a modification in the standard PSO algorithm to handle discrete values, such as characters in DNA sequences. They used both consensus and position-specific weight matrix representations in their algorithm to take advantage of the efficiency of the former and the accuracy of the latter. Many real motifs contain gaps; to address this issue, their method models gaps explicitly and provides a solution to find gapped motifs without any detailed knowledge of gaps.

Lee and Wang (2011) developed an SOM (Kohonen 2001)-based extraction algorithm (SOMEA) to discover overrepresented motifs in DNA datasets. SOMEA seek to use SOM to project k-mers (i.e., a subsequence with length k of DNA sequences) onto a two-dimensional lattice of nodes. Through this projection, input patterns (i.e., k-mers) with closely related features are projected onto the same or adjacent nodes on the map. Hence, the complex similarity relationships of the high-dimensional input sequence space become apparent on the map. Analysis of selected nodes therefore can reveal potential patterns (i.e., motifs) in the dataset.

Other Methods

Hu et al. (2005) introduced the ensemble approach for motif finding to improve the prediction accuracy of the motif-finding algorithms. They developed a clustering-based ensemble algorithm named EMD (Hu et al. 2006) for motif discovery by combining multiple predictions from multiple runs of one or more base component algorithms. The potential of an EMD algorithm lies in the fact that it could take advantage of superb predictions of every component algorithm. The authors used five component algorithms, namely, AlignACE, BioProspector, MDScan (Liu et al. 2002), MEME, and MotifSampler in their study. They tested their algorithm on a benchmark dataset generated from *Escherichia coli* RegulonDB. The EMD algorithm achieved 22.4 % improvement in terms of the nucleotide level prediction accuracy over the best stand-alone component algorithm.

Chakravarty et al. (2007) proposed an ensemble learning method, SCOPE, that is based on the assumption that transcription factor binding sites belong to one of three broad classes of motifs: nondegenerate, degenerate, and gapped motifs. SCOPE employs a unified scoring metric to combine the results from three motif-finding algorithms each aimed at the discovery of one of these classes of motifs.

Sandve et al. (2008) proposed a discrete approach to composite motif discovery (Compo) that supports rich modeling of composite motifs and a realistic background model. Compo can return either an ordered list of motifs, ranked according to the general significance measure, or a Pareto front corresponding to a multi-objective evaluation on sensitivity, specificity, and spatial clustering.

Marschall and Rahmann (2009) proposed an exact motif discovery method for a practically relevant space of IUPAC-generalized string patterns, using the p-value, with respect to a Markov model as the measure of overrepresentation. The key characteristics of their method can be observed threefold. First, they used a compound Poisson approximation for the null distribution of the number of motif occurrences. Second, they defined two p-value scores for overrepresentation, one based on the total number of motif occurrences and the other based on the number of sequences in a collection with at least one occurrence. Third, their method exploits monotonic properties of the compound Poisson approximation and is by orders of magnitude faster than exhaustive enumeration of IUPAC strings.

Prediction Accuracy of Motif Identification Algorithms

There are several prediction accuracy measures for evaluating motif discovery algorithms (Sinha et al. 2004; Liu et al. 2002; Thijs et al. 2002). Many of them are derived from the accuracy definitions for evaluating gene predictions (Burset and Guigo 1996; Rogic et al. 2001). Most of the algorithm uses the parameters precision (specificity) and recall (sensitivity) as prediction accuracy measures. The accuracy of a prediction can be measured by comparing the predicted motifs with the true motifs.

To measure the accuracy of the algorithms at nucleotide level, the following values for calculating accuracy metrics at the nucleotide level are defined for each target binding site with overlapping predicted binding sites in an input sequence:

1. nTP (true positive) – the number of target binding site positions predicted as binding site positions
2. nTN (true negative) – the number of nontarget binding site positions predicted as non-binding site positions
3. nFP (false positive) – the number of nontarget binding site positions predicted as binding site positions
4. nFN (false negative) – the number of target binding site positions predicted as non-binding site positions

Precision or specificity over a pair of target/predicted binding sites is defined as the number of predicted sites that are true sites divided by the total number of predicted sites:

$$nS_p = \frac{nTP}{nTP + nFP} \tag{8.1}$$

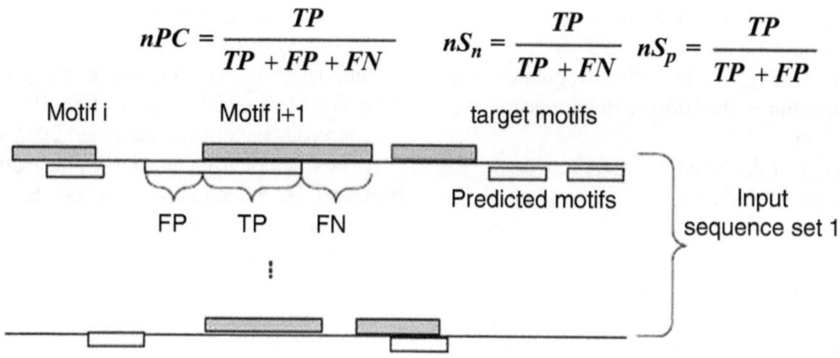

Fig. 8.3 Measures of prediction accuracy at the nucleotide levels. Accuracy scores over an input sequence set are the average accuracy scores over all its sequences (Hu et al. 2005)

Recall or sensitivity is the number of predicted sites that are true sites divided by the total number of true sites:

$$nS_n = \frac{nTP}{nTP + nFN} \tag{8.2}$$

In order to capture both specificity and sensitivity in a single accuracy measurement, the nucleotide level performance coefficient (nPC) is used as follows (Pevzner and Sze 2000; Tompa et al. 2005):

$$nPC = \frac{nTP}{nTP + nFP + nFN} \tag{8.3}$$

According to this definition, the nPC value ranges over (0, 1) with the perfect prediction being the value of 1. Compared with the correlation coefficient (CC) (Burset and Guigo 1996; Rogic et al. 2001), nPC has several benefits: it is straightforward to interpret, and practically it also tells the experimental biologists the probable range where the true binding sites are located around the predicted positions. The F-score or harmonic mean is also used as the overall accuracy measurement. Compared with geometric or arithmetic mean, it tends to penalize more the imbalance of sensitivity and specificity. The F-score is defined as (Fig. 8.3)

$$F = \frac{2 \times S_n \times S_p}{S_n + S_p} \tag{8.4}$$

References

Bailey TL, Elkan C (1995a) The value of prior knowledge in discovering motifs with MEME. Proc ISMB 1995:21–29

Bailey TL, Elkan C (1995b) Unsupervised learning of multiple motifs in biopolymers using expectation maximization. Mach Learn 21:51–80

Bailey TL, Gribskov M (1998) Combining evidence using p-values: application to sequence homology searches. Bioinformatics 14:48–54

Bailey TL, Williams N, Misleh C, Li WW (2006) MEME: discovering and analyzing DNA and protein sequence motifs. Nucleic Acids Res 34:W369–W373

Bailey TL et al (2009) MEME SUITE: tools for motif discovery and searching. Nucleic Acids Res 37(Web Server issue):W202–W208. doi:10.1093/nar/gkp335

Blanchette M, Tompa M (2002) Discovery of regulatory elements by a computational method for phylogenetic footprinting. Genome Res 12:739–748

Blanco E, Farr´e D, Alb`a MM, Messeguer X, Guig´o R (2006) ABS: a database of Annotated regulatory Binding Sites from orthologous promoters. Nucleic Acids Res 34(1):D63–D67

Boucher C, Church P, Brown D (2007) A graph clustering approach to weak motif recognition. Proc WABI 2007:149–160

Boyer LA, Lee TI et al (2005) Core transcriptional regulatory circuitry in human embryonic stem cells. Cell 122(6):947–956

Bucher P (1990) Weight matrix description for four eukaryotic RNA polymerase II promoter element derived from 502 unrelated promoter sequences. J Mol Biol 212:563–578

Buhler J, Tompa M (2002) Finding motifs using random projections. J Comput Biol 9:225–242

Bulyk ML, Johnson PL, Church GM (2002) Nucleotides of transcription factor binding sites exert interdependent effects on the binding affinities of transcription factors. Nucleic Acids Res 30:1255–1261

Burset M, Guigo R (1996) Evaluation of gene structure prediction programs. Genomics 34:353–367

Bussemaker HJ, Li H, Siggia ED (2001) Regulatory element detection using correlation with expression. Nat Genet 27:167–171

Carmack CS, McCue LA, Newberg LA, Lawrence CE (2007) PhyloScan: identification of transcription factor binding sites using cross-species evidence. Algorithms Mol Biol 2:1

Carvalho AM, Oliveira AL (2011) GRISOTTO: a greedy approach to improve combinatorial algorithms for motif discovery with prior knowledge. Algorithms Mol Biol 6:13

Carvalho AM, Freitas AT, Oliveira AL, Sagot MF (2006) An efficient algorithm for the identification of structured motifs in DNA promoter sequences. IEEE/ACM Trans Comput Biol Bioinform 3(2):126–140

Chakravarty A, Carlson JM, Khetani RS, Gross RH (2007) A novel ensemble learning method for de novo computational identification of DNA binding sites. BMC Bioinformatics 8:249. doi:10.1186/1471-2105-8-249

Chan T, Li G, Leung K, Lee K (2009) Discovering multiple realistic TFBS motifs based on a generalized model BMC. Bioinformatics 10:321. doi:10.1186/1471-2105-10-321

Chengwei L, Jianhua R (2010) A particle swarm optimization-based algorithm for finding gapped motifs. BioData Min 3:9

Chin FYL, Leung CM (2005) Voting algorithms for discovering long motifs. Proc APBC 2005:261–271

Chin FYL, Leung CM (2006) An efficient algorithm for string motif discovery. Proc APBC 2006:79–88

Chin F, Leung HCM (2008) DNA motif representation with nucleotide dependency. IEEE/ACM Trans Comput Biol Bioinform 5:110–119

Cliften PF, Hillier LW, Fulton L, Graves T, Miner T, Gish WR, Waterston RH, Johnston M (2001) Surveying *Saccharomyces* genomes to identify functional elements by comparative DNA sequence analysis. Genome Res 11:1175–1186

Das MK, Dai H-K (2007) A survey of DNA motif finding algorithms. BMC Bioinformatics 8(7):S21

Davila J, Balla S, Rajasekaran S (2007) Fast and practical algorithms for planted (l, d) motif search. IEEE/ACM Trans Comput Biol Bioinform 4(4):544–552

Elnitski L, Jin VX, Farnham PJ, Jones SJ (2006) Locating mammalian transcription factor binding sites: a survey of computational and experimental techniques. Genome Res 16(12):1455–1464

Eskin E, Pevzner P (2002) Finding composite regulatory patterns in DNA sequences. Bioinformatics 18(1): S354–S363

Ettwiller L, Paten B, Ramialison M, Birney E, Wittbrodt J (2007) Trawler: de novo regulatory motif discovery pipeline for chromatin immunoprecipitation. Nat Methods 4:563–565

Evans PA, Smith A, Wareham HT (2003) On the complexity of finding common approximate substrings. Theor Comput Sci 306(1–3):407–430

Frith MC, Saunders NFW, Kobe B, Bailey TL (2008) Discovering sequence motifs with arbitrary insertions and deletions. PLoS Comput Biol 4:e1000071

Hertz GZ, Hartzell GW, Stormo GD (1990) Identification of consensus patterns in unaligned DNA sequences known to be functionally related. Bioinformatics 6:81–92

Hongwei H, Zhenhua Z, Vojislav S, Lifang L (2010) Optimizing genetic algorithm for motif discovery. Math Comput Model 52(11–12):2011–2020

Horak CE, Snyder M (2002) ChIP-chip: a genomic approach for identifying transcription factor binding sites. Methods Enzymol 350(469–483):2002

Hu J, Li B, Kihara D (2005) Limitations and potentials of current motif discovery algorithms. Nucleic Acids Res 33:4899–4913

Hu J, Yang YD, Kihara D (2006) EMD an ensemble algorithm for discovering regulatory motifs in DNA sequences. BMC Bioinformatics 7:342

Sandve GK, Drablos F (2006) A survey of motif discovery methods in an integrated framework. Biol Direct 1:11. doi:10.1186/1745-6150-1-11

Kohonen T (2001) Self-organizing maps, vol 30, 3rd edn, Springer series in information sciences. Springer, Berlin, Heidelberg, New York

Kolchanov NA, Podkolodnaya OA et al (2000) Transcription Regulatory Regions Database (TRRD): its status in 2000. Nucleic Acids Res 28(1):298–301

Lawrence CE, Reilly AA (1990) An expectation maximization (EM) algorithm for the identification and characterization of common sites in unaligned biopolymer sequences. Proteins 7:41–51

Lawrence CE, Altschul SF, Boguski MS, Liu JS, Neuwald AF et al (1993) Detecting subtle sequence signals: a Gibbs sampling strategy for multiple alignment. Science 262:208–214

Lee NK, Wang D (2011) SOMEA: self-organizing map based extraction algorithm for DNA motif identification with heterogeneous model. BMC Bioinformatics 12(1):S16

Li L (2009) GADEM: a genetic algorithm guided formation of spaced dyads coupled with an EM algorithm for motif discovery. J Comput Biol 16(2):317–329. doi:10.1089/cmb.2008.16TT

Liang S (2003) cWINNOWER algorithm for finding fuzzy DNA motifs. IEEE Compu Soc Bioinform Conf 2003:260–265

Liu XS, Brutlag DL, Liu JS (2002) An algorithm for finding protein-DNA binding sites with applications to chromatin-immunoprecipitation microarray experiments. Nat Biotechnol 20:835–839

Liu F et al (2004) FMGA: finding motifs by genetic algorithm. In proceedings of the fourth IEEE symposium on bioinformatics and bioengineering, Taichung, Taiwan, pp 459–466

Liu D, Xiong X, DasGupta B, Zhang H (2006) Motif discoveries in unaligned molecular sequences using self-organizing neural network. IEEE Trans Neural Netw 17:919–928

Lockhart D, Winzeler E (2000) Genomics, gene expression and DNA arrays. Nature 405:827–836

Marinescu VD, Kohane IS, Riva A (2005) MAPPER: a search engine for the computational identification of putative transcription factor binding sites in multiple genomes. BMC Bioinformatics 6:79

Marsan L, Sagot M (2000) Algorithms for extracting structured motifs using a suffix tree with an application to promoter and regulatory site consensus identification. J Comput Biol 7(3–4):345–362

Marschall T, Rahmann S (2009) Efficient exact motif discovery. Bioinformatics 25:i356–i364. doi:10.1093/bioinformatics/btp188

Matys V, Fricke E et al (2003) TRANSFAC: transcriptional regulation, from patterns to profiles. Nucleic Acids Res 31(1):374–378

McCue L, Thompson W, Carmack C, Ryan M, Liu J, Derbyshire V, Lawrence C (2001) Phylogenetic footprinting of transcription factor binding sites in proteobacterial genomes. Nucleic Acids Res 29:774–782

Odom DT, Dowell RD et al (2006) Core transcriptional regulatory circuitry in human hepatocytes. Mol Syst Biol 2:0017

Pavesi G, Mauri G, Pesole G (2001) An algorithm for finding signals of unknown length in DNA sequences. Bioinformatics 17(1):S207–S214

Pavesi G, Mereghetti P, Mauri G, Pesole G (2004) Weeder web: discovery of transcription factor binding sites in a set of sequences from co-regulated genes. Nucleic Acids Res 32:W199–W203

Pevzner PA, Sze SH (2000) Combinatorial approaches to finding subtle signals in DNA sequences. Proc Int Conf Intell Syst Mol Biol 8:269–278

Pisanti N, Carvalho AM, Marsan L, Sagot MF (2006) RISOTTO: fast extraction of motifs with mismatches. In Proceedings of LATIN'06, Vol 3887 of LNCS. Springer, pp 757–768

Rajasekaran S, Balla S, Huang CH (2005) Exact algorithms for planted motif problems. J Comput Biol 12(8):1117–1128

Ren B, Robert F et al (2000) Genome-wide location and function of DNA binding proteins. Science 290(5500):2306–2309

Rogic S, Mackworth AK, Ouellette FB (2001) Evaluation of gene-finding programs on mammalian sequences. Genome Res 11:817–832

Roth FP, Hughes JD, Estep PW, Church GM (1998) Finding DNA regulatory motifs within unaligned noncoding sequences clustered by whole-genome mRNA quantitation. Nat Biotechnol 16:939–945

Sagot M (1998) Spelling approximate repeated or common motifs using a suffix tree. Lect Notes Comput Sci 1380:111–127

Sandelin A, Alkema W, Engstrom P, Wasserman WW, Lenhard B (2004) JASPAR: an open-access database for eukaryotic transcription factor binding profiles. Nucleic Acids Res 32:D91–D94

Sandve G, Abul O, Drablos F (2008) Compo: composite motif discovery using discrete models. BMC Bioinformatics 9:527. doi:10.1186/1471-2105-9-527

Schneider TD, Stephens RM (1990) Sequence logos: a new way to display consensus sequences. Nucleic Acids Res 18:6097–6100

Sharon E, Lubliner S, Segal E (2008) A feature-based approach to modeling protein-DNA interactions. PLoS Comput Biol 4:e1000154

Siddharthan R, Siggia ED, Nimwegen E (2005) PhyloGibbs: a Gibbs sampling motif finder that incorporates phylogeny. PLoS Comput Biol 1:534–556

Sinha S, Tompa M (2003) YMF: a program for discovery of novel transcription factor binding sites by statistical overrepresentation. Nucleic Acids Res 31:3586–3588

Sinha S, Blanchette M, Tompa M (2004) PhyME: a probabilistic algorithm for finding motifs in sets of orthologous sequences. BMC Bioinformatics 5:170

Smith AD, Sumazin P, Das D, Zhang MQ (2005) Mining ChIP-chip data for transcription factor and cofactor binding sites. Bioinformatics 21(1):i403–i412

Stormo GD (2000) DNA binding sites: representation and discovery. Bioinformatics 16:16–23

Sze S, Lu S, Chen J (2004) Integrating sample-driven and pattern driven approaches in motif finding. Proc WABI 2004:438–449

Thijs G, Marchal K, Lescot M, Rombauts S, DeMoor B, Rouze P, Moreau Y (2002) A Gibbs sampling method to detect overrepresented motifs in the upstream regions of coexpressed genes. J Comput Biol 9:447–464

Thompson JD, Higgins DG, Gibson TJ (1994) CLUSTALW improving the sensitivity of progressive multiple sequence alignment through sequence weighting, position specific gap penalties and weight matrix choice. Nucleic Acids Res 22:4673

Tompa M (2001) Identifying functional elements by comparative DNA sequence analysis. Genome Res 11:1143–1144

Tompa M, Li N, Bailey TL, Church GM, De Moor B et al (2005) Assessing computational tools for the discovery of transcription factor binding sites. Nat Biotechnol 23:137–144

Vavouri T, Elgar G (2005) Prediction of cisregulatory elements using binding site matrices—the successes, the failures and the reasons for both. Curr Opin Genet Dev 15:395–402

Vilo J, Brazma A, Jonassen I, Robinson A, Ukonnen E (2000) Mining for putative regulatory elements in the yeast genome using gene expression data. In: In proceedings of the eighth international conference on intelligent systems for molecular biology. AAAI Press, San Diego, pp 384–394

Vlieghe D, Sandelin A, Bleser P, Vleminckx K, Wasserman W, Roy F, Lenhard B (2006) A new generation of JASPAR, the open-access repository for transcription factor binding site profiles. Nucleic Acids Res 34(Database issue):D95–D97

Wang T, Stormo GD (2003) Combining phylogenetic data with coregulated genes to identify regulatory motifs. Bioinformatics 19:2369–2380

Wang C, Xie J, Craig BA (2006) Context dependent models for discovery of transcription factor binding sites. Stat Methodol 3:55–68

Wasserman WW, Sandelin A (2004) Applied bioinformatics for the identification of regulatory elements. Nat Rev Genet 5:276–287

Wray GA, Hahn MW, Abouheif E, Balhoff JP, Pizer M, Rockman MV, Romano LA (2003) The evolution of transcriptional regulation in eukaryotes. Mol Biol Evol 20(9):1377–1419

Zaslavsky E, Singh M (2006) A combinatorial optimization approach for diverse motif finding applications. Algorithms Mol Biol 1:13. doi:10.1186/1748-7188-1-13

Zhang Y, Zaki MJ (2006a) SMOTIF: efficient structured pattern and profile motif search. Algorithms Mol Biol 1:22. doi:10.1186/1748-7188-1-22

Zhang Y, Zaki MJ (2006b) EXMOTIF: efficient structured motif extraction. Algorithms Mol Biol 1:21. doi:10.1186/1748-7188-1-21

Zhang S, Xu M, Li S, Su Z (2009) Genome-wide de novo prediction of cis-regulatory binding sites in prokaryotes. Nucleic Acids Res 37(10):e72

Zhang S, Li S et al (2010) Simultaneous prediction of transcription factor binding sites in a group of prokaryotic genomes. BMC Bioinformatics 11:397

Zhao F, Xuan Z, Liu L, Zhang MQ (2005) TRED: a Transcriptional Regulatory Element Database and a platform for in silico gene regulation studies. Nucleic Acids Res 33(Database issue):D103–D107

Zhu J, Zhang MQ (1999) SCPD: a promoter database of the yeast *Saccharomyces cerevisiae*. Bioinformatics 15:607–611

Chitin Deacetylase: Characteristic Molecular Features and Functional Aspects

9

Nidhi Pareek, V. Vivekanand, and R.P. Singh

Abstract

Chitosan has a broad and impressive array of applications in diverse industrial sectors, like pharmaceutics (drug delivery), gene delivery, tissue engineering, food and cosmetics industry, water treatment, and agriculture. To date, majority of the chitosan is produced from thermo-alkaline deacetylation of chitin from crustacean shells. The process is incompatible as it leads to variability in the product properties, increased cost of production, and environmental concerns. Functional properties and in turn industrial applicability of chitosan depend on its degree of deacetylation; hence, a controlled biological process needs to be developed so as to realize the commercial value of the product. Chitin deacetylase (CDA) is the key enzyme employed for bioconversion of chitin to chitosan. It catalyzes deacetylation of *N*-acetyl-D-glucosamine residues under mild reaction conditions and results into production of novel superior-quality chitosan. The enzyme-aided production is a vital step towards the chitosan production in the green chemistry realm as the chemical process is engraved with a number of limitations and bottlenecks. Apart from being used in bioconversion reactions, CDA has a number of biological roles, namely, formation of spore wall in *Saccharomyces cerevisiae* and vegetative cell wall in *Cryptococcus neoformans*, responsible for pathogenesis of plant pathogenic fungi, and utilization of chitin in marine ecosystems.

N. Pareek
Department of Biotechnology, Indian Institute of Technology Roorkee, Roorkee 247667, India

Department of Chemistry, Umeå University, SE-90187 Umeå, Sweden

V. Vivekanand
Department of Biotechnology, Indian Institute of Technology Roorkee, Roorkee 247667, India

Protein Engineering and Proteomics Group, Department of Chemistry, Biotechnology and Food Sciences, Norwegian University of Life Sciences, Ås 1430, Norway

R.P. Singh (✉)
Department of Biotechnology, Indian Institute of Technology Roorkee, Roorkee 247667, India
e-mail: rpsbsfbs@iitr.ernet.in

P. Shukla and B.I. Pletschke (eds.), *Advances in Enzyme Biotechnology*,
DOI 10.1007/978-81-322-1094-8_9, © Springer India 2013

Keywords
Chitin deacetylase • Chitin • Chitosan • Deacetylation • Bioconversion

Introduction

One of the major objectives of present age of biotechnology is headed towards production and application of a range of bio-based value-added products. Biopolymers, i.e., cellulose, xylan, lignin, chitin, tannin, and pectin, are considered as the major resource materials for a number of industrial sectors due to their renewable and biodegradable nature. Among all the polymers, in spite of its abundance in nature, chitin is the most underutilized one due to its high degree of crystallinity and insolubility in most of thesolvents (Ruiz-Herrera 1978). Chitosan, the *N*-deacetylated derivative of chitin has enormous commercial potential due to its characteristically superior properties like biodegradability, biocompatibility, solubility, and non-toxicity (Kurita 2006). It is present in nominal amounts in animal biomass, in the shells or cuticles of many crustaceans, and also in the fungal cell wall; it is therefore mainly derived from chitin by chemical or biocatalytic alkaline deacetylation process. Chemical route utilizes larger amounts of concentrated alkali that in turn eventually leads into environmental deterioration (Chang et al. 1997). Bioconversion to chitosan is assisted by a member of carbohydrate esterase family 4, i.e., chitin deacetylase (CDA, EC 3.5.1.41). The enzyme catalyzes the removal of acetyl groups from the nascent chitin chain via a multiple-attack mechanism, resulting into a polymer consisting of both glucosamine and *N*-acetylglucosamine monomers (Tsigos et al. 2000). CDA-assisted chitosan production had initially begun as a thought and now considered as a worthwhile, promising technology of the future. Intense research and developmental activities are under way to develop a robust enzyme preparation that can be explored further as an eco-friendly approach for production of chitosan with desired properties.

Chitin Deacetylase

Source

The enzyme has been detected in an array of organisms including fungi, bacteria, and insects, among which fungal deacetylases are widely explored (Tsigos et al. 2000; Meens et al. 2001; Zhao et al. 2010a) (Table 9.1). The physiological role of CDAs in microbes is primarily concerned with the production of chitosan, a cell wall component, along with the pathogenesis of plant pathogenic fungi.

Deacetylases of Fungal Origin

A number of attempts have been made for CDA production from fungi mainly from *Mucor rouxii* (Davis and Bartnicki-Garcia 1984; Kafetzopoulos et al. 1993a), *Colletotrichum lindemuthianum* (Tsigos and Bouriotis 1995; Tokuyasu et al. 1996; Shrestha et al. 2004), *Absidia coerulea* (Gao et al. 1995), *Aspergillus nidulans* (Alfonso et al. 1995), *Gongronella butleri* (Maw et al. 2002a, b), *Metarhizium anisopliae* (Nahar et al. 2004), *Rhizopus nigricans* (Jeraj et al. 2006), *Scopulariopsis brevicaulis* (Cai et al. 2006), *Mortierella* sp. DY-52 (Kim et al. 2008), *Rhizopus circinans* (Gauthier et al. 2008), *Flammulina velutipes* (Yamada et al. 2008), *Absidia corymbifera* (Zhao et al. 2010b), etc. Fungi are known to produce both extra- and intracellular deacetylases during different periods owing to their specific biological roles, namely, development of fungal cell wall, modifying the hyphal chitin of plant pathogenic fungi to aid in pathogenesis by protecting it from the plant resistance system.

Deacetylases of Yeast Origin

Apart from these fungal strains, some yeast species, namely, *Saccharomyces cerevisiae* (Martinou et al. 2002) and *Schizosaccharomyces pombe* (Matsuo et al. 2005), are also known to produce CDAs during sporulation, which may be involved in the synthesis of spore wall.

Table 9.1 Characteristic features of chitin deacetylase from microorganisms

Organism	MW (kD)	pI	Opt. pH	Opt. temp (°C)	Acetate inhibition	Metal ion activation/ inhibition	References
Mucor rouxii ATCC 24905	75	3	4.5	50	Yes	–	Kafetzopoulos et al. (1993a)
Aspergillus nidulans CECT 2544	27	2.7	7.0	50	Yes	–	Alfonso et al. (1995)
Colletotrichum lindemuthianum DSM 63144	150	–	8.5	50	No	Co^{+2} (*Ac.*)	Tsigos and Bouriotis (1995)
Colletotrichum lindemuthianum ATCC 56676	31.5	3.7	11.5–12.0	60	Yes	Co^{+2} (*Ac.*)	Tokuyasu et al. (1996)
Vibrio alginolyticus H-8	48, 46	3.3, 3.5	8.5, 8.0	45, 40	–	Ag^{+}, Hg^{+2} (*Ac.*)	Ohishi et al. (2000)
Thermus caldophilus	45	–	7.5	80	–	Mn^{+2}, Co^{+2}, Fe^{+3} (*Ac.*); Cu^{+2} (*In.*)	Shin et al. (1999)
Metarhizium anisopliae	70, 37, 26	2.6, 3.8, 4.1	8.5–8.8	–	No	–	Nahar et al. (2004)
Mucor circinelloides	–	–	4.5	50	–	–	Amorim et al. (2005)
Scopulariopsis brevicaulis	55	–	7.5	55	–	–	Cai et al. (2006)
Mortierella sp. DY-52	50, 59	–	5.5	60	–	Co^{+2}, Ca^{+2} (*Ac.*)	Kim et al. (2008)
Absidia corymbifera DY-9	–	–	6.5	55	Yes	Co^{+2} Ca^{+2}, Mg^{+2} (*Ac.*)	Zhao et al. (2010b)

Ac activation, *In* inhibition

Deacetylases of Bacterial Origin

Among the bacterial strains, members belonging to family Vibrionaceae, widely distributed in all oceanic and estuarine waters, are considered as the major CDA producers (Ferguson and Gooday 1996), where they are involved in the chitin metabolism (Hunt et al. 2008). Most of the *Vibrio* strains are known to produce chitin oligosaccharide deacetylase (COD), which in turn produces a characteristic inducer for chitinase production for chitin catabolism (Hirano et al. 2009). COD from *Vibrio parahaemolyticus* and *Vibrio cholerae* was purified and characterized by Kadokura et al. (2007a, b) and Li et al. (2007), respectively. Hyperthermophilic archaeon *Thermococcus kodakaraensis* KOD1 also possesses a deacetylase, involved in their chitin catabolic pathway (Tanaka et al. 2003). *Bacillus amyloliquefaciens* was also known to produce CDA (Zhou et al. 2010).

Deacetylases of Insect Origin

Occurrence of CDAs is not only restricted to microbial population but also detected in insect species. *Trichoplusia ni* (Guo et al. 2005), *Helicoverpa armigera* (Campbell et al. 2008), and *Mamestra configurata* (Toprak et al. 2008) are reported to have CDAs in their midgut peritrophic matrix. In *Drosophila melanogaster* two CDA-like proteins, CDA1 and CDA2 (serpentine and vermiform), have critical roles in shaping the tracheal tubes as well as regulating the structural properties of epidermal cuticle by influencing the structure and orientation of chitin fibrils (Luschnig et al. 2006; Wang et al. 2006). Apart from the above-mentioned species, *Anopheles gambiae*, *Apis mellifera*, and *Tribolium castaneum* (Dixit et al. 2008; Arakane et al. 2009; Noh et al. 2011) are reported to produce CDAs. In spite of the presence of CDAs in a large number of insect species, their roles are still unresolved.

Classification

CDAs (EC 3.5.1.41) are the members of carbohydrate esterase family 4 (CE-4s) as defined in the CAZY database (http://afmb.cnrs-mrs.fr/~cazy/CAZY). According to the Henrissat classification, the family also includes rhizobial NodB chitooligosaccharide deacetylases, peptidoglycan *N*-acetylglucosamine deacetylases (EC 3.5.1.104), acetyl xylan esterases (EC 3.1.1.72), and xylanases A, C, D, E (EC 3.5.1.8) (Coutinho and Henrissat 1999). Members of this family share a conserved region in their primary structure named the "NodB homology domain" or "polysaccharide deacetylase domain" (Caufrier et al. 2003; Gauthier et al. 2008). All five members of this family catalyze the hydrolysis of either *N*-linked acetyl groups from *N*-acetylglucosamine residues or *O*-linked acetyl groups from *O*-acetylxylose residues of their substrates, namely, chitin, NodB factors, peptidoglycan, and acetyl xylan. Insect CDA-like proteins are classified into five groups based on phylogenetic analysis and the presence of additional motifs. Group I include CDA1 and isoforms of CDA2, containing a polysaccharide deacetylase-like catalytic domain, a chitin-binding peritrophin-A domain (ChBD), and a low-density-lipoprotein receptor class A domain (LDLa). Group II is composed of CDA3 proteins having similar domain organization as group I CDAs, but with substantially different sequences. Group III includes CDA4s, which have only the ChBD domain. Group IV comprises of CDA5s, which are the largest CDAs because of a very long intervening region separating the ChBD and catalytic domains. Group V contains divergent group of proteins containing only a catalytic domain, lacking the *N*-terminal ChBD and the LDLa domain (Dixit et al. 2008).

Multiplicity

Multiplicity corresponds to existence of multiple forms of the similar enzymes or isozymes, which catalyzes the same reaction but differ in their amino acid sequence and physicochemical properties, such as molecular weight, isoelectric point, and kinetic constants. Such multiple forms of CDAs were observed in many fungal and bacterial species. In *Uromyces viciae-fabae*, five isoforms of CDA (12.7–48.1 kDa) were produced during penetration of fungus through leaf stomata (Deising and Siegrist 1995). Three CDA isoforms were also reported by Trudel and Asselin (1990) in *Mucor racemosus* with molecular mass of 26, 30, and 64 kDa. Similarly, *C. lindemuthianum*, *M. anisopliae*, and *R. nigricans* are also known to produce two, three, and four CDA isozymes, respectively. Among bacterial species, *Vibrio alginolyticus* H-8 produced two extracellular CDA isoforms (Ohishi et al. 2000). Genomes of *Tribolium castaneum*, *Drosophila melanogaster*, *Anopheles gambiae*, and *Apis mellifera* contain 9, 6, 5, and 5 genes, encoding proteins with a CDA motif. The presence of alternative exons in two of the *Tribolium* CDA genes, CDA2 and CDA5, leads into protein diversity further due to alternative splicing. However, these CDA genes are quite different in terms of tissue specificity and developmental patterns of expression. Possible reasons for this multiplicity may be due to gene duplication, differential *m*RNA processing, posttranslational modifications such as glycosylation and autoaggregation. Multiple CDAs can also be the product from different alleles of the same gene, i.e., allozymes.

Substrate Specificity

Substrate specificity of *M. rouxii* CDA was first evaluated by Araki and Ito (1975) and they observed the enzyme activity for glycol chitin and chitooligomers with degree of polymerization greater than two. Deacetylase activity of the enzyme depends on the number of monosaccharide units in the substrate. The values of kinetic parameters, namely, k_{cat}, V_{max}, and k_{cat}/K_m, were observed to increase along with the degree of polymerization of chitin oligomers, while the values of Michaelis constant (K_m) decreased with the increasing polymerization of glucosamine moieties. Caufrier et al. (2003) had analyzed the activity of CDA from *M. rouxii* and acetyl xylan esterase from *Streptomyces lividans* on acetyl xylan, peptidoglycan, and soluble chitin as the substrates. The enzymes were observed to be active on acetyl xylan and soluble chitin while inactive on peptidoglycan. This might be attributed to the difference in the sequence and structure of the catalytic domain of the enzymes. One such difference is the absence of the disulfide bond, tethering the *N*-terminal and *C*-terminal ends from the homologous peptidoglycan deacetylases from *Streptococcus pneumoniae* and *Bacillus subtilis*, while observed to be conserved in CDA from *M. rouxii* and *C. lindemuthianum* and also in *S. lividans* acetyl xylan esterase (Blair et al. 2005, 2006; Taylor et al. 2006).

Catalytic Mechanism

Mode of catalysis by CDAs on both chitin polymers and oligomers had been studied by several research groups. Catalytic action of CDAs strongly depends on the degree of polymerization of the substrate. Tokuyasu et al. (1997) studied the deacetylation of chitooligosaccharides (degree of polymerization (DP), 2–4) by purified *C. lindemuthianum* CDA using FAB-MS (fast atom bombardment mass spectrometry) and ^{1}H NMR spectroscopy and concluded that *N, N′, N″, N‴*-tetraacetylchitotetraose and *N, N′, N″*-triacetylchitotriose were deacetylated to corresponding chitosan oligomers. But *N, N′*-diacetylchitobiose was deacetylated on either of the glucosamine residues to yield a unique compound, i.e., 2-acetamido-4-*O*-(2-amino-2-deoxy-*β*-D-glucopyranosyl)-2-deoxy-D-glucose [GlcN-GlcNAc]. *M. rouxii* CDA deacetylates water-soluble partially deacetylated chitosans (DP, 30) following a multiple-attack mechanism with a degree of multiple attack of at least three (Martinou et al. 1998). The polarity of CDA was preferentially towards the reducing end and it follows an endo-type mechanism with no preferential attack at any sequence in the chitosan chain. Deacetylation reaction was not detected at the nonreducing end of the chain. The relative rate of enzymatic deacetylation increased linearly with the increasing fraction of acetylated units. Deacetylation of *N*-acetylchitooligosaccharides (DP 2–7) by *Mucor rouxii* ATCC 24905 CDA was studied by Tsigos et al. (1999) employing an exo-splitting system. The extent of deacetylation depends on the length of the substrate. The enzyme could not effectively deacetylate chitin oligomers with a degree of polymerization less than three. Only tetra-*N*-tetraacetylchitotetraose and penta-*N*-pentaacetylchitopentaose were fully deacetylated by the enzyme while the reducing end residues of *N*-acetylchitotriose, *N*-acetylchitohexaose, and *N*-acetylchitoheptaose always remained intact. The enzyme initially removes an acetyl group from the nonreducing end residue and further catalyzes the hydrolysis of the next acetamido group in progressive fashion. This report was in contrast with the findings of Martinou et al. (1998) in which the enzyme was observed to be more active towards reducing end.

C. lindemuthianum ATCC 56676 CDA was observed to follow a multiple chain mechanism to remove the acetyl groups (Tokuyasu et al. 2000). The structural analysis of deacetylation products of $(GlcNAc)_4$ suggested that the enzyme has four subsites (−2, −1, 0, +1), in which subsite 0 is the catalytic subsite. Reaction rate analysis of partially deacetylated substrates showed that subsite −2 strongly recognizes the *N*-acetyl group of the GlcNAc residue of the substrate. Hekmat et al. (2003) had also suggested the presence of four enzyme subsites (−2 to +1) in the *C. lindemuthianum* CDA. Steady-state kinetic analyses for the initial deacetylation reaction of $(GlcNAc)_{2–6}$ by *C. lindemuthianum* ATCC 56676 elucidated that the kinetic parameters, i.e., K_m and K_{cat}/K_m, depends on the degree of polymerization of the substrate, while K_{cat} is independent. A more detailed insight of the catalytic action of *C. lindemuthianum* was outlined by Blair et al. (2006) via coupled structural and biochemical analysis. The reaction proceeds via generation of a tetrahedral oxyanion intermediate following a nucleophilic attack to the carbonyl carbon of the substrate, the charge of which is stabilized by the oxyanion hole generated by the backbone nitrogen of Tyr145 and zinc. Transfer of a proton from the water molecule to the catalytic base Asp49 leads to the generation of a nucleophile to attack the substrate carbonyl carbon. Further, His206 protonates the reaction intermediate on the nitrogen as it breaks down, generating a free amine and the acetate as the product.

Cloning of Chitin Deacetylase Gene

Till date CDA genes from various sources have been isolated, cloned, and expressed into suitable homologous as well as heterologous hosts. Kafetzopoulos et al. (1993b) isolated and characterized a cDNA for CDA from *M. rouxii* ATCC 24905. Two CDA genes (CDA1 and CDA2) of *S. cerevisiae* were cloned and expressed into the plasmid pSK and sequenced (Mishra et al. 1997). Both of these deacetylases are expressed constitutively during sporulation, since spore wall of the yeast contains chitosan, which along with the dityrosine layer have a role in the spore protection in extreme conditions. CDA1- and CDA2-deficient mutants were observed to be more sensitive towards chemical and environmental challenges. CDA gene from *C. lindemuthianum* ATCC 56676 was overexpressed in *Escherichia coli* as a fusion protein using an expression vector pQE60 (Tokuyasu et al. 1999). The CDA open reading frame (ORF) consists of two regions: one from the start codon (ATG) encoding a deduced preprodomain of 27 amino acids and the other encoding a mature CDA of 221 amino acids. Ohishi et al. (2000) cloned a deacetylase encoding gene DA1 from *V. alginolyticus* H-8 and sequenced using a shotgun approach. The ORF of the gene starts at base 256 and ends at the base 1,536. This 1,281-nucleotide-long sequence encoded a 427-amino-acid-long enzyme. A putative Shine-Dalgarno (SD) sequence, ACGA, was found upstream to the start codon, ATG.

R. nigricans B 154 CDA was cloned by Jeraj et al. (2006) using yeast expression vector pFL61 in *E. coli* DH5α. Sequence analysis divulged an open reading frame of 1,341 nucleotides encoding a complete 447-amino-acid protein. CDA gene from *C. lindemuthianum* UPS9 was isolated and cloned in *Pichia pastoris* as a tagged protein with six added terminal histidine residues. The specific activity of the purified protein was 72 Umg^{-1} (Shrestha et al. 2004). A complete CDA cDNA from *M. racemosus* was cloned and sequenced by RT-PCR and RACE with conserved primers (Xia-Yun et al. 2007). Predicted three-dimensional structure of the gene had outlined the whole CDA functional domain and a polysaccharide deacetylase domain. The cloned 1,506 bp CDA gene from *M. racemosus* included a 67 bp 5′-untranslated region, an open reading frame of 1,344 bp, and a 95 bp 3′-untranslated region including a tailing site AATAAA. The gene coded for a 448-amino-acid protein and consisted of core nucleotides encoding a polysaccharide deacetylase conserved domain, which had 144 amino acids and covered 32 % of the entire sequence in the middle part. The structural domain of CDA deduced from the primary sequence was very similar to that of other species. A 75 kDa CDA from *R. circinans* was cloned in *P. pastoris* expression system using cDNA library, and 4.8-fold increment was achieved in the production level (Gauthier et al. 2008). *Fv-pda*, a gene encoding CDA, was isolated from *F. velutipes* during fruiting body development and expressed in *P. pastoris* (Yamada et al. 2008). The *fv-pda* open reading frame comprises 250 amino acid residues and is interrupted by 10 introns. The recombinant FV-PDA was observed to effectively catalyze the deacetylation of chitin oligomers (dimer to pentamer), glycol chitin and colloidal chitin. cDNA amplification of nine *Tribolium* CDA genes were carried out using gene-specific primers, designed from the available expressed sequence tags (ESTs) or from ORFs in GLEAN predictions from the *Tribolium* shotgun genome sequences.

Crystal Structure

The structure of CDA from *C. lindemuthianum* (*Cl*CDA, Blair et al. 2006) is the first and till now the only structure among CDAs that is known. It is a metalloenzyme and consists of a single catalytic domain, similar to the deformed $(\beta/\alpha)_8$ fold in other CE-4 family members (Blair and Van Aalten 2004; Blair et al. 2005). The active site cleft of the protein is formed from the *C*-terminal ends of β-strands 2,4,5,7, and 8 of the $(\beta/\alpha)_8$ barrel and includes five distinct sequence motifs (MT1-MT5) conserved in the family (Blair et al. 2006). The catalytic subsite contains a zinc-binding triad consisting of two histidines (His104, His108) from motif 2 and an aspartic acid (Asp50) from motif 1. In addition, a loop is present between strand β3 and helix α2, supporting Trp79 that protrudes into the catalytic cleft. Two intramolecular disulfide linkages (Cys38-Cys237, Cys148-Cys152) are known to stabilize the structure. Sequence alignment suggests that one disulfide (Cys38-Cys237), tethering the *N*- and *C*-terminal ends of the structure, is conserved in the fungal CDA from *M. rouxii* and other CE-4 family members.

Salient Structural Features

Among the carbohydrate active enzymes, CDAs are relatively less studied in terms of structural features. *Cl*CDA is a member of the carbohydrate esterases family 4 (CE-4s) which include several members that share the "NodB homology domain" (Caufrier et al. 2003). CDA, acetyl xylan esterase, rhizobial NodB, and peptidoglycan deacetylases are involved in deacetylating chitin, xylan, the nonreducing end GlcNAc from short chitooligosaccharides for synthesis of Nod factors, and *N*-acetylmuramic acid (MurNAc) or GlcNAc residues of the disugar repeats to modify bacterial cell wall peptidoglycan, respectively. In spite of sharing a homologous catalytic domain, significant topological differences are observed between the deacetylases group, i.e., *Cl*CDA, *Sp*PgdA (*Streptococcus pneumoniae* peptidoglycan deacetylase A), and *Bs*PdaA (*Bacillus subtilis* peptidoglycan deacetylase), via structural analysis. The intramolecular disulfide bonds and an extended loop between strand β3 and helix α2 are present in *Cl*CDA while absent from the *Sp*PgdA and *Bs*PdaA. The N/C termini are present on the same side of the barrel in *Cl*CDA while they are located at opposite ends in *Sp*PgdA and BsPdaA. Apart from these, the conserved catalytic subsite lined by the His-His-Asp residues has only two metal coordinating residues in *Bs*PdaA (Blair et al. 2006).

Functional Aspects

Bioconversion to Chitosan

A major objective for developing CDA has been to replace the harsh chemical process for conversion of chitin into chitosan, the process as considered is not an environment-friendly one. The enzymatic deacetylation would be able to provide chitosans with defined levels of deacetylation, as like the chemical process, it does not proceed in a random fashion and permit the partially deacetylated chitosans obtained from the chemical process to undergo the desired degree of deacetylation. The enzymatic deacetylation of various chitinous substrates was investigated by Aye et al. (2006) using the CDA isolated from *Rhizopus oryzae*. Chitin was observed to be a poor substrate for the enzyme, but reprecipitated chitin was moderately better. *F. velutipes* CDA catalyzes deacetylation of *N-acetyl-chitooligomers*, from dimer to pentamer, glycol chitin and colloidal chitin (Yamada et al. 2008). Chitosan with lower degree of deacetylation (28 % and 42 %) can be further deacetylated with *M. rouxii* deacetylase (Martinou et al. 1995, 1997a, b). Yield and rate of deacetylation was observed to be more with amorphous chitin substrates. Martinou et al. (2002) had attempted deacetylation of glycol chitin, chitin-50, and *N*-acetylchitooligosaccharides using cobalt-activated CDA from *S. cerevisiae* and concluded that the enzyme requires at least two *N*-acetyl-D-GLUCOSAMINE residues for catalysis, exhibiting maximum activity on hexa-*N*-acetylchitohexaose. CDAs from *C. lindemuthianum* (Kauss and Bausch 1988; Tsigos and Bouriotis 1995; Tokuyasu et al. 1997), *A. nidulans* (Alfonso et al. 1995), *S. brevicaulis* (Cai et al. 2006), and *Mortierella* sp. (Kim et al. 2008) also exhibited similar deacetylation kinetics towards chitin and its oligomers. Crude CDA from *C. lindemuthianum* was active on partially deacetylated chitin (chitin with 50 %, 65 %, 70 %, and 82 % degree of deacetylation) (Shrestha et al. 2004).

Crystallinity and insolubility of chitin makes it a poor substrate for enzymatic deacetylation. The degree of crystallinity of the chitin must be reduced to enable enzymes to access the internal polysaccharide structure. Pretreatment of chitin by either physical or chemical means prior to enzymatic reaction was necessary to increase the substrate accessibility to the enzyme (Martinou et al. 1997a, b; Win et al. 2000). Nearly 90 % deacetylation of superfine chitin generated using $CaCl_2.2H_2O$/methanol solvent system was achieved using CDA from *A. coerulea* and *C. lindemuthianum* (Win and Stevens 2001). Beaney et al. (2007) also modified chitin by physical or chemical methods followed by deacetylating the same with extracellular deacetylase from *C. lindemuthianum.* Modifications of the chitin led to decreased crystallinity of the substrate with simultaneous increase in enzymatic deacetylation. It was observed that the dissolution and drying methods used in modifying the chitin had significant impact on the final efficiency of the enzymatic deacetylation reaction. Jaworska et al. (2009) found that immobilization of CDA to diethylaminoethyl cellulose via divinyl sulfone led to high activity and stability towards various chitin and chitosans.

Chitin Catabolism

Marine ecosystems contain huge quantities of chitin, with an annual production of $>10^{11}$ metric tons. This bulk amount of chitin is recycled by the members of the family Vibrionaceae, capable of using chitin as the sole carbon source. Chitin sensing, attachment, and degradation are the three steps involved in the chitin utilization by *Vibrio furnissii* (Bassler et al. 1989, 1991). Interactions between *V. cholerae* and chitin occur at multiple hierarchical levels in the environment and include cell metabolic and physiological responses, e.g., chemotaxis, cell multiplication, induction of competence, biofilm formation, commensal and symbiotic relationship with higher organisms, cycling of nutrients, as well as pathogenicity for humans and aquatic animals (Pruzzo et al. 2008). Three sets of differentially regulated genes, i.e., a $(GlcNAc)_2$ catabolic operon, two extracellular chitinases, a chitoporin, make *V. cholerae* to be able to utilize chitin (Meibom et al. 2004). Induction of *V. cholerae* with $(GlcNH_2)_2$ or crab shells resulted into the production of COD, hydrolyzes the *N*-acetyl group attached to the penultimate GlcNAc unit (Li et al. 2007). *V. alginolyticus* was observed to produce a deacetylase specific for $(GlcNAc)_2$, but inactive with higher oligosaccharides (Ohishi et al. 2000). COD is considered as an essential part of the chitin catabolic cascade of marine bacteria (Jung et al. 2008) due to its ability to produce heterodisaccharide GlcNAc-GlcN, a unique inducer for chitinase production. Chitin utilization among several bacterial species, namely, *V. parahaemolyticus* KN1699 (Kadokura et al. 2007a), *V. furnissii* (Bassler et al. 1991), *T. kodakaraensis* KOD1 (Tanaka et al. 2003), *Serratia marcescens* (Watanabe et al. 1997), *S. lividans* (Miyashita et al. 2000), and *Streptomyces coelicolor* (Saito et al. 2000, 2007), is found to be associated with the expression of COD.

Synthesis of Cell Wall/Spore Wall

Expression of CDA among fungal strains is thought to be related to various stages of growth to execute different functions. In *M. rouxii* and *A. coerulea*, CDA was localized near the periplasmic space in the mycelia and contributed to formation of chitosan in the cell wall from nascent chitin synthesized by the action of chitin synthetase (Davis and Bartnicki-Garcia 1984; Ruiz-Herrera and Martinez-Espinoza 1999). Even though enzymology and cytology of chitin biosynthesis in fungi have been extensively studied, very little information exists on the correlation between CDA and chitosan biosynthesis (Hunt et al. 2008).

S. cerevisiae requires chitin as an essential component for vegetative growth; however, for spore wall formation both chitin synthesis and chitin deacetylation are necessary. Chitin is synthesized by three chitin synthases, Chs1, Chs2, and Chs3, in *S. cerevisiae*, and its conversion to chitosan by either Cda1 or Cda2 had allowed the second layered structure of the spore wall next to the outer dityrosine layer to retain its structural rigidity and resistance to various stresses (Mishra et al. 1997; Christodoulidou et al. 1999). Cda2p is the predominant deacetylase and performs most of the deacetylation task, while Cda1p contributes to the proper ascospore wall assembly (Martinou et al. 2002, 2003). In addition, a *cda1*$^+$ encoded CDA in a fission yeast *S. pombe* was identified and required for proper spore formation (Matsuo et al. 2005). Four CDAs, namely,

Cda1, Cda2, Cda3, and Fpd1, have been identified from *C. neoformans*, responsible for chitosan synthesis which is an important component of the vegetative cell wall and helps to maintain cell integrity and aids in bud separation (Baker et al. 2007).

Plant-Pathogen Interaction

CDA aids pathogenesis of plant pathogenic fungi, namely, wheat stem rust fungus *Puccinia graminis* f. sp. *tritici* and the broad bean rust fungus *U. fabae*, and the causative agents of anthracnose, *Colletotrichum graminicola* and *C. lindemuthianum*, by performing dual roles, i.e., protection of penetrating fungal hyphae from being lysed by secretory plant chitinases by transforming the superficial cell wall chitin into chitosan and decrease the activity of chitin oligomers. Plants, when attacked by fungal pathogens, activate an elaborate defense system consisting of chemically and physically performed resistance factors and have induced resistance reactions to evade the pathogenesis. Chitin oligomers play a vital role in eliciting the plant-defense mechanisms, namely, callose formation, lignification, and synthesis of coumarin derivatives. Plant endo-type chitinases produce the chitin oligomers by degrading the fungal chitin. Fungal pathogens escape the plant hydrolases by partially deacetylating the exposed chitin polymers during the initial growth via action of CDA (Kauss et al. 1983; Walker-Simmons et al. 1984; Vander et al. 1998; Gueddari et al. 2002; Hekmat et al. 2003). Chitin-binding domain (CBM14) of avirulence protein Avr4 of *Cladosporium fulvum* contributes to its pathogenicity by protecting the fungal cell wall against hydrolysis by plant chitinases during infection (van den Burg et al. 2006).

Biocontrol Agent

The role displayed by CDAs during pathogenesis of plant pathogenic fungi makes them a crucial target in the biological control of plant pathogenic fungi (Brosson et al. 2005; Das et al. 2006; Baker et al. 2007) and insect pests (Nahar et al. 2004). Inhibition of CDA would result into hydrolysis of fungal cell wall by plant chitinases; thus, the control of the plant pathogenic fungi becomes feasible (Tokuyasu et al. 1996). In the biological control of insect pests, CDAs were observed to have dual roles. The enzyme initiates the pathogenesis of the insect-pathogenic fungus by softening the insect cuticle to aid the mycelial penetration and it also alters the fungal cell wall to protect it from insect chitinase.

Conclusion and Future Perspectives

Chitosan is a biopolymer with immense commercial potential. Its functional characteristics correlate with its structural features. Enzyme-aided production of chitosan implying CDA possesses ecological and economic benefits over conventional production approaches. Various research groups are involved in studying the production and biochemical and molecular characterization of the enzyme but implication of bioconversion at large scale needs intensive screening of novel CDA hyper-producers that are able to produce the enzyme with higher catalytic efficacy and lower inhibition by end product, i.e., acetate. Attempts should be made to further improve the catalytic activity and stability of the enzyme by cloning and characterizing the corresponding molecular domains and then targeted mutagenesis of the corresponding sequences in the CDA gene. Apart from this, an efficient pretreatment approach needs to be developed to decrystallize the chitin structure to improve not only the substrate accessibility but also the rate of enzymatic deacetylation.

References

Alfonso C, Nuero OM, Santamaria F, Reyes F (1995) Purification of a heat-stable chitin deacetylase from *Aspergillus nidulans* and its role in cell wall degradation. Curr Microbiol 30:49–54

Amorim RVS, Ledingham WM, Fukushima K, Campos-Takaki GM (2005) Screening of chitin deacetylase from Mucoralean strains (Zygomycetes) and its relationship to cell growth rate. J Ind Microbiol Biotechnol 32:19–23

Arakane Y, Dixit R, Begum K, Park Y, Specht CA, Merzendorfer H, Kramer KJ, Muthukrishnan S, Beeman RW (2009) Analysis of functions of the chitin deacetylase gene family in *Tribolium castaneum*. Insect Biochem Mol Biol 39:355–365

Araki Y, Ito E (1975) A pathway of chitosan formation in *Mucor rouxii*. Eur J Biochem 55:71–78

Aye KN, Karuppuswamy R, Ahamed T, Stevens WF (2006) Peripheral enzymatic deacetylation of chitin and reprecipitated chitin particles. Bioresour Technol 97:577–582

Baker LG, Specht CA, Donlin MJ, Lodge JK (2007) Chitosan, the deacetylated form of chitin, is necessary for cell wall integrity in *Cryptococcus neoformans*. Eukaryot Cell 6:855–867

Bassler BL, Gibbons PJ, Roseman S (1989) Chemotaxis to chitin oligosaccharides by *Vibrio furnissii*, a chitinivorous marine bacterium. Biochem Biophys Res Commun 161:1172–1176

Bassler BL, Gibbons PJ, Yu C, Roseman S (1991) Chitin utilization by marine bacteria: chemotaxis to chitin oligosaccharides by *Vibrio furnissii*. J Biol Chem 266:24268–24275

Beaney PD, Gan Q, Magee TRA, Flealy M, Lizardi-Mendoza J (2007) Modification of chitin properties for enzymatic deacetylation. J Chem Technol Biotechnol 82:165–173

Blair DE, Van Aalten DM (2004) Structures of *Bacillus subtilis* PdaA, a family 4 carbohydrate esterase, and a complex with *N*-acetyl-glucosamine. FEBS Lett 570:13–19

Blair DE, Schuttelkopf AW, Macrae JI, van Aalten DM (2005) Structure and metal-dependent mechanism of peptidoglycan deacetylase, a streptococcal virulence factor. Proc Natl Acad Sci USA 102:15429–15434

Blair DE, Hekmat O, Schuttelkopf AW, Shrestha B, Tokuyasu K, Withers SG, van Aalten DMF (2006) Structure and mechanism of chitin deacetylase from the fungal pathogen *Colletotrichum lindemuthianum*. Biochemistry 45:9416–9426

Brosson D, Kuhn L, Prensier G, Vivarès CP, Texier C (2005) The putative chitin deacetylase of *Encephalitozoon cuniculi*: a surface protein implicated in microsporidian spore-wall formation. FEMS Microbiol Lett 247:81–90

Cai J, Yang J, Du Y, Fan L, Qiu Y, Li J, Kennedy JF (2006) Purification and characterization of chitin deacetylase from *Scopulariopsis brevicaulis*. Carbohydr Polym 65:211–217

Campbell PM, Cao AT, Hines ER, East PD, Gordon KHJ (2008) Proteomic analysis of the peritrophic matrix from the gut of the caterpillar, *Helicoverpa armigera*. Insect Biochem Mol Biol 38:950–958

Caufrier F, Martinou A, Dupont C, Bouriotis V (2003) Carbohydrate esterase family 4 enzymes: substrate specificity. Carbohydr Res 338:687–692

Chang KLB, Tsai G, Lee J, Fu W-R (1997) Heterogeneous N-deacetylation of chitin in alkaline solution. Carbohydr Res 303:327–332

Christodoulidou A, Briza P, Ellinger A, Bouriotis V (1999) Yeast ascospore wall assembly requires two chitin deacetylase isozymes. FEBS Lett 460:275–279

Coutinho PM, Henrissat B (1999) Carbohydrate-active enzymes: an integrated database approach. In: Gilbert HJ, Davis GJ, Henrissat B, Svensson B (eds) Recent advances in carbohydrate bioengineering. Royal Society of Chemistry, Cambridge, UK, pp 3–12

Das S, Dellen KV, Bulik D, Magnelli P, Cui J, Head J, Robbins PW, Samuelson J (2006) The cyst wall of *Entamoeba invadens* contains chitosan (deacetylated chitin). Mol Biochem Parasitol 148:86–92

Davis LL, Bartnicki-Garcia S (1984) Chitosan synthesis by the tandem action of chitin synthetase and chitin deacetylase from *Mucor rouxii*. Biochemistry 23:1065–1073

Deising H, Siegrist J (1995) Chitin deacetylase activity of the rust *Uromyces viciae-fabae* is controlled by fungal morphogenesis. FEMS Microbiol Lett 127:207–212

Dixit R, Arakane Y, Specht CA, Richard C, Kramer KJ, Beeman RW, Muthukrishnan S (2008) Domain organization and phylogenetic analysis of proteins from the chitin deacetylase gene family of *Tribolium castaneum* and three other species of insects. Insect Biochem Mol Biol 38:440–451

Ferguson MJL, Gooday GW (1996) Environmental recycling of chitin. In: Muzzarelli RAA (ed) Chitin enzymology. Atec, Grottammare, pp 393–396

Gao XD, Katsumoto T, Onodera K (1995) Purification and characterization of chitin deacetylase from *Absidia coerulea*. J Biochem 117:257–263

Gauthier C, Clerisse F, Dommes J, Versali MJ (2008) Characterization and cloning of chitin deacetylases from *Rhizopus circinans*. Protein Expr Purif 59:127–137

Gueddari NEE, Rauchhaus U, Moerschbacher BM, Deising HB (2002) Developmentally regulated conversion of surface-exposed chitin to chitosan in cell walls of plant pathogenic fungi. New Phytol 156:103–112

Guo W, Li G, Pang Y, Wang P (2005) A novel chitin-binding protein identified from the peritrophic membrane of the cabbage looper, *Trichoplusia ni*. Insect Biochem Mol Biol 35:1224–1234

Hekmat O, Tokuyasu K, Withers SG (2003) Subsite structure of the endo-type chitin deacetylase from a Deuteromycete, *Colletotrichum lindemuthianum*: an investigation using steady-state kinetic analysis and MS. Biochem J 374:369–380

Hirano T, Kadokura K, Ikegami T, Shigeta Y, Kumaki Y, Hakamata W, Oku T, Nishio T (2009) Heterodisaccharide 4-*O*-(*N*-acetyl-beta-D-glucosaminyl)-D-glucosamine is a specific inducer of chitinolytic enzyme production in Vibrios harboring chitin oligosaccharide deacetylase genes. Glycobiology 19:1046–1053

Hunt DE, Gevers D, Vahora NM, Polz MF (2008) Conservation of the chitin utilization pathway in the Vibrionaceae. Appl Environ Microbiol 74:44–51

Jaworska MM, Bryjak J, Liesiene J (2009) A search of an optimal carrier for immobilization of chitin deacetylase. Cellulose 16:261–270

Jeraj N, Kunic B, Lenasi H, Breskvar K (2006) Purification and molecular characterization of chitin deacetylase from *Rhizopus nigricans*. Enzyme Microb Technol 39:1294–1299

Jung BO, Roseman S, Park JK (2008) The central concept for chitin catabolic cascade in marine bacterium, *Vibrios*. Macromol Res 16:1–5

Kadokura K, Rokutani A, Yamamoto M, Ikegami T, Sugita H, Itoi S, Hakamata W, Oku T, Nishio T (2007a) Purification and characterization of *Vibrio parahaemolyticus* extracellular chitinase and chitin oligosaccharide deacetylase involved in the production of heterodisaccharide from chitin. Appl Microbiol Biotechnol 75:357–365

Kadokura K, Sakamoto Y, Saito K, Ikegami T, Hirano T, Hakamata W, Oku T, Nishio T (2007b) Production of a recombinant chitin oligosaccharide deacetylase from *Vibrio parahaemolyticus* in the culture medium of *Escherichia coli* cells. Biotechnol Lett 29:1209–1215

Kafetzopoulos D, Martinou A, Bouriotis V (1993a) Bioconversion of chitin to chitosan: purification and characterization of chitin deacetylase from *Mucor rouxii*. Proc Natl Acad Sci USA 90:2564–2568

Kafetzopoulos D, Thireos G, Vournakis JN, Bouriotis V (1993b) The primary structure of a fungal chitin deacetylase reveals the function for two bacterial gene products. Proc Natl Acad Sci USA 90:8005–8008

Kauss H, Bausch B (1988) Chitin deacetylase from *Colletotrichum lindemuthianum*. In: Woods WA, Kellogg ST (eds) Methods in enzymology. Academic, San Diego, pp 518–523

Kauss H, Jeblick W, Young DH (1983) Chitin deacetylase from the plant pathogen *Colletotrichum lindemuthianum*. Plant Sci 28:231–236

Kim YJ, Zhao Y, Oh KT, Nguyen VN, Park RD (2008) Enzymatic deacetylation of chitin by extracellular chitin deacetylase from a newly screened *Mortierella* sp. DY-52. J Microbiol Biotechnol 18:759–766

Kurita K (2006) Chitin and chitosan: functional biopolymers from marine crustaceans. Marine Biotechnol 8:203–226

Li XB, Wang LX, Wang XS, Roseman S (2007) The chitin catabolic cascade in the marine bacterium *Vibrio cholerae*: characterization of a unique chitin oligosaccharide deacetylase. Glycobiology 17:1377–1387

Luschnig S, Bätz T, Armbruster K, Krasnow MA (2006) *serpentine* and *vermiform* encode matrix proteins with chitin binding and deacetylation domains that limit tracheal tube length in *Drosophila*. Curr Biol 16:186–194

Martinou A, Kafetzopoulos D, Bouriotis V (1995) Chitin deacetylation by enzymatic means: monitoring of deacetylation process. Carbohydr Res 273:235–242

Martinou A, Tsigos I, Bouriotis V (1997a) Enzymatic deacetylation of chitooligosaccharides. In: Muzzarelli RAA, Peter MG (eds) Chitin handbook. Atec, Grottammare, pp 191–194

Martinou A, Tsigos I, Bouriotis V (1997b) Preparation of chitosan by enzymatic deacetylation. In: Muzzarelli RAA, Peter MG (eds) Chitin handbook. Atec, Grottammare, pp 501–506

Martinou A, Bouriotis V, Stokke BT, Vårum KM (1998) Mode of action of chitin deacetylase from *Mucor rouxii* on partially *N*-acetylated chitosans. Carbohydr Res 311:71–78

Martinou A, Koutsioulis D, Bouriotis V (2002) Expression, purification, and characterization of a cobalt-activated chitin deacetylase (Cda2p) from *Saccharomyces cerevisiae*. Protein Expr Purif 24:111–116

Martinou A, Koutsiouli D, Bouriotis V (2003) Cloning and expression of chitin deacetylase gene (chitin deacetylase 2) from *Saccharomyces cerevisiae* in *Escherichia coli*. Purification and characterization of the cobalt-dependent recombinant enzyme. Enzyme Microb Technol 32:757–763

Matsuo Y, Tanaka K, Matsuda H, Kawamukai M (2005) *cda1+*, encoding chitin deacetylase is required for proper spore formation in *Schizosaccharomyces pombe*. FEBS Lett 579:2737–2743

Maw T, Tan TK, Khor E, Wong SM (2002a) Complete cDNA sequence of chitin deacetylase from *Gongronella butleri* and its phylogenetic analysis revealed clusters corresponding to taxonomic classification of fungi. J Biosci Bioeng 93:376–381

Maw T, Tan TK, Khor E, Wong SM (2002b) Selection of *Gongronella butleri* strains for enhanced chitosan yield with UV mutagenesis. J Biotechnol 95:189–193

Meens J, Schreiber C, Deshpande MV (2001) Screening of marine fungi for new chitin deacetylase enzymes. In: Muzzarelli RAA (ed) Chitin enzymology. Atec, Grottammare, pp 533–540

Meibom KL, Li XBB, Nielsen AT, Wu CY, Roseman S, Schoolnik GK (2004) The *Vibriocholerae* chitin utilization program. Proc Natl Acad Sci USA 101:2524–2529

Mishra C, Semino CE, Mccreath KJ, De La Vega H, Jones BJ, Specht CA, Robbins PW (1997) Cloning and expression of two chitin deacetylase genes of *Saccharomyces cerevisiae*. Yeast 13:327–336

Miyashita K, Fujii T, Saito A (2000) Induction and repression of a *Streptomyces lividans* chitinase gene promoter in response to various sources. Biosci Biotechnol Biochem 64:39–43

Nahar P, Ghormade V, Deshpande MV (2004) The extracellular constitutive production of chitin deacetylase in *Metarhizium anisopliae*: possible edge to entomopathogenic fungi in the biological control of insect pests. J Invertebr Pathol 85:80–88

Noh MY, Muthukrishnan S, Kramer KJ, Beeman RW, Arakane Y (2011) Expression profiles and functional analysis of genes encoding chitin deacetylases, extracellular matrix-modifying proteins in *Tribolium castaneum*. Entomol Res. doi:10.1111/j.1748-5967.2011.00425.x

Ohishi H, Murase K, Ohta T, Etoh H (2000) Cloning and sequencing of the deacetylase gene from *Vibrio alginolyticus* H-8. J Biosci Bioeng 90:561–563

Pruzzo C, Vezzulli L, Colwell RR (2008) Global impact of *Vibrio cholerae* interactions with chitin. Environ Microbiol 10:1400–1410

Ruiz-Herrera J (1978) The distribution and quantitative importance of chitin in fungi. In: Muzzarelli RA, Pariser ER (eds) In proceedings of the Ist international conference on chitin and chitosan. Massachusetts Institute of Technology, Cambridge, MA, pp 11–21

Ruiz-Herrera J, Martinez-Espinoza AD (1999) Chitin biosynthesis and structural organization in vivo. In: Jolles

P, Muzzarelli RAA (eds) Chitin and chitinases. Birkhauser Verlag, Basel, pp 39–53

Saito A, Ishizaka M, Francisco PB, Fujii T, Miyashita K (2000) Transcriptional co-regulation of five chitinase genes scattered on the *Streptomyces coelicolor* A3(2) chromosome. Microbiology 146:2937–2946

Saito A, Shinya T, Miyamoto K, Yokoyama T, Kaku H, Minami E, Shibuya N, Tsujibo H, Nagata Y, Ando A, Fujii T, Miyashita K (2007) The *dasABC* gene cluster, adjacent to *dasR*, encodes a novel ABC transporter for the uptake of *N*, *N′*-diacetylchitobiose in *Streptomyces coelicolor* A3(2). Appl Environ Microbiol 73:3000–3008

Shrestha B, Blondeau K, Stevens WF, Hegarat FL (2004) Expression of chitin deacetylase from *Colletotrichum lindemuthianum* in *Pichia pastoris*: purification and characterization. Protein Expr Purif 38:196–204

Shin H-J, Kim M, Lee DS (1999) Purification and characterization of *N*-acetylglucosamine 6-phosphate deacetylase from *Thermus caldophilus*. J Biosci Bioeng 88:319–322

Tanaka T, Fukui T, Atomi H, Imanaka T (2003) Characterization of an Exo-β-D-glucosaminidase Involved in a novel chitinolytic pathway from the hyperthermophilic archaeon *Thermococcus kodakaraensis* KOD1. J Bacteriol 185:5175–5181

Taylor EJ, Gloster TM, Turkenburg JP, Vincent F, Brzozowski AM, Dupont C, Shareck F, Centeno MSJ, Prates JAM, Puchart V, Ferreira LMA, Fontes CMGA, Biely P, Davies GJ (2006) Structure and activity of two metal-ion dependent acetyl xylan esterases involved in plant cell wall degradation reveals a close similarity to peptidoglycan deacetylase. J Biol Chem 281:10968–10975

Tokuyasu K, Ohnishi-Kameyama M, Hayashi K (1996) Purification and characterization of extracellular chitin deacetylase from *Colletotrichum lindemuthianum*. Biosci Biotechol Biochem 60:1598–1603

Tokuyasu K, Ono H, Ohnishi-Kameyama M, Hayashi K, Mori Y (1997) Deacetylation of chitin oligosaccharides of dp 2–4 by chitin deacetylase from *Colletotrichum lindemuthianum*. Carbohydr Res 303:353–358

Tokuyasu K, Ohnishi-Kameyama M, Hayashi K, Mori Y (1999) Cloning and expression of chitin deacetylase gene from a deuteromycete, *Colletotrichum lindemuthianum*. J Biosci Bioeng 87:418–423

Tokuyasu K, Mitsutomi M, Yamaguchi I, Hayashi K, Mori Y (2000) Recognition of chitooligosaccharides and their *N*-acetyl groups by putative subsites of chitin deacetylase from a deuteromycete, *Colletotrichum lindemuthianum*. Biochemistry 39:8837–8843

Toprak U, Baldwin D, Erlandson M, Gillott C, Hou X, Coutu C, Hegedus DD (2008) A chitin deacetylase and putative insect intestinal lipases are components of the *Mamestra configurata* (Lepidoptera: Noctuidae) peritrophic matrix. Insect Mol Biol 17:573–585

Trudel J, Asselin A (1990) Detection of chitin deacetylase activity after polyacrylamide gel electrophoresis. Anal Biochem 189:249–253

Tsigos I, Bouriotis V (1995) Purification and characterization of chitin deacetylase from *Colletotrichum lindemuthianum*. J Biol Chem 270:26286–26291

Tsigos I, Zydowicz N, Martinou A, Domard A, Bouriotis V (1999) Mode of action of chitin deacetylase from *Mucor rouxii* on *N*-acetylchitooligosaccharides. Eur J Biochem 261:698–705

Tsigos I, Martinou A, Kafetzopoulos D, Bouriotis V (2000) Chitin deacetylases: new, versatile tools in biotechnology. Trends Biotechnol 18:305–312

Van den Burg HA, Harrison SJ, Joosten MHAJ, Vervoort J, de Wit PJGM (2006) *Cladosporium fulvum* Avr4 protects fungal cell walls against hydrolysis by plant chitinases accumulating during infection. Am Phytopathol Soc 19:1420–1430

Vander P, Varum KM, Domard A, Gueddari NEE, Moersschbacher BM (1998) Comparison of the ability of partially N-acetylated chitosans and chitooligosaccharides to elicit resistance reactions in wheat leaves. Plant Physiol 118:1353–1359

Walker-Simmons M, Jin D, West CA, Hadwiger CA, Ryan CA (1984) Comparison of proteinase inhibitor-inducing activities and phytoalexin elicitor activities of a pure fungal endopolygalacturonase, pectic fragments, and chitosans. Plant Physiol 76:833–836

Wang S, Jayaram SA, Hemphälä J, Senti KA, Tsarouhas V, Jin H, Samakovlis C (2006) Septate-junction-dependent luminal deposition of chitin deacetylases restricts tube elongation in the *Drosophila* Trachea. Curr Biol 16:180–185

Watanabe T, Kimura K, Sumiya T, Nikaidou N, Suzuki K, Suzuki M, Taiyoji M, Ferrer S, Regue M (1997) Genetic analysis of the chitinase system of *Serratia marcescens* 2170. J Bacteriol 179:7111–7117

Win NN, Stevens WF (2001) Shrimp chitin as substrate for fungal chitin deacetylase. Appl Microbiol Biotechnol 57:334–341

Win NN, Pengju G, Stevens WF (2000) Deacetylation of chitin by fungal enzymes. In: Peter MG, Domard A, Muzzarelli RAA (eds) Advances in chitin science, vol 4. University of Postdam, Postdam, pp 55–62

Xia-Yun J, Shu-Ming Z, Pei-Gen Z (2007) Cloning and sequence analysis of complete cDNA of chitin deacetylase from *Mucor racemosus*. Chin J Agric Biotechnol 4:167–172

Yamada M, Kurano M, Inatomi S, Taguchi G, Mitsuo M, Shimosaka M (2008) Isolation and characterization of a gene coding for chitin deacetylase specifically expressed during fruiting body development in the basidiomycete *Flammulina velutipes* and its expression in the yeast *Pichia pastoris*. FEMS Microbiol Lett 289:130–137

Zhao Y, Park R-D, Muzzarelli AA (2010a) Chitin deacetylases: properties and applications. Mar Drugs 8: 24–46

Zhao Y, Kim YJ, Oh KT, Van Nam Nguyen VN, Ro Dong Park RD (2010b) Production and characterization of extracellular chitin deacetylase from *Absidia corymbifera* DY-9. J Korean Soc Appl Biol Chem 53:119–126

Zhou G, Zhang H, He Y, He L (2010) Identification of a chitin deacetylase producing bacteria isolated from soil and its fermentation optimization. Afr J Microbiol Res 4:2597–2603

Role of Enzymes and Proteins in Plant-Microbe Interaction: A Study of *M. oryzae* Versus Rice

10

Jahangir Imam, Mukund Variar, and Pratyoosh Shukla

Abstract

The wall interface between rice and *Magnaporthe oryzae* plays an important role in the outcome of their interactions, i.e., resistance or susceptibility. A number of enzymes and proteins are involved in both external and internal interactions. The blast fungus secretes many enzymes which help in the plant cell wall degradation and the entry of fungus into the plant cell which results in the development of disease. To restrict the growth and development of blast fungus, the rice plants have also developed many defense mechanisms like generation of defense substances and hydrogen peroxide catalysis by the production of some enzymes in plant cells. These enzymes occur frequently in many isoforms and help in plant defense. Proteins also participate in the defense against blast fungus attack. These proteins are called as pathogenesis-related proteins (PRs). PR proteins have activities of both proteins and hydrolytic enzymes. Chitinase and β-1,3-glucanase are the most common PR proteins which can hydrolyze major components of blast fungal cell walls, chitin and β-1,3-glucan, respectively.

Keywords

Rice • *Magnaporthe oryzae* • Xylanase • Cutinase • PR proteins • Resistance

J. Imam
Biotechnology Laboratory, Central Rainfed Upland Rice Research Station (CRRI), Hazaribagh 825301, Jharkhand, India

Enzyme Technology and Protein Bioinformatics Laboratory, Department of Microbiology, Maharshi Dayanand University, NH10, Rohtak 124001, Haryana, India

M. Variar
Biotechnology Laboratory, Central Rainfed Upland Rice Research Station (CRRI), Hazaribagh 825301, Jharkhand, India

P. Shukla (✉)
Enzyme Technology and Protein Bioinformatics Laboratory, Department of Microbiology, Maharshi Dayanand University, NH10, Rohtak 124001, Haryana, India
e-mail: pratyoosh.shukla@gmail.com

P. Shukla and B.I. Pletschke (eds.), *Advances in Enzyme Biotechnology*,
DOI 10.1007/978-81-322-1094-8_10, © Springer India 2013

Introduction

The interaction between plants and microbial pathogens is among the most complex phenomena in biology. Different aspects of interaction specificity and defense mechanisms of plants against potential fungal pathogens have received great attention in the last few years. Generally plants have two levels of defense mechanisms according to their function, structural (constitutive) and biochemical (active). The structural compounds present the first line of defense against invading pathogens by forming mechanical barriers or by forming preformed chemical substances. Biochemical processes participate in active defense reaction of plants against pathogens (Lebeda et al. 1999).

The primary walls of plant cells are pivotal battlegrounds between microbial pathogens and their hosts. Microbial pathogens secrete an array of cell wall-degrading enzymes (CWDEs) and other enzymes capable of breaking the plant cell wall and causing infection (Wu et al. 2006). Interaction between cells of *Magnaporthe oryzae* and rice involves a complex of biological influences which lead to rice blast disease. Pathogenesis in general and the initial infection steps in particular may be viewed as a sequence of discrete, critical events. In *Magnaporthe oryzae-rice* pathogenesis, CWDEs as well as other enzymes play a crucial role and involve both external and internal interactions. Proteins participating in defense mechanisms after the fungal attack are generally called pathogenesis-related proteins (PR proteins). Initially, it was assumed that PR proteins are devoid of any enzymatic activity, but Legrand et al. (1987) detected chitinase activity in four members of group 3 tobacco PRs and later on established β-1,3-glucanase activity in four members of group 2 tobacco PRs. PRs also have antifungal effect and show stronger accumulation in resistant than susceptible plants. There is also a high level of constitutive expression of PR proteins in naturally resistant plants (Edreva 2005).

Role of Enzymes in *Magnaporthe oryzae*-Rice Interaction

The rice blast fungus secretes a battery of enzymes which facilitate its colonization in the plant tissue. These enzymes are mainly cell wall-degrading enzymes (CWDEs) like xylanases, cutinases, and other enzymes. *Magnaporthe oryzae* also secretes metabolic enzymes like trehalase that helps in plant tissue colonization (Foster et al. 2003). On the other hand, rice plant synthesizes enzymes and proteins as their defense against the blast fungus. This chapter focuses mainly on enzymes and proteins from plant pathogens which have been extensively studied and characterized in *Magnaporthe oryzae*-rice interaction.

Xylanases

Xylan is the predominant hemicellulose component in plant cell walls and the second most abundant polysaccharide in nature (Subramaniyan and Prema 2002). Xylan is a heteropolysaccharide having a backbone of β-1,4-linked xylopyranose units, with groups of acetyl, 4-O-methyl-D-glucuronosyl, and α-arabinofuranosyl residues linked to the backbone (Subramaniyan and Prema 2002). The complete degradation of xylan in plant requires the activity of complex hydrolytic enzymes with diverse mode of action (Beg et al. 2001). Out of the many xylanolytic enzymes, endo-β-1,4-xylanase is the most important, which is required to cleave the main xylan backbone chain (Biely and Tenkanen 1998). The many xylan-degrading enzymes secreted by fungi are one of their components of offensive arsenal (Belien et al. 2006). Many plant pathogenic fungi secrete endoxylanases when grown in the presence of host cell walls (Cooper et al. 1988; Lehtinen 1993; Ruiz et al. 1997; Wu et al. 1997; Giesbert et al. 1998; Carlile et al. 2000; Hatsch et al. 2006).

The recently published *Magnaporthe oryzae* genome sequence unveiled the possible presence of as many as 20 xylanase genes, which encodes six glycoside hydrolase family 10 (GH10), 5 GH11, and 9 GH43 members (Dean et al. 2005). The high level of redundancy is an indication that xylanase activity is essential for the vitality of *Magnaporthe oryzae*, either saprophytically or pathogenetically or both (Wu et al. 2006). *M. oryzae* secretes several isoforms of endo-β-1, 4-xylanase, and these isoforms act as pathogenicity factors (Wu et al. 1997). Experiment proves that deletion of one or two xylanase genes did not abolish endoxylanase activity (Wu et al. 1997) and no detectable effect on virulence is observed. The presence of multiple endoxylanase and β-xylosidase genes in *M. oryzae* may be the reason why mutants in individual xylanase genes remain pathogenic (Apel-Birkhold and Walton 1996; Wegener et al. 1999; Gomez-Gomez et al. 2002). One evidence that supports an important role for xylanases in the pathogenicity of *M. oryzae* is that when cultured rice cells were treated with commercial xylanase, it causes cell death (Ishii 1998). Many fungi produce glycanases which facilitate colonization of plant tissue (eg. galacturonases, xylanases and glucanases) that fragments plant cell wall polysaccharides that are generated by these glycanases, provide the fungus with a carbon source but also elicit the plant defence response (Wu et al. 1997). The purification, cloning, and characterization of two xylanases from *M. oryzae* are steps towards analyzing the role of xylanase in the interaction of *M. oryzae* with its rice host. One early study reported two types of xylanases with different pH optima from *M. oryzae* (Sumizu et al. 1961). Till now at least 17 putative xylanases in the genome of *M. oryzae* have been identified. Six of them (Xyl 1–6) have been partially characterized (Wu et al. 1997). Xyl 1, Xyl 3, and Xyl 4 encode class XI endo-β-xylanases, while Xyl 2, Xyl 5, and Xyl 6 encode class 10 endo-β-xylanases (Wu et al. 1997). Knockout studies suggest that Xyl 1, Xyl 4, and Xyl 5 are pathogenicity factors, while Xyl 2 may have a role in initiating the host plant defense responses.

Cutinases

The cuticle which is present over all parts of the aerial plant presents the first physical barrier to pathogen entry and infection. The main structural component of the plant cuticle is cutin which occurs as a hydrophobic cutin network of esterified hydroxyl and epoxy fatty acids which are n-C_{16} and n-C_{18} types intermingled with wax (Kolattukudy 2001; Lequeu et al. 2003; Nawrath 2006). *Magnaporthe oryzae* uses direct method of penetration, i.e., through cuticle. The penetration through cuticle requires both physical pressure (Howard et al. 1991; Bechinger et al. 1999) and enzymatic degradation by extracellular cutinases (Skamnioti et al. 2007). Cutin monomers promote germ tube and appressorium differentiation on chemically inert surfaces in *M. oryzae* (Gilbert et al. 1996; DeZwaan et al. 1999) and showed enhanced resistance to infection by *M. oryzae* (Schweizer et al. 1994). Sweigard et al. (1992a, b) cloned and identified cutin-degrading enzyme from *M. oryzae*. They named it as *CUTINASE1* (*CUT1*) gene. They showed that this gene is expressed when cutin is the sole carbon source but not when carbon source is cutin and glucose together or glucose alone. Dean et al. (2005) revealed seven more members of the cutinase family in *M. oryzae* genomes and 16 putative cutinases in genome sequence release five. Such large number of cutinases in *M. oryzae* genome reflects functional redundancy or varying specificity of these enzymes (Skamnioti et al. 2007). Appressoria formation in *M. oryzae* occurs either in hydrophobic surfaces or in the presence of soluble host cutin monomers but not on hydrophilic surfaces (Lee and Dean 1994; Gilbert et al. 1996), and the cutin monomers alone are sufficient to induce appressorium differentiation (Choi and Dean 1997). The plasma membrane protein Pth11p functions at the cell cortex as an upstream effector of appressorium differentiation in response to soluble plant cutin monomers (DeZwaan et al. 1999). Skamnioti et al. (2007) identified a specific *M. oryzae* cutinase,

CUTINASE2 (*CUT2*), which showed a dramatic uplift in transcription during appressorium maturation and penetration. They proposed that *CUT2* is an upstream activator of the cAMP/PKA and DAG/PKC signaling pathways that directs appressorium formation and infection growth in *M. oryzae*. *CUT2* mutant shows reduced extracellular serine esterase and plant cutin-degrading activity and attenuated pathogenicity on rice. Exogenous application of synthetic cutin monomers, cAMP and DAG, restores the morphological and pathogenicity defects of the *Cut2* mutant to wild-type levels. *CUT2* plays no part in spore or appressorium adhesion or in appressorial turgor generation, but mediates the formation of penetration peg (Skamnioti and Gurr 2008). The cutin monomer ligand released by *CUT2* is perceived by one of the G-protein-coupled receptors (GPCRs) in *M. oryzae* (Kulkarni et al. 2005). Overall, *CUT2* is required for surface sensing leading to correct germ lining differentiation, penetration, and full virulence in *M. oryzae* (Skamnioti et al. 2007).

Trehalases

Trehalase is a glycoside hydrolase enzyme that catalyzes the conversion of trehalose to glucose. Trehalose is a nonreducing disaccharide commonly found in all eukaryotic cells except mammals as storage carbohydrate (Arguelles 2000). The disaccharide is hydrolyzed into two molecules of glucose by the enzyme trehalase. Trehalose mobilization may be involved by many virulence-associated functions in *M. oryzae* like germination of conidia, development of infected cells on the leaf surface, and subsequent plant tissue colonization (Foster et al. 2003). In *M. oryzae*, breakdown of trehalose requires two trehalases: a neutral trehalase encoded by a gene *NTH1* and a novel trehalase encoded by *TRE1* which is required for mobilization during spore germination, but dispensable for pathogenicity (Foster et al. 2003). Neutral trehalase *NTH1* is regulated by protein phosphorylation-dephosphorylation in *M. oryzae*. *NTH1* has phosphorylation site for cAMP-dependent protein kinase (PKA) and a putative Ca^{2+} binding site. This reveals that *NTH1* is a regulated protein. The gene for trehalase (*NTH1*) in *M. oryzae* is expressed during its sporulation, plant infection, and in response to environmental stress (Foster et al. 2003). The mutant for Δnth1 gene in *M. oryzae* showed slow proliferation of invasive hyphae as compared to its wild type. This concludes that *NTH1* is required by *M. oryzae* to generate severe blast symptoms (Foster et al. 2003). Three putative *TRE1* products (trehalase encoding) are 33 % similar to human and mouse TreA trehalase but are distinct from both acidic and neutral trehalases from fungi (Foster et al. 2003). *TRE1*-encoded trehalase is required both for growth on trehalose and mobilization of intracellular trehalose in *M. oryzae*.

Catalase-Peroxidases

The generation of reactive oxygen species (ROS) such as superoxide anion radicals (O_2^-), hydroxyl radicals ($OH^{\cdot}$), and hydrogen peroxide (H_2O_2) in plant cells is one of the most rapid and drastic defense reactions activated following pathogen attack (Doke 1983; Lamb and Dixon 1997). Catalase enzyme is present in peroxisomes and catalyzes the hydrogen peroxide into water and oxygen, while peroxidase enzyme catalyzes the oxidation of substrates like phenol and its derivative with the help of hydrogen peroxide. Both the enzymes are the competitor of each other because they both use the same substrate. Class III peroxidases (*POXs*) provide resistance to plants against blast disease infection, but there is no clear-cut evidence of *POX* as self-defense for plants at the molecular level (Sasaki et al. 2004). *POX* genes constitute a multigene family, and the redundant expression of many *POX* genes against pathogen attack and environmental stresses may guarantee its necessities in self-defense (Sasaki et al. 2004). The *M. oryzae* genome contains two true heme catalases, catalase A (*CATA*) and catalase B (*CATB*), plus two bifunctional catalase-peroxidase genes, catalase-peroxidase A (*CPXA*) and catalase-peroxidase B (*CPXB*) (Skamnioti et al. 2007). In vitro, the *CATA* expression varied little with time and H_2O_2 concentration, whereas *CATB* transcript abundance showed a moderate increase with increasing H_2O_2 concentration. In vivo also there is upregulation of *CATB* gene at the time of penetration of the host by *M. oryzae*. Skamnioti et al. (2007) showed that *CATB* plays a part in strengthening the fungal cell wall and not in the detoxification of host-produced H_2O_2. Tanabe et al. (2011) in a gene knockout experiment showed that *CPXB* is the major gene encoding the secretory catalase and confers resistance to H_2O_2 in *M. oryzae* hyphae. Their results suggest that *CPXB* plays a role in fungal defense against H_2O_2 accumulated in the epidermal cells of rice at the early stage of infection but not in pathogenicity of *M. oryzae* (Tanabe et al. 2011).

Role of Proteins in *Magnaporthe oryzae*-Rice Interaction

The small secreted proteins play an important and decisive role in plant pathogenesis. Generally these proteins are less than 200 amino acid residues. These small secreted proteins from *M. oryzae* into rice lead to disease symptom development. As a defense, the rice plant also produces proteins against the fungus. Depending on their function during the defense response, proteins can be grouped into three classes. The first class of proteins is structural proteins that participate in strengthening and repairing of the cell wall or modification of the properties of the extracellular matrix. The second class of proteins exhibits direct antimicrobial activities or catalyzes the synthesis of antimicrobial compounds (Lebeda et al. 1999). The third class comprises of proteins, which function in plant defense is not well known (Schoeltens et al. 1991).

Pathogenesis-Related (PR) Proteins

Pathogenesis-related (PR) proteins are encoded by host plants in response to pathological or situations of nonpathogenic origin. Antoniw et al. (1980) coined the term "pathogenesis-related proteins" (PRs). To be included among the PRs, a protein has to be newly expressed upon infection but not necessarily in all pathological conditions. A unifying nomenclature for PRs was proposed based on their grouping into families sharing amino acid sequences, serological relationships, and enzymatic or biological activity (Van Loon et al. 1994; Van Loon and Van Strien 1999) (Table 10.1).

The classified PR proteins are grouped into two subclasses on the basis of acidic and basic subclass. The acidic subclass proteins are generally secreted to the extracellular spaces, and basic subclass proteins are transported to the vacuole by C-terminal end signal sequence (Takeda et al. 1991; Koiwa et al. 1994; Sato et al. 1995). The expression of basic PR proteins is constitutive and independent of pathogen infection in some organs like roots, seedling, and cultured cells (Agrios 1997). There are two criteria on the basis of which new families are included in PR proteins. The first is that the protein must be induced by a pathogen in tissues that do not normally express it, and the second is that the induced expression must occur in at least two different plant-pathogen combinations or expression in a single plant-pathogen combination must be confirmed independently in different laboratories. These are low molecular weight proteins (6–43KDa), stable at pH<3, can be extracted biochemically, are thermostable and most importantly highly resistant to protease. Till now the presence of PR proteins is established in almost all parts of plants like leaves, stems, roots, and flowers. Five to ten percent of total leaf proteins account for PR proteins (Van Loon and Van Strien 1999). NMR reveals α-β-α sandwich structure which provides compactness to the structure of PR proteins and possibly helps in the resistance to protease (Fernandez et al. 1997).

PR proteins have been well studied as a major defense response in several dicot plants, both in R gene-mediated resistance and in SAR. The roles of PR genes in disease resistance have been suggested by the tight correlation between expression levels of PR genes and disease resistance and by the observation of enhanced disease resistance in the transgenic plants overexpressing certain PR genes (Song and Goodman 2001).

Table 10.1 Recommended classifications and properties of families of Pathogenesis-Related Proteins (*PRs*) (Sels et al. 2008)

S.No.	Family	Type member	Typical size (KDa)	Properties	Proposed microbial target	Original reference
1	PR-1	Tobacco PR-1a	15	Antifungal	Unknown	Antoniw et al. (1980)
2	PR-2	Tobacco PR-2	30	β-1,3-Glucanase	β-1,3-Glucan	Antoniw et al. (1980)
3	PR-3	Tobacco P, Q	25–30	Chitinase (Class I,II,IV,V,VI,VII)	Chitin	Van Loon (1982)
4	PR-4	Tobacco R	15–20	Chitinase class I,II	Chitin	Van Loon (1982)
5	PR-5	Tobacco S	25	Thaumatin-like	Membrane	Van Loon (1982)
6	PR-6	Tomato inhibitor I	8	Proteinase-inhibitor	_a	Green and Ryan (1972)
7	PR-7	Tomato P_{69}	75	Endoproteinase	_a	Vera and Conejero (1988)
8	PR-8	Cucumber chitinase	28	Chitinase class III	Chitin	Metraux et al. (1988)
9	PR-9	Lignin-forming peroxidase	35	Peroxidase	_a	Lagrimini et al. (1987)
10	PR-10	Parsley PR-1	17	'Ribonuclease-like'	_a	Somssich et al. (1986)
11	PR-11	Tobacco class V chitinase	40	Chitinase class I	Chitin	Melchers et al. (1994)
12	PR-12	Radish Ps-AFP3	5	Defensin	Membrane	Terras et al. (1995)
13	PR-13	*Arabidopsis* THI2.1	5	Thionin	Membrane	Epple et al. (1995)
14	PR-14	Barley LTP4	9	Lipid-transfer protein	Membrane	Garcia-Olmedo et al. (1995)
15	PR-15	Barley OxOa (germin)	20	Oxalate oxidase	_a	Zhang et al. (1995)
16	PR-16	Barley OxOLP	20	'Oxalate oxidase-like'	_a	Wei et al. (1998)
17	PR-17	Tobacco PRp27	27	Unknown	_a	Okushima et al. (2000)

-[a]No in vitro antimicrobial activity reported

Some PR proteins have activities of hydrolytic enzymes including chitinase and β-1,3-glucanase, which can hydrolyze major components of fungal cell walls, chitin and β-1,3-glucan, respectively. Hydrolysis of these fungal cell wall constituents leads to the inhibition of the growth of several fungi in vitro (Punja 2006). Genes encoding chitinase or β-1,3-glucanase from rice and microbes have been extensively used in generation of transgenic rice resistant to fungal pathogens (Punja 2006). Transgenic plants constitutively expressing the *Gns1* gene, encoding a β-1,3-glucanase, accumulated Gns1 protein up to 0.1 % of total soluble protein in leaves. The *Gns1*-overexpressing transgenic plants developed many resistant-type lesions on the inoculated leaf, accompanying earlier activation of defense genes *PR-1* and *PBZ1*, when inoculated with virulent *M. oryzae* (Nishizawa et al. 2003). Transgenic plants which constitutively expressed a rice class I chitinase gene, *Cht-2* or *Cht-3*, showed significant resistance against two races of *M. oryzae* (Nishizawa et al. 1999). Interestingly, hydrolytic enzymes of microbial origin have also been demonstrated to be effective in engineering rice disease resistance against fungal pathogens. Ninety percent of transgenic rice plants expressing *ChiC* had higher resistance against *M. oryzae* than non-transgenic plants. Disease resistance in the transgenic plants was correlated with the *ChiC* expression levels (Itoh et al. 2003). Three genes, *ech42*, *nag70*, and *gluc78*, encoding hydrolytic enzymes, from a biocontrol fungus *Trichoderma atroviride*, were introduced in single or in combinations into rice. *Gluc78-overexpressing* transgenic plants showed enhanced resistance to *M. oryzae*, while transgenic plants overexpressing the *ech42* gene encoding for an endochitinase increased resistance to *Rhizoctonia solani*, resulting in a reduction of 62 % in the sheath blight disease index (Liu et al. 2004; Shah et al. 2008).

Conclusion

It can thus be inferred that enzymes and proteins play an integral role in the interaction of *M. Oryzae* with rice and vice versa. A detailed study of the interaction will further help us understand the mechanism of fungal invasion and how plant defenses are activated in response to the attack and give insight to subsequent changes in response in case of any deviation in the normal mechanism/mode of fungal virulence. Research till now has already elucidated on the fact that infection initiation requires some prerequisites (as in the case of cutinases) and that each enzyme is specific for a particular component of the plant with varying specificity not only among the classes of enzyme but also between members of the classes. There exist a lot of possibilities on discovering the biochemical pathways and the genes involved with the activity of each enzyme and in turn comprehend the plant response.

Continued efforts in this research area will also help us understand better how the fungus subverts the first line of defense and further overcomes the different levels of defense. An analytical approach to the interaction will also throw light on the plant response elicited towards pathogen invasion and why knowledge of the plant defense mechanism is important, notwithstanding the fungal defense evoked on plant response mechanisms.

Resistant varieties as in the case of transgenic rice or induction of overexpression of chitinases and β-1,3-glucanases for increased plant defense in transgenic plants have already been developed. There still exists a lot of scope in the vastly unexplored territory of transgenic plants. One can also foray in the development of an antidote to the enzymes breaking the first line of defense and being responsible for invasion. The study of this plant-microbe interaction is thus of interest not only to the academician but also the researcher.

Acknowledgements This chapter is a part of NAIP-C4, ICAR work. The authors thank the National Agriculture Innovative Project-C4 for supporting the work. The authors also express the sincere thanks to Miss Neha Nancy Toppo for her valuable inputs in the conclusion.

References

Agrios GN (1997) Plant pathology, 4th edn. Academic, San Diego

Antoniw JF, Ritter CE, Pierpoint WS, Van Loon LC (1980) Comparison of three pathogenesis-related proteins from plants of two cultivars of tobacco infected with TMV. J Gen Virol 47:79–87

Apel-Birkhold PC, Walton JD (1996) Cloning, disruption, and expression of two endo-β-1,4Xylaanse genes, *XYL2* and *XYL3* from *Cochliobolus carbonum*. Appl Environ Microbiol 62:4129–4135

Arguelles JC (2000) Physiological roles of trehalose in bacteria and yeasts: a comparative analysis. Arch Microbiol 174:217–224

Bechinger C, Giebel KF, Schnell M, Leiderer P, Deising HB, Bastmeyer M (1999) Optical measurements of invasive forces exerted by appressoria of a plant pathogenic fungus. Science 285:1896–1899

Beg Q, Kapoor M, Mahajan L, Hoondal G (2001) Microbial xylanase and their industrial applications: a review. Appl Microbiol Biotechnol 56:326–338

Belien T, Van Campenhout S, Robben J, Volckaert G (2006) Microbial endoxylanases: effective weapons to breach the plant cell-wall barrier or, rather, triggers of plant defense systems? Mol Plant Microbe Interact 19:1072–1081

Biely P, Tenkanen M (1998) Enzymology of hemicelluloses degradation. In: Kubicek CP, Harman GE, Ondik KL (eds) *Trichoderma* and *Gliocladium*: basic biology, taxonomy and genetics. Taylor and Francis, London, pp 25–47

Carlile AJ, Bindschedler LV, Bailey AM, Bowyer P, Clarkson JM, Cooper RM (2000) Characterization of SNP1, a cell wall-degrading trypsin, produced during infection by *Stagonospora nodorum*. Mol Plant Microbe Interact 13:538–550

Choi W, Dean R (1997) The adenylate cyclase gene MAC1 of *Magnaporthe oryzae* controls appressorium formation and other aspects of growth and development. Plant Cell 9:1973–1983

Cooper R, Longman D, Campbell A, Henry M, Lees P (1988) Enzymic adaptation of cereal pathogens to the monocotyledonous primary wall. Physiol Mol Plant Pathol 32:33–47

Dean R et al (2005) The genome sequence of the rice blast fungus *Magnaporthe oryzae*. Nature 434:980–986

DeZwaan TM, Carroll AM, Valent B, Sweigard JA (1999) *Magnaporthe oryzae Pth11p* is a novel plasma membrane protein that mediates appressorium differentiation in response to inductive substrate cues. Plant Cell 11:2013–2030

Doke N (1983) Involvement of superoxide anion generation in the hypersensitive response of potato tuber tissues to infection with an incompatible race of *Phytophthora infestans* and to the hyphal wall components. Physiol Plant Pathol 23:345–357

Edreva A (2005) Pathogenesis-related proteins: research progress in the last 15 years. Gen Appl Plant Physiol 31:105–124

Epple P, Apel K, Bohlmann H (1995) An *Arabidopsis thaliana* thionin gene is inducible via a signal transduction pathway different from that for pathogenesis-related proteins. Plant Physiol 109:813–820

Fernandez C, Scyperski T, Bruyere T, Ramage P, Mosinger E, Wutrich K (1997) NMR solution structure of the pathogenesis-related protein p 14a. J Mol Biol 266:576–593

Foster AJ, Jenkinson JM, Talboit NJ (2003) Trehalose synthesis and metabolism are required at different stages of plant infection by *Magnaporthe oryzae*. Eur Mol Biol Organ J 22:225–235

Garcia-Olmedo F, Molina A, Segura A, Moreno M (1995) The defensive role of nonspecific lipid-transfer proteins in plants. Trends Microbiol 3:72–74

Giesbert S, Lepping H, Tenberge K, Tudzynski P (1998) The xylanolytic system of *Claviceps purpurea*: cytological evidence for secretion of xylanases in infected rye tissue and molecular characterization of two xylanase genes. Phytopathology 88:1020–1030

Gilbert RD, Johnson AM, Dean RA (1996) Chemical signal responsible for appressorium formation in the rice blast fungus *Magnaporthe oryzae*. Physiol Mol Plant Pathol 48:335–346

Gomez-Gomez E, Ruiz-Roldan MC, Di Pietro A, Roncero MI, Hera C (2002) Role in pathogenesis of two endo-β-1,4-Xylanase genes from the vascular wilt fungus *Fusarium oxysporum*. Fungal Genet Biol 35:213–222

Green TR, Ryan CA (1972) Wound-induced proteinase inhibitor in plant leaves: a possible defense mechanism against insects. Science 175:776–777

Hatsch D, Phalip V, Petkovski E, Jeltsch JM (2006) *Fusarium graminearum* on plant cell wall: no fewer than 30 xylanase genes transcribed. Biochem Biophys Res Commun 345:959–966

Howard RJ, Ferrari MA, Roach DH, Money NP (1991) Penetration of hard substrates by a fungus employing enormous turgor pressures. Proc Natl Acad Sci USA 88:11281–11284

Ishii S (1998) Factors influencing protoplast viability of suspension-cultured rice cells during isolation process. Plant Physiol 88:26–29

Itoh Y, Takahashi K, Takizawa H, Nikaidou N, Tanaka H, Nishihashi H, Watanabe T, Nishizawa Y (2003) Family 19 chitinase of *Streptomyces griseus* HUT6037 increases plant resistance to the fungal disease. Biosci Biotechnol Biochem 67:847–855

Koiwa H, Sato F, Yamada Y (1994) Characterization of accumulation of tobacco PR-5 proteins by IEF-immunoblot analysis. Plant Cell Physiol 35:821–827

Kolattukudy PE (2001) Polyesters in higher plants. Adv Biochem Eng Biotechnol 71:1–49

Kulkarni RD, Thon MR, Pan HQ, Dean RA (2005) Novel G-protein-coupled receptor-like proteins in the plant pathogenic fungus *Magnaporthe grisea*. Genome Biol 6:R24.

Lagrimini LM, Burkhart W, Moyer M, Rothstein M (1987) Molecular cloning of complementary DNA encoding the lignin-forming peroxidase from tobacco: molecular analysis and tissue-specific expression. Proc Natl Acad Sci USA 84:7542–7546

Lamb C, Dixon RA (1997) The oxidative burst in plant disease resistance. Annu Rev Plant Physiol Plant Mol Biol 48:251–275

Lebeda A, Jancova D, Luhova L (1999) Enzymes in fungal plant pathogenesis. Phyton (Austria) 39:51–56

Lee YH, Dean R (1994) Hydrophobicity of contact surface induces appressorium formation in *Magnaporthe oryzae*. FEMS Microbiol Lett 115:71–76

Legrand M, Kauffmann S, Geoffroy P, Fritig B (1987) Biological function of pathogenesis-related proteins: four tobacco pathogenesis related proteins are chitinase. Proc Natl Acad Sci USA 84:6750–6754

Lehtinen U (1993) Plant-cell wall degrading enzymes of *Septoria nodorum*. Physiol Mol Plant Pathol 43:121–134

Lequeu J, Fauconnier ML, Chammai A, Bronner R, Blee E (2003) Formation of plant cuticle: evidence for the occurrence of the peroxygenase pathway. Plant J 36:155–164

Liu M, Sun ZX, Zhu J, Xu T, Harman GE, Lorito M (2004) Enhancing rice resistance to fungal pathogens by transformation with cell wall degrading enzyme genes from *Trichoderma atroviride*. J Zhejiang Univ Sci 5:133–136

Melchers LS, Apotheker-de Groot M, Van der Knaap JA, Ponstein AS, Sela-Buurlage MB, Bol JF, Cornelissen BJC, Van den Elzen PJM, Linthorst HJM (1994) A new class of tobacco chitinases homologous to bacterial exochitinases displays antifungal activity. Plant J 5:469–480

Metraux JP, Streit L, Staub T (1988) A pathogenesis-related protein in cucumber is a chitinase. Physiol Mol Plant Pathol 33:1–9

Nawrath C (2006) Unravelling the complex network of cuticular structure and function. Curr Opin Plant Biol 9:281–287

Nishizawa Y, Nishio Z, Nakazono K, Soma M, Nakajima E, Ugaki M, Hibi M (1999) Enhanced resistance to blast (*Magnaporthe grisea*) in transgenic japonica rice by constitutive expression of rice chitinase. Theor Appl Genet 99:383–390

Nishizawa Y, Saruta M, Nakazono K, Nishio Z, Soma M, Yoshida T, Nakajima E, Hibi T (2003) Characterization of transgenic rice plants over-expressing the stress-inducible beta-glucanase gene *Gns1*. Plant Mol Biol 51:143–152

Okushima Y, Koizumi N, Kusano T, Sano H (2000) Secreted proteins of tobacco cultured BY2 cells: identification of a new member of pathogenesis-related proteins. Plant Mol Biol 42:479–488

Punja ZK (2006) Recent developments toward achieving fungal disease resistance in transgenic plants. Can J Plant Pathol 28:298–308

Ruiz MC, DiPietro A, Roncero MIG (1997) Purification and characterization of an acidic endo-beta-1,4-xylanase from the tomato vascular pathogen *Fusarium oxysporum* f sp. *lycopersici*. FEMS Microbiol Lett 148:75–82

Sasaki K, Iwai T, Hiraga S, Kuroda K, Seo S, Mitsuhara I, Miyasaka A, Iwano M, Ito H, Matsui H, Ohashi Y (2004) Ten rice peroxidases redundantly respond to multiple stresses including infection with rice blast fungus. Plant Cell Physiol 45:1442–1452

Sato F, Koiwa H, Sakai Y, Kato N, Yamada Y (1995) Synthesis and secretion of tobacco neutral PR-5 protein by transgenic tobacco and yeast. Biochem Biophys Res Commum 211:909–913

Schoeltens-Toma IMJ, Joosten MHA, De Wit PJGM (1991) Appearance of pathogen-related proteins in plant hosts: relationships between compatible and incompatible interactions. In: Cole GT, Hoch HC (eds) The fungal spore and disease initiation in plants and animals. Plenum Press, New York/London, pp 247–265

Schweizer P, Jeanguenat A, Mdossinger E, Metraux JP (1994) Plant protection by free cutin monomers in two cereal pathosystems. In: Daniels MJ, Downie JA, Osbourn AE (eds) Advances in molecular genetics of plant-microbe interaction. Springer Publication, The Netherlands, pp 371–374

Sels J, Mathys J, De Coninck BM, Cammue BP, De Bolle MF (2008) Plant pathogenesis-related (PR) proteins: a focus on PR peptides. Plant Physiol Biochem 46:941–950

Shah JM, Raghupathy V, Veluthambi K (2008) Enhanced sheath blight resistance in transgenic rice expressing an endochitinase gene from *Trichoderma virens*. Biotechnol Lett. doi:10.1007/s10529-008-9856-5

Skamnioti P, Gurr SJ (2008) Cutinase and hydrophobin interplay, a herald for pathogenesis? Plant Signal Behav 3:248–250

Skamnioti P, Henderson C, Zhang Z, Robinson Z, Gurr SJ (2007) A novel role of catalase B in the maintenance of fungal cell wall integrity during host invasion in the rice blast fungus *Magnaporthe oryzae*. Mol Plant Microbe Interact 20:568–580

Somssich IE, Schmelzer E, Bollmann J, Hahlbrock K (1986) Rapid activation by fungal elicitor of genes encoding "pathogenesis-related" proteins in cultured parsley cells. Proc Natl Acad Sci USA 83:2427–2430

Song F, Goodman RM (2001) Molecular biology of disease resistance in rice. Physiol Mol Plant pathol 59:1–11

Subramaniyan S, Prema P (2002) Biotechnology of microbial xylanases: enzymology, molecular biology, and application. Crit Rev Biotechnol 22:33–64

Sumizu K, Yoshikawa M, Tanaka S (1961) Studies on xylanase of *Pyricularia oryzae*. J Biochem 50:538–543

Sweigard JA, Chumley FG, Valent B (1992a) Disruption of a *Magnaporthe oryzae* cutinase gene. Mol Gen Genet 232:183–190

Sweigard JA, Chumley FG, Valent B (1992b) Cloning and analysis of *Cut1*, a cutinase gene from *Magnaporthe oryzae*. Mol Gen Genet 232:174–182

Takeda S, Sato F, Ida K, Yamada Y (1991) Nucleotide sequence of a cDNA for osmotin-like protein from cultured tobacco cells. Plant Physiol 97:844–846

Tanabe S, Ishii-Minami N, Saitoh KI, Otake Y, Kaku H, Shibuya N, Nishizawa Y, Minami E (2011) The role of catalase-peroxidase secreted by *Magnaporthe oryzae* during early infection of rice. Mol Plant Microbe Interact 24:163–171

Terras FR, Eggermont K, Kovaleva V, Raikhel NV, Osborn RW, Kester A, Rees SB, Vanderleyden J, Cammue BP, Broekaert WF (1995) Small cysteine-rich antifungal proteins from radish: their role in host defense. Plant Cell 7:573–588

Van Loon LC (1982) Regulation of changes in proteins and enzymes associated with active defense against virus infection. In: Wood RKS (ed) Active defense mechanisms in plants. Plenum Press, New York, pp 247–273

Van Loon LC, Van Strien EA (1999) The families of pathogenesis-related proteins, their activities, and comparative analysis of PR-1 type proteins. Physiol Mol Plant Pathol 55:85–97

Van Loon LC, Pierpoint WS, Boller T, Conejero V (1994) Recommendations for naming plant pathogenesis-related proteins. Plant Mol Biol Report 12:245–264

Vera P, Conejero V (1988) Pathogenesis-related proteins of tomato P-69 as an alkaline endoproteinase. Plant Physiol 87:58–63

Wegener S, Ransom RF, Walton JD (1999) A unique eukaryotic β-xylosidase gene from the phytopathogenic fungus *Cochliobolus carbonum*. Microbiology 145:1089–1095

Wei Y, Zhang Z, Anderson CH, Schmelzer E, Gregersen PL, Collinge DB, Smedegaard-Petersen V, Thordal-Christensen H (1998) An epidermis/papilla-specific oxalate oxidase like protein in the defence response of barley attacked by the powdery mildew fungus. Plant Mol Biol 36:101–112

Wu SC, Ham KS, Darvill AG, Albersheim P (1997) Deletion of two endo-β-1,4-Xylanase genes reveals additional isozymes secreted by the rice blast fungus. Mol Plant Microbe Interact 10:700–708

Wu SC, Helley JE, Luttig C, Fernekes LM, Gutierrez-Sanchez G, Darvill AG, Albersheim P (2006) Identification of an *endo*-β-1,4-D-Xylanase from *Magnaporthe oryzae* by gene knockout analysis, purification, and heterologous expression. Am Soc Microbiol 72:986–993

Zhang Z, Collinge DB, Thordal-Christensen H (1995) Germin-like oxalate oxidase, a H_2O_2-producing enzyme, accumulates in barley attacked by the powdery mildew fungus. Plant J 8:139–145

11 Industrial Enzyme Applications in Biorefineries for Starchy Materials

Vipul Gohel, Gang Duan, and Vimal Maisuria

Abstract

This chapter reviews recent advances in technology developments in biorefinery industries through enzymatic approaches where various starchy materials have been used as feedstock for biofuel and various syrup productions. It further discusses the enzymes discovery, industrial challenges, and how enzymatic-based approaches help different industries to develop environmentally sustainable and cost-effective solutions by making industrial process into more simplified without compromising the product and by-product yields and their qualities.

Keywords

Biorefinery • Enzymatic process • Ethanol • Speciality starch syrup • Corn wet milling • Starchy grain feedstock

Introduction

Recent biotechnological advances to enhance the manufacturing performance of bio-based chemical products for a wide range of applications in many industries have resulted in a surge of bioprospecting approaches (http://ec.europa.eu/enterprise/policies/innovation/files/lead-market-initiative/bio_based_products_taksforce_report_en.pdf). Bio-based chemical products such as enzymes, emulsifiers, and plastics provide an excellent opportunity to reverse the trends through the creation of a new generation of renewable, environmentally sustainable products (Chandel et al. 2007).

(Advances in biotechnology have resulted in a revealing impact on bio-based industrial enzymes.) A large number of commercial enzymes used for various applications, from grain processing to the textile industry, are produced through large-scale microbial fermentation process. Bio-based enzymes have found extensive application in the food, feed and beverage, pharmaceutical, detergent, and textile industries and in recent times also as analytical agents (http://www.usda.gov/oce/reports/energy/

V. Gohel (✉) • V. Maisuria
Genencor®, a Danisco Division, DuPont Industrial Biosciences, Danisco (India) Pvt. Ltd.,
Plot No -46, Roz-Ka-Meo Industrial Area, Sohna, Tehsil NUH, Sohna, Gurgaon 122 103, India
e-mail: vipul.gohel@dupont.com

G. Duan
Genencor (China) Bio-Products Co. Ltd.,
102, Mei Li Road, 214028 Wuxi,
People's Republic of China
e-mail: gang.duan@dupont.com

P. Shukla and B.I. Pletschke (eds.), *Advances in Enzyme Biotechnology*,
DOI 10.1007/978-81-322-1094-8_11, © Springer India 2013

BiobasedReport2008.pdf). The largest market in the enzyme sector, accounting for 59 % of sales, consists of industrial use enzymes, including those used in the starch, food and animal feed, beverage, detergent, textile, leather, pulp, and paper industries. The remaining 41 % of sales are accounted for by the personal care and pharmaceutical industries (Modilal et al. 2011). Food enzymes, including enzymes that are employed in the dairy, brewing, wine and juice, fats and oils, and baking industries, account for the second largest segment with 17 % of the market share. Finally, feed enzymes that are used in animal feeds account for approximately 10 % of the enzyme market share (Chandel et al. 2007; Maurer 2004; Gupta et al. 2002; Sivaramakrishnan et al. 2006; Herrera 2004).

Enzymes have significant advantages over chemical catalysts in that they are derived from natural resources (animal, plant, microbial) and exhibit very high specificity under various reaction conditions, such as pH, temperature, and aqueous and nonaqueous environment. They are easily biodegradable, thereby reducing the risk of environmental pollution and providing an eco-friendly and sustainable solution for today's industries in a variety of process applications (Buchholz et al. 2005; Aehle 2007; Polaina and MacCabe 2007; Olempska-Beer et al. 2006).

The industrial enzyme business is steadily growing worldwide due to enhanced production technologies, engineered enzyme with novel properties, and discovery of new application fields. The global market for industrial enzymes, which is estimated to be at about $3.3 billion (₹14,904.4 crore), is attracting large investments throughout the world (http://www.bccresearch.com/report/enzymes-industrial-applications-bio030f.html; http://www.marketwire.com/press-release/industrial-enzymes-market-estimated-at-33-billion-in-2010-1395348.htm). In India, a developing country, the bio-industrial market is estimated to be at about ₹625.94 crore in 2010–2011, with a growth rate of 10.98 % in comparison to 2009–2010 (₹564 crore). This segment in India is forecast to grow at a compounded annual growth rate (CAGR) of 15 % until 2015 (http://biospectrumindia.ciol.com/content/BSTOP20/112062612.asp). The major worldwide enzyme producers are Genencor (now a part of DuPont Industrial Biosciences) and Novozymes A/S (www.wiley-vch.de/books/biopoly/pdf_v07/vol07_04.pdf) (Sutherland 2000).

Bio-based sustainable solutions are becoming important for food and energy security due to limited availability and increasing demand with ever-increasing population (Gohel et al. 2006). Grain processing is the biggest component in the organized food sector consuming over 40 % of the total value of all enzymes (http://www.apind.gov.in/Library/Note%20fp.pdf). Grain-processing industries include milling of rice, wheat, maize, barley, millets, sorghum, finger millet, and pulses to grind them into fine flour; malting by germinating seeds; and extracting soluble carbohydrates, proteins, vegetable oils, and fibers for use in the food and livestock feed sectors (http://www.ebrd.com/downloads/policies/environmental/grain.pdf). At the same time, the demand for these starchy feedstocks for ethanol production has increased as an alternative transportation energy source and for use in recreational consumption (Mussatto et al. 2010; Szulczyk et al. 2010). To fulfill the competing demands of the food, feed, and energy sectors, the focus of the starch-processing industries has been to develop efficient processes to either maximize energy (starch) availability in the grains using a biotechnological approach or improve starch utilization in value-added starch derivatives through sustainable bio-based enzymatic solutions in order to produce cheaper high-sweetening agents such as glucose, maltose, fructose, and specialty syrups. Sweetener production is based on acid or enzymatic hydrolysis of starch extracted mainly from corn through wet milling. The residual cornstarch is used as feedstock for ethanol production through yeast fermentation in which starch is converted into fermentable sugars using industrial enzymes and the steep liquor generated through wet milling as a fermentation booster. Recent industrial trends show a shift to the production of sweeteners and ethanol, both potable and fuel ethanol from a variety of starch sources such as rice, millet, sorghum, and wheat through dry-milling process. Currently, many different

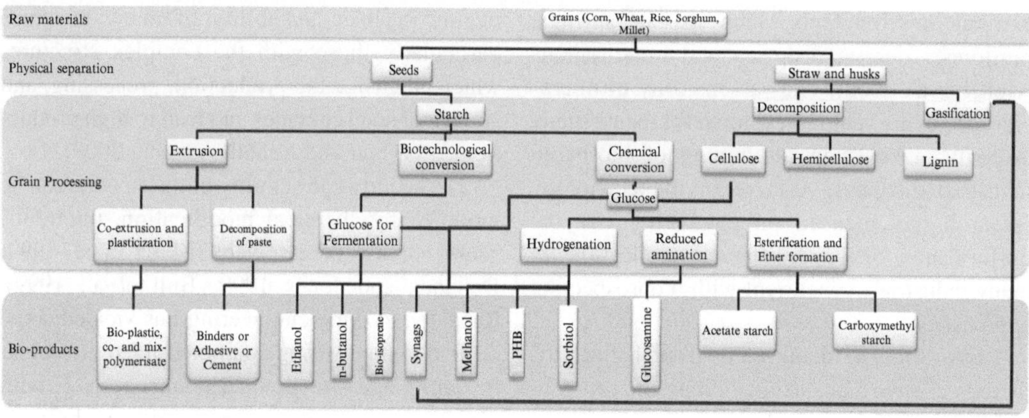

Fig. 11.1 Grain-based biorefinery processes and their bio-products

starch sources have become key industrial raw materials apart from being the major ingredient in the human diet over centuries as the major source of daily caloric intake (Tharanathan 2005).

Ethanol from various sugar substrates such as molasses and sugarcane juice has become a major biofuel, used to replace gasoline (Gough et al. 1997). India produces ethanol mostly from feedstock molasses, a byproduct of sugar manufacturing, unlike Brazil where ethanol is produced directly from sugarcane juice. The trend of replacing gasoline is expected to continue worldwide and increase at a rate almost one billion gallons annually (Gopinathan and Sudhakaran 2009). To maintain this accelerating momentum, continuous innovations in various technologies for ethanol production from different starch sources other than sugarcane juice and molasses are necessary, in addition to bringing down enzyme cost with increasing production volumes.

A whole-crop biorefinery process has a unique advantage, because it consumes the entire crop to obtain useful bio-products. Several raw materials such as wheat, rye, triticale, and maize can be utilized as feedstock input in a whole-crop biorefinery (Fernando et al. 2006). The process is initiated by mechanical separation of biomass into various components that are then treated separately. For example, seeds can be utilized directly after grinding to meal or can be converted to starch, followed by (i) extrusion, (ii) plasticization, (iii) chemical conversion, and (iv) biotechnological conversion to ethanol via glucose fermentation process (Fig. 11.1).

With the rapid growth of biorefineries, there is a pressing need for environmentally sustainable and cost-effective enzyme-based solutions. A plethora of technologies have been developed to screen and discover the potential robust enzymes for application under stringent industrial conditions. Technology has evolved from conventional screening to the use of protein engineering and direct evaluation approaches.

Screening and Discovery of Industrial Enzymes

To identify suitable enzymes for an application, the traditional process has been to screen microorganisms either from naturally occurring environmental samples or from known cultures (Yeh et al. 2010; Warnecke and Hess 2009). With recent advances in biotechnology, it is possible to isolate novel enzymes using bioprospecting approaches. Bioprospecting, also known as biodiversity prospecting, is the exploration of wild species of organisms for commercially valuable biochemical and genetic resources (Gohel et al. 2006).

In general, bioprospecting is a search for unique and robust bioactive compounds including novel enzymes existing in or produced by microorganisms, animal, and plant species found in extreme environments, such as hot springs, rainforests, deserts, deep sea, and arctic regions (Gohel et al. 2006). The vast majority of microbial species are yet to be explored for their genetic diversity, a key step for mining of bio-based industrial enzymes. Also recent biotechnology research has focused on enhancing enzyme performance for a wide range of applications in many industries, concurrent with bioprospecting approaches.

Looking at the depth of microbial diversity, there is always a chance of finding microorganisms that produce novel enzymes with better properties and suitable for commercial exploitation. The multitude of physicochemically diverse habitats has challenged microbes to develop appropriate molecular adaptations for survival (Oberoi et al. 2001). Microbial diversity is a major resource for biotechnological products and processes (Bull et al. 1992). Natural biodiversity among microbes is a vast but little-developed resource for biotechnological innovation. The biosphere is dominated by microorganisms, yet to date most microbes in nature have not been studied. This is mainly due to the fact that historically the only way to reliably characterize a microorganism was by isolation, purification, and fermentation, to specify its biochemical and physiological features on the "macroscopic" level of a pure culture (Gupta et al. 2002).

Many methodologies have been developed for discovering enzymes including conventional screening of environmental isolates and cultures, genetic engineering of existing molecules, environmental gene screening or genomic database mining, protein engineering of an existing enzyme, gene shuffling, and metagenomics (Yeh et al. 2010; Martin et al. 2009; Warnecke and Hess 2009; Böttcher and Bornscheuer 2010). These search processes have greatly benefitted enzyme manufacturing by providing innovative bio-based sustainable solutions. Improvements in enzyme technology include reduction in manufacturing cost of industrial enzymes and their purification, enzymes with high specificity (molecular and chiral), high turnover number, very high activity under mild conditions, and biodegradability. These advantages are offset, however, by the intrinsic instability of many enzymes particularly those with the complex structures which challenge the production, processing, and storage of such enzymes and lead to high production costs (Iyer and Ananthanarayan 2008).

The activity and stability of the enzyme can be modified by chemical modification, immobilization, or use of a solvent (Marrs et al. 1999; DeSantis and Jones 1999; Bull et al. 1999). Recently, enzyme engineering has yielded some desired mutants using computer-aided molecular modeling and site-directed mutagenesis or with directed (molecular) evolution techniques (Fig. 11.2). These techniques have helped in improving the catalytic efficiency of certain processes, reduce the formation of unusable by-products, or stabilize a biocatalyst or protein for prolonged activity under different process conditions. Within a decade, directed evolution has emerged as a standard methodology for protein engineering, used either as a complementary method or in combination with rational protein design. Directed evolution technology for protein designing has focused on meeting the demands for industrially applicable biocatalysts with the desired level of chemoselectivity, regioselectivity, and stereoselectivity, as well as the ability to perform under various process parameters (i.e., high substrate concentrations, solvents, temperatures, long-term stability) (Böttcher and Bornscheuer 2010). This fascinating area of protein design will no doubt be at the heart of future developments in the enzyme industry.

In contrast, rational design generally requires structural information regarding the enzyme and of the associations between sequence, structure, and mechanism/function, a very information-intensive effort. In past decades, it has been possible to predict how to increase enzyme activity, substrate specificity, and stability by employing molecular modeling tools, even in the absence of structural data for an enzyme, using the structure of a homologous enzyme as a model in many cases. Amino acid substitutions are often selected by comparisons of homologous sequences, depending on the purpose of the mutagenesis. However, the resulting molecules have to be carefully evaluated, because minor sequence

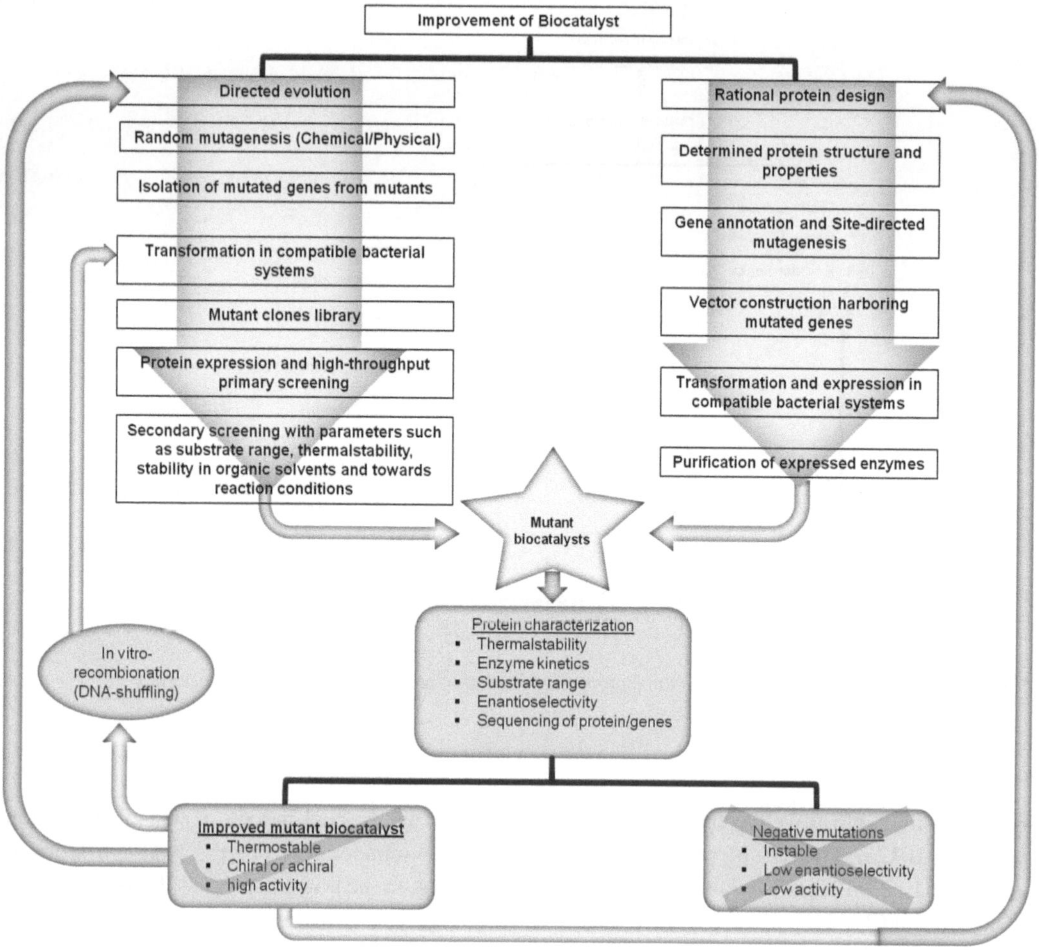

Fig. 11.2 An approach toward improvement of biocatalyst

changes by a single point mutation may cause significant structural disturbance, resulting in collateral negative mutations (Fig. 11.2). Therefore, comparisons of the three-dimensional structures of mutant and wild-type enzymes are essential to verify that a single mutation was really site-directed. Genetic decoding of the enzyme(s) of interest, documenting a suitable (usually microbial) expression system, and developing a sensitive detection system are prerequisites for both direct evolution and rational protein design (Bornscheuer and Pohl 2001). With these discovery and screening methods, enzymes can be developed in short periods of time to meet the needs of biotechnological industries in creating the robust process. The next following section explains how these new enzymes are applied in different industries to enable sustainable solutions in the VUCA (volatility, uncertainty, complexity, and ambiguity) world.

Industrial Enzymes' Application in Ethanol Industries for Starchy Materials

Worldwide, ethanol producers use different starchy grains as feedstock, such as rice, millet, corn, sorghum, sweet sorghum, wheat, potato, sweet potato, cassava, rye, triticale, barley, and

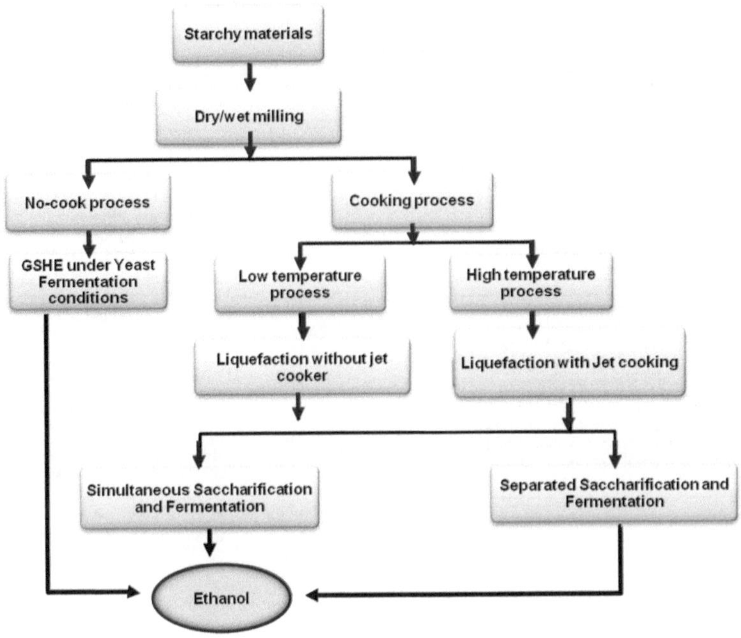

Fig. 11.3 Conventional and nonconventional processes for ethanol production using starchy materials

tapioca, for potable as well as fuel ethanol (Gohel and Duan 2012a). The type of grain is selected by industry based on local availability and its price (Kim and Dale 2004; Gopinathan and Sudhakaran 2009). Industrial enzymes used to convert starches into fermentable sugars include amylases, glucoamylases, pullulanases, and proteases (Gohel and Duan 2012a), mostly sourced from bacteria, fungi, and plants.

Most ethanol producers follow the conventional process for conversion of grain starches into fermentable sugars and ethanol (Suresh et al. 1999; Gibreel et al. 2009; Nikolić et al. 2010; Linko et al. 1983). The conventional process consists of either of a four-step process with milling (wet or dry), liquefaction, saccharification, and fermentation (Fig. 11.3) or a three-step process with the saccharification step omitted by carrying out simultaneous saccharification and fermentation in one tank (Fig. 11.3). Due to recent developments in enzyme technology, higher temperatures are no longer required to break down starch. These technologies are known as low-temperature process and no-cook process of ethanol production. The low-temperature process works without jet cooking. In the no-cook process (Gohel and Duan 2012b), the liquefaction and saccharification processes take place simultaneously, and it does not involve cooking of the starchy materials. The no-cook process eliminates steam consumption entirely, while the low-temperature process saves about 50 % steam compared to conventional ethanol production.

Today, most fuel ethanol is produced from corn using either the dry-grinding (67 %) or the wet-milling (33 %) process (Bothast and Schlicher 2005). The key distinction between wet-mill and dry-grind facilities is the focus of resource use. In a dry-grind plant, the business focus is on maximizing the return on capital exclusively based on the number of gallons of ethanol produced. In a wet-mill plant, capital investments allow for the separation of other valuable grain components in the grain before fermentation to ethanol (Bothast and Schlicher 2005).

Wet-Milling Process: *Mainly Corn*

Wet milling is a complex, capital- and energy-intensive industrial process by which starch-containing grain is hydrated and separated into starch, fiber, germ, and proteins or gluten. Corn is the grain mainly used in the wet-milling process. The germ is separated from the kernel, and corn oil extracted from the germ. The remaining germ meal is added to the fiber and hull which results in corn gluten feed. Gluten is separated to produce corn gluten meal, a high-protein animal feed. The residual starch is subjected to wet milling to produce ethanol, involving preparation of starch solution, followed by fermentation of the fermentable sugars into ethanol. Wet-mill facilities are true "biorefineries," producing a number of high-value products (Shapouri et al. 2004; Bothast and Schlicher 2005).

These wet millers are engaged in producing a variety of starch-based sweeteners to maximize profitability apart from adding to their profit margins through sale of by-products. However, due to massive capital investments and operating costs of wet milling, new manufacturing units are mainly based on dry-milling process. Recently, a number of enzyme-based innovative technologies have helped simplify the wet-milling process (Singh and Johnston 2002; Ramírez et al. 2009).

Enzymes in the Corn-Steeping Process

In the wet-milling process, steeping of corn is the most time-consuming step. The toxicity of sulfur dioxide (SO_2) causes an additional problem, as it poses an environmental and health concern, and there is growing pressure from environmental agencies to find alternatives to SO_2-based wet-milling processes (Johnston and Singh 2001; Singh and Johnston 2002). It requires at least 48–52 h of incubation of corn with SO_2 (about 2,000–2,500 ppm) in the presence of 0.5–2 % lactic acid to soften the hard corn kernels which then swell up. This lactic acid is usually produced by bacteria belonging to the genus *Lactobacillus* during steeping. The steeping process is sometimes prolonged to 72 h because of poor corn quality or improper soaking with SO_2. Wet milling uses steeping at different temperatures ranging from ambient temperature to 50 °C to sustain 45–50 % moisture in the corn grains. Steeping is a diffusion-driven process. During steeping, chemicals and water diffuse into the corn kernel through the grain tip and cap and cross the tube cells of the pericarp into the kernel crown and finally into the endosperm. Apart from softening the corn, SO_2 also plays a vital role in releasing starch from the endosperm, breaking down the disulfide bonds holding the protein matrix that surrounds the starch particles. This increases the starch yield. Reducing steeping process time helps the manufacturer to increase plant capacity and reduce energy and SO_2 consumption associated with prolonged steeping time; these are the three leading challenges for wet millers to sustain profit margins (Johnston and Singh 2004).

Many chemical approaches have been developed for reducing the steeping time, all of which either require major capital investments for existing process modification or require pretreatment of corn kernel which increases energy consumption and environmental safety concerns due to increasing pollution. To address these challenges, Johnston and Singh (2004) demonstrated an enzymatic process in which maize kernels were soaked in water with SO_2 in the steeping tank for about 8–10 h so that the germ gets fully hydrated and tensile enough to resist breakdown when the corn is coarsely grounded and treated with acid protease in separate reactor. This process removes the diffusion barriers for protease penetration into the corn endosperm to break down the protein substrate. After enzymatic treatment, the corn was milled using the conventional wet-milling process. Degermination milling was proposed by Johnston and Singh (2004) using coarse grinding (also called first grinding) with a larger gap between the grinding stones than what is currently used by wet millers, in order to significantly improve germ recovery. Apart from proteases, Johnston and Singh (2004) also studied various carbohydrate enzymes including β-glucanase, cellulase, and xylanase in the steeping process, which were found to produce significantly lower starch yields compared to pepsin, acid protease, and bromelain. An increase in incubation time with the three proteases resulted in reduction of total fiber content. These proteases not only reduced the stepping process time but also reduced the residual protein content in the final starch fraction. The benefits of this enzymatic wet-milling process included a drastic reduction or elimination in the use of SO_2 during steeping and also reduced the steeping time by 70 % compared to the conventional corn wet-milling steeping process (Johnston and Singh 2001, 2004; Ramírez et al. 2009; Steinke and Johnson 1991). DuPont has a product for enzymatic steeping in the wet-milling process, which it has marketed as PROSTEEP™, a fungal protease.

Maize comprises phytate, which to a large extent ends up in the corn steep liquor and constitutes an undesirable component. Corn kernel is treated with phytase under steeping conditions in the presence of SO_2 to eliminate or greatly reduce the phytate content in corn steep liquor. This further helps in reducing the steeping time and also by facilitating the separation of starch from fiber and gluten, resulting in higher starch and gluten yields as well as lower energy consumption (Shetty et al. 2010).

Enzymes in Gluten Filtration

In conventional wet-milling process, gluten dewatering and drying of corn consumes almost 26 % of the total energy, next only to the starch-drying process (32 %) (http://www.energystar.gov/ia/business/industry/LBNL-52307.pdf). To reduce this energy consumption and improve filtration efficiency, DuPont Genencor Science has powered the production of cellulase and hemicellulase enzymes. After germ and fiber separation, the next stage is to separate gluten and starch with 5–6 % protein. This mixture is known as mill starch, which is processed through a de-gritting cyclone to remove any foreign particles or sand to prevent damage or blocking of centrifuge nozzles. After de-gritting, the mill starch having 6–8°Be (Baumé) is concentrated to specific Be (Baumé) of 10–12° (Blanchard 1992). The concentrated starch is passed through primary separators, a kind of centrifuge systems to separate starch and gluten factions based on density differences (1.5 g/cm^3 for starch vs. 1.1 g/cm^3 for gluten particles) (Singh and Johnston 2004).

Primary centrifuges consist of a rotating bowl in which a stack of conical discs are separated by a distance of 0.4–1.0 mm, depending on the density of particles to be separated. On the periphery of the rotating bowl, there are 6–12 nozzles. The mill starch enters the rotating bowl from the top or the bottom. Due to centrifugal force, the heavier starch particles are forced toward the periphery of the rotating bowl and exit through the nozzles as underflow. The starch slurry coming out from the primary separator has a protein content of 2–4 % and a specific gravity of 1.160–1.198 (35–42 % dry solids). Lighter gluten particles move up between the discs and exit out as overflow. Gluten slurry from the primary centrifuges comes out at a concentration of 15–30 g/L (2–4 oz/gal) and contains about 68–75 % protein (db). Routine maintenance of centrifuges is required to optimize performance (Blanchard 1992; Singh and Johnston 2004).

The gluten slurry is concentrated from 15–30 g/L (2–4 oz/gal) to 150–165 g/L (20–22 oz/gal) by using another nozzle-bowl gluten thickener (GT) (Blanchard 1992). In this process, cellulase and hemicellulase enzymes are added to accelerate the dewatering of gluten with the prevailing process of rotary vacuum iterations (Fig. 11.4). These enzymes are marketed by DuPont as OPTIFLOW® RC 2.0, produced by controlled fermentation of *Trichoderma reesei*. The enzymes hydrolyze the thin plate material and fine fiber, resulting in lower gluten cake moisture and improved dewatering characteristics which result in reduced drying energy cost by 5–25 %. This process also increases the gluten filtration capacity from 20 % to 25 %, reduces filtration cloth wash, and decreases gluten cake recycle. These technologies, however, are sensitive to multiple factors such as corn quality, upstream processing, filter dryer operations, and design.

The cornstarch separated through the wet-milling process is used either for producing starch derivatives, sweeteners, or ethanol/biofuel by enzymatic liquefaction and saccharification, followed by yeast fermentation in case of ethanol/biofuel production.

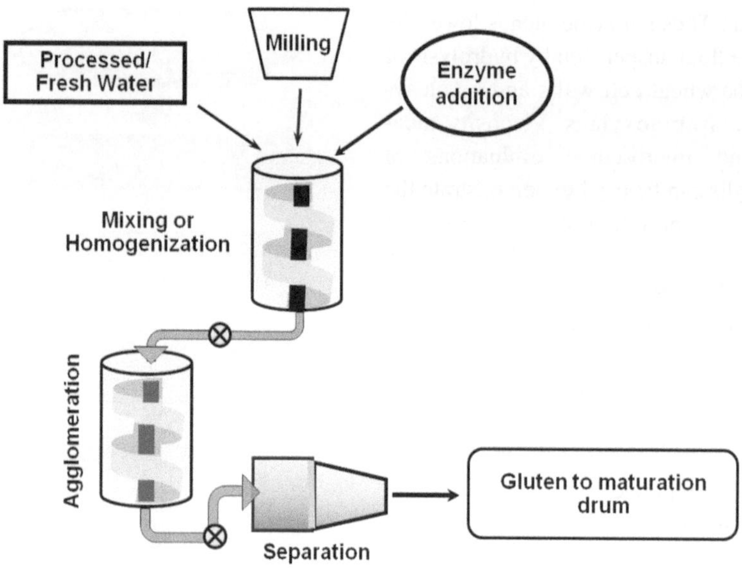

Fig. 11.4 Enzymatic gluten separation processes used in wheat starch industries

Enzymes to Separate Gluten from Wheat Starch

Manufacture of starch and gluten from wheat is typically done in a dry-milling operation (Sugden 1997). The disintegration of milled wheat particles into A- or B-starch fractions and gluten, as well as their separation by hydrocyclones and decanters, is affected by the interaction with non-starch polysaccharides like arabinoxylans and β-glucans (Sayaslan 2004). Multicomponent blends of endo-β-1,4-glucanases and endo-β-1,4-xylanases derived from controlled fermentation of *Trichoderma reesei* (SPEZYME® CP and GC 220) improve the starch yield and gluten purity (Fig. 11.5). The enzymes are added at concentrations of 0.1–0.3 kg per ton of milled wheat. The enzymes used should be free from α-amylase or protease activity. The level of exo-beta-xylosidase should be low to limit the formation of monosaccharides, which might have a negative effect due to the reactivity of xylose. A novel enzyme preparation (GC 220) has been developed which exhibits high endo-β-glucanase and endo-xylanase activity but is virtually free of

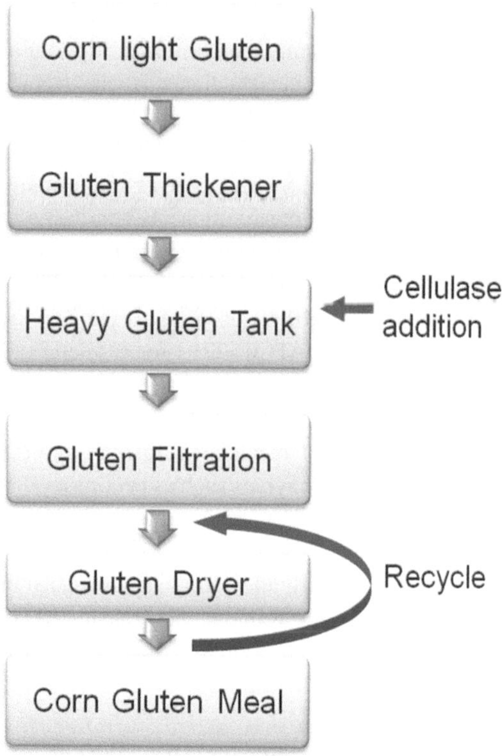

Fig. 11.5 Industrial enzymes used in corn gluten filtration processes

exo-xylosidase. These enzyme blends lower the viscosity of the flour suspension by hydrolysis of β-glucans in the wheat cell walls and by cleavage of soluble arabinoxylans. Viscosity measurements and microscopic evaluations of stained cell walls can be used to demonstrate the mode of action of these hydrolases. The impact of added enzymes on starch yield and gluten quality will be discussed in detail.

In wheat starch processing, utilization of xylanase results in reduced viscosity of slurry along with improved starch-gluten separation and higher relative yields compared to the conventional process (Christophersen et al. 1997). Similar results have been reported in pilot plant experiments with hemicellulases (Weegels et al. 1992).

Enzyme for Cassava Starch Separation

Cassava (*Manihot esculenta* Crantz) is a perennial tuber plant widely grown in many tropical countries including Nigeria and is one of the most important commercial crops. It is the third largest source of carbohydrates in the tropics after rice and corn and a staple for over 600 million people (FAO 2002). There are sweet and bitter cassavas. Sweet cassavas are normally used for human consumption. Bitter cassavas with higher starch content are used as animal feed or processed for industrial use (Vessia 2007). Dzogbefia et al. (1999) used pectinase enzyme from *S. cerevisiae* ATCC52712 for extraction of starch from cassava.

The extraction of starch from cassava involves the combination of mechanical rasping and the use of hydrolytic enzymes such as pectinases that disintegrates the pulp releasing the starch. Usage of enzyme saves the energy costs involved in starch extraction by alleviating the intensity of mechanical rasping (Rahman and Rakshit 2003). The pectolytic enzymes break down the pectate-composed cell wall leading to release of starch granules.

Commercial pectases have been documented to enhance cassava starch extraction (Sriroth et al. 2000; Daiuto et al. 2005). In a modified method, the combined application of pectinase, xylanase, cellulase, and protease manufactured by DuPont increased cassava starch extraction by about 11 % (Fig. 11.6). In addition, more water was released from the mash after enzyme treatment, which resulted in lower water content of the pulp and fiber residue after separating starch. Therefore, the pulp and fiber could be dried easily while also reducing water consumption (Duan et al. 2010b).

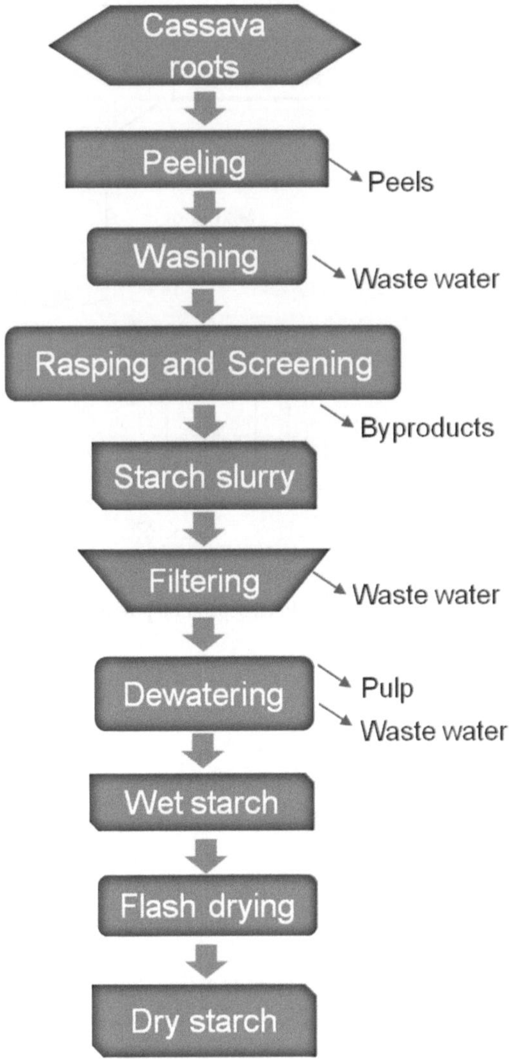

Fig. 11.6 Enzymatic processes for separation and processing of cassava starch

Dry Milling of Starchy Materials

Dry milling is cost-effective and also less capital intensive. Due to this, most of the newly built ethanol manufacturing units are coming up with dry-milling technology (https://engineering.purdue.edu/~lorre/16/Midwest%20Consortium/DM%20DescManual%2042006-1.pdf). Starchy grains such as rice, millet, corn, sorghum, cassava, and wheat are the most commonly used grains worldwide due to their ample availability and relatively low price (Gohel and Duan 2012a). Milling is done to reduce grain kernel to small particle sizes for distribution in solutions. Two types of milling devices are used by industry, roller and hammer mills (Baron et al. 2005). Roller mills are used in many potable ethanol plants because of the use of grains with high husk content, e.g., rye. Hammer mills are most commonly used by industries producing non-potable ethanol. The grain is fed into a chamber with hammers rotating at high speed to grind the grain. Screens between 6/64 and 3/16 in. are used to control particle size.

Milling is an essential step for efficient hydrolysis by enzymes since it directly impacts the physicochemical reaction between insoluble grain flour and enzyme. Large particle size impedes gelatinization, resulting in lower enzyme efficiency with corresponding loss in yield up to 10 %. Hence, particle size should be such that 92–94 % of the flour passes through US standard 20-mesh sieves, while very finely ground grain particles (<200 mesh) may increase yield due to efficient enzymatic action on the substrate. However, it may hamper mixing of the flour with water and cause problems in centrifugation, evaporation, and drying, with the result of reduced DDGS yield (distilled dry grains with solubles), a by-product of dry millers. Additionally, the milling process should have temperature controls to limit the milled flour temperature to 40 °C to avoid the formation of nonconvertible starch. To prevent the need for aggressive dry milling, "DuPont Genencor Science" has powered the development of a fungal protease, which supports by mashing the grain under the hydrolytic conditions.

After dry milling, the grain flour is subjected to liquefaction, saccharification, series of filtrations (charcoal treatment followed by ion-exchange process), and evaporation to concentrate the syrup to more than 60 % Bx. (Brix). For ethanol production the grain flour is subjected to liquefaction followed by sequential saccharification and fermentation, or simultaneous saccharification and fermentation (SSF).

Liquefaction

Liquefaction is the first and most important step in starch processing for ethanol production.

Conventional Liquefaction Process

The liquefaction process is variable, usually processed in three to four tanks in the ethanol industries, involving a slurry tank (55–65 °C), primary liquefaction (85–95 °C), followed by jet cooking (105–120 °C), and secondary liquefaction (85–98 °C). Alternatively, a three-step process involves the use of slurry tank (55–65 °C), jet cooking (105–120 °C), and liquefaction (85–98 °C) (Fig. 11.7). In the starch-processing industries, liquefaction is carried out in a three-step process, slurry tank process (40–45 °C), jet cooking (105–108 °C), and dextrinization process (92–98 °C) (Fig. 11.8) (Gohel and Duan 2012b).

In the slurry tank, vigorous agitation is required to keep the mash mixed at pH 5.5–5.8 at a temperature of 60–65 °C for a holding period of 20–30 mins to produce 4–6 DE mash at tank outlet. DE is equal to the percentage of bonds hydrolyzed in the starch. A DE of 100 indicates that all bonds between the glucosyl units have been hydrolyzed (Rong et al. 2009; Chaplin and Bucke 1990). In jet cooking, the mash is heated by mixing with high-pressure steam in a jet, to gelatinize the starch and prepare it for easier enzymatic breakdown. It is then agitated under pressure for several minutes before flash cooling. Although some of the thermostable α-amylase activity is reduced during this stage, it remains adequately effective in reducing the viscosity at the high temperature. Many factors impact the jet cooking process including temperature, pH, mash solids, retention time, and shear (back pressure, breaking up of granules). The initial and final

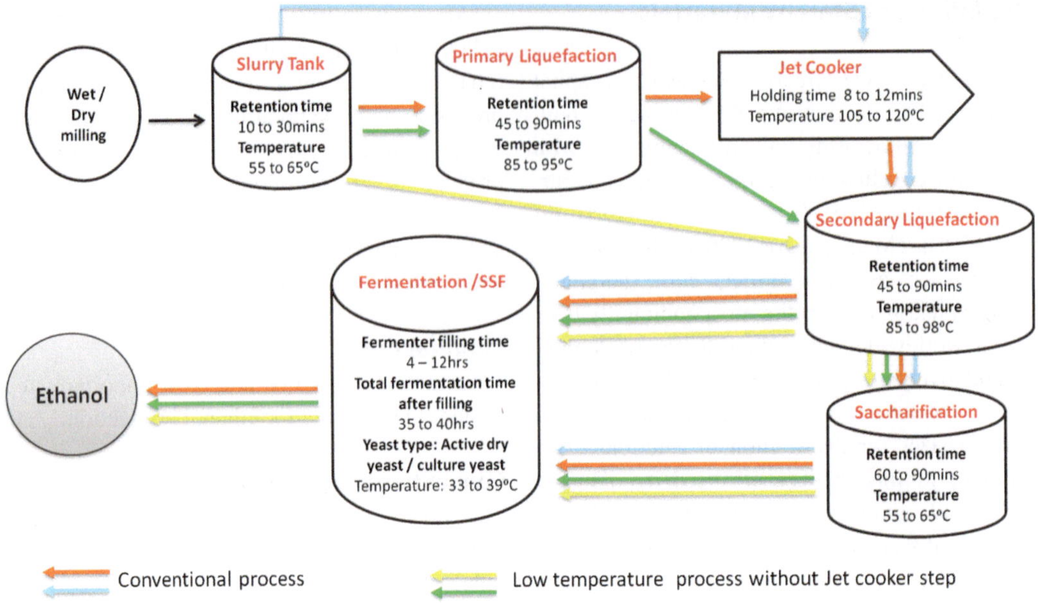

Fig. 11.7 Industrial enzymatic processes for ethanol production

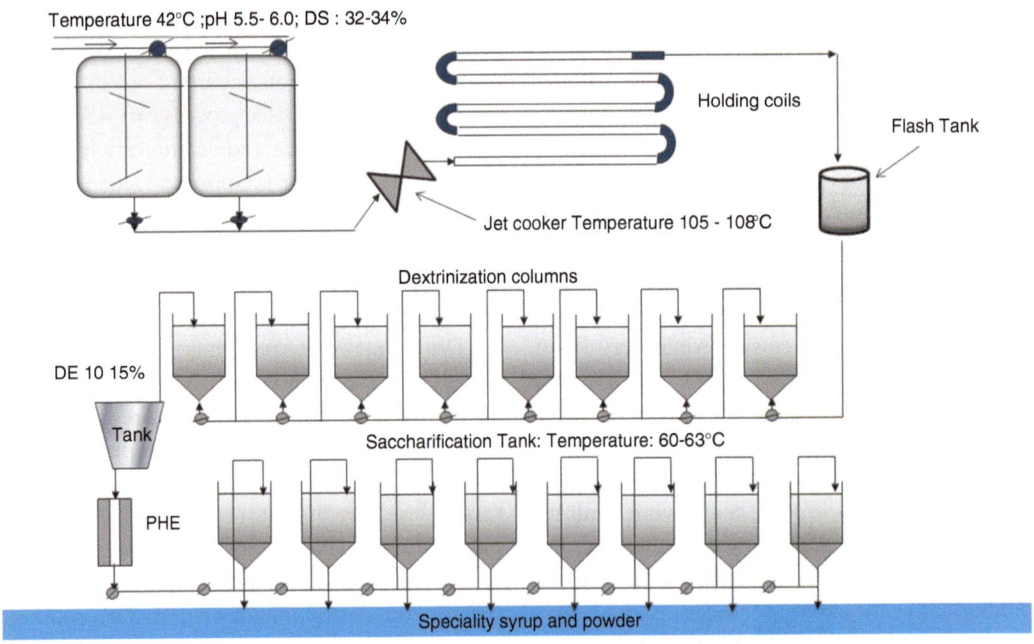

Fig. 11.8 Industrial enzyme-based carbohydrate processing for glucose syrup production

liquefaction results in partial and complete solubilization of starch, respectively, by hydrolyzing the starch into low-molecular-weight glucose chains called dextrins and generating a final liquefact DE of 12–14 % with no precipitate in the mash coolers. DE is dextrose equivalent, a measure of the reducing sugars in solution. Hence, it also measures the degree of starch hydrolysis (Rong et al. 2009; Chaplin and Bucke 1990).

The liquefaction process for about 1.5–2 h results in complete gelatinization and dextrinization of pure starch or raw starchy grain substrate. A shortened liquefaction process time (0.5 or 1 h) produces inefficient hydrolysis into glucose polymers (Montesinos and Navarro 2000; Neves et al. 2006). An increase in the dosage of α-amylase reduces the liquefaction process time but may not produce the proper substrate for saccharification required for the production of various sweeteners or for ethanol production in which glucoamylase and pullulanase enzymes produce fermentable sugars in yeast fermentation. This is because α-amylase produces shorter 1-6 linkage branches of glucose polymers which reduce the yield of the desired end products (Gohel and Duan 2012a; Roy and Gupta 2004; Kłosowski et al. 2010).

The key functional characteristics of starch that are critical for the biorefinery industry are granular swelling, gelatinization, and pasting. Granular swelling facilitates starch liquefaction, which should be followed by gelatinization and dextrinization. Gelatinization causes the intermolecular bonds of starch molecules to hydrolyze in the presence of water and heat. This irreversibly dissolves the starch granule. Penetration of water increases randomness in the general granule structure and decreases the number and size of crystalline regions. Crystalline regions do not allow water entry. Heat causes such regions to become diffuse, so that the chains begin to separate into an amorphous form (Jenkins and Donald 1998).

During gelatinization, water acts as a plasticizer. Water is first absorbed in the amorphous space of starch leading to swelling during heating, and the water is then transmitted through connecting molecules to crystalline regions (Cauvain and Young 2001). Water enters tightly bound amorphous regions of double helical structures to swell the amylopectin, thus causing crystalline structures to melt and break free. Stress caused by this swelling phenomenon eventually interrupts the structure and allows the leaching of amylose molecules into the surrounding water (Englyst et al. 1992). In industry, a jet cooker is used to gelatinize starch by mixing starch slurry with steam under pressure (Van der Maarel et al. 2002).

The thermostable α-amylase, an endo-acting enzyme, acts faster on starch granules during swelling and randomly hydrolyzes α-1, 4-glucosidic bonds to reduce the peak viscosity of gelatinized starch, producing soluble dextrins and oligosaccharides (Tester et al. 2004). Thermostable α-amylase also plays a key role as a thinning agent (Aiyer 2005). After gelatinization, partial hydrolysis to produce dextrin has to be sustained without interruption by ensuring that the slurry does not cool down to avoid the formation of crystalline-stable starch molecules which resist further degradation by α-amylase. The formation of such crystallized starch is called retrograding (Aiyer 2005; Englyst et al. 1992; Zeleznak and Hoseney 1986). For pure starch liquefaction after gelatinization, the dextrinization process temperature should be between 90 °C and 98 °C in a closed dextrinization tank after passing through the flash tank.

Higher Dry Solid (DS) Liquefaction

Utility consumption involves energy, electricity, water, and temperature changes, i.e., cooling and heating. Thus, water and energy (steam and cooling generated with water) are highly used commodities in the processing industries. Water scarcity and environmental regulations on water effluents are a major concern currently in these industries (Mann and Liu 1999; Karuppiah et al. 2008). To become more sustainable and cost-effective, industries constantly look for ways to control the use of these inputs. Due to these limitations, industries are pushing their plants to run on higher DS (dry solids) from 15–20 % to 25–35 % to save water and save up to 40–50 % energy consumption. In this process, the maximum backset or thin slop is recycled through liquefactions (Pohit et al. 2009; Grafelman and Meagher 1995; Gohel and Duan 2012a). The use of concentrated dry solids requires appropriate processes for plant optimization depending on the feedstock. For example, whole wheat grain feedstock contains fibers, hemicellulase, and pentosans other than starch, increasing the dry solid concentration which increases viscosity. Thermostable α-amylase alone cannot reduce the viscosity generated through non-starch polysaccharides. To reduce such viscosity, it requires a mixture of cellulase, hemicellulase, and xylanase added with α-amylase before or during liquefaction (Crabb and Shetty 1999).

In industrial processes, the concentration of dry solids in the starch suspension subjected to liquefaction is usually 25–35 %, which results in very high viscosity following gelatinization (Aiyer 2005; De Cordt et al. 1994). Many thermostable α-amylases are commercially produced from different microorganisms (Klibanov 1983). The efficacy of α-amylase is measured by the speed and uniformity with which it reduces the peak viscosity, when used at a suitable concentration in the slurry at 105–107 °C. This translates into fewer problems in the jet cooker or liquefaction tank for mixing slurry and, in turn, lowers power consumption and more rapid liquefaction. Use of split dosages of α-amylase is a common practice among ethanol producers to generate adequate fermentable sugars for yeast propagation in the SSF process to keep up a steady production of ethanol. Many ethanol producers also use an online dosing pump or periodic enzyme feeding in the continuous liquefaction process.

The thermostability of α-amylase varies with the starchy substrate used and the quality of feedstock used in liquefaction. Reduced thermostability of α-amylase in prolonged liquefaction may result in poor starch digestion and therefore interfere with filtration of the resultant liquefact due to turbidity (Rosendal et al. 1979). To overcome this, researchers have identified many factors that reduce the digestibility of grain protein and starches. Grain containing phytic acid and tannin reduce the thermostability of α-amylase enzyme, resulting in reduced starch digestibility under high-temperature liquefaction (due to reacting with proteins, including hydrolyzing enzymes). However, at lower temperatures, the impact of tannin can be minimized (Cawley and Mitchell 1968; Yan et al. 2009). DuPont Genencor Science has pioneered the development of low-temperature and no-cook processes, which are effective in reducing negative impact of tannin in the grain-based ethanol process (Gohel and Duan 2012b).

DuPont-Genencor Science has developed a thermostable phytase enzyme (marketed as GC 980) to be used with α-amylase to enhance its thermostability during liquefaction of grains containing phytic content, such as sorghum, rice, millet, and corn. Other thermostable α-amylases produced from different genetically modified *Bacillus licheniformis* are marketed as SPEZYME® XTRA, SPEZYME® FRED, SPEZYME® ALPHA, and CLEARFLOW® AA, which function within the pH range of 5.0–7.0 and at temperatures above 85 °C. Liu et al. (2008) have reported that α-amylase is sensitive to acidic medium and loses its hydrolytic activity. α-Amylase functions optimally at 90 °C and pH 6.0 (Liu et al. 2008). Previously, the liquefaction step in cornstarch hydrolysis was performed at 85 °C and pH 6.0 (Mojović et al. 2006). Optimum α-amylase action to produce reducing sugar in continuous enzymatic hydrolysis was obtained at pH 6.0 and 30 °C. The amount of reducing sugar produced from sago starch was 0.464 g/L. It is documented that 5.9 % (w/v) fermentable sugar is produced from 25 % DS corn with RSH (raw starch hydrolyzing) added at 2.5 kg per MT of corn during liquefaction at 48 °C for 2 h, while 3.70 % (w/v) fermentable sugar produced from 30 % DS cassava starch treated with SPEZYME® XTRA at 90 °C applied at 0.66 kg per MT of starch (Shanavas et al. 2011). After liquefaction, SPEZYME® ALPHA treated 30 % and 35 % Indian broken rice produced about 4–5 % and 6–7 % fermentable sugars, respectively. The same enzyme in the presence of 30 % and 35 % pearl millet produced 3–4 % and 5–6 % fermentable sugars, respectively (Gohel and Duan 2012a).

Liquefaction Without Jet Cooking

Steam generation is expensive due to limited availability of fossil fuel (Mussatto et al. 2010; Szulczyk et al. 2010). Hence, many ethanol producers have already started looking at alternatives to reduce or omit the jet cooker step which cooks the starchy materials at high temperature and pressure. For this, manufacturers use 30–40 %

higher doses of α-amylase in liquefaction with the rest of the process remaining the same (Fig. 11.7). This process reduces about 30–50 % of steam consumption. However, research is continuing to develop an efficient α-amylase with persistent activity while cooking at or above gelatinization temperature.

Saccharification and Fermentation: Sequential or Concurrent

After liquefaction, the liquefact is subjected to saccharification at 55–65 °C (Figs. 11.3 and 11.7). Saccharification is a widely used process in the production of almost all sweeteners and of ethanol. Saccharification yields about 96 % glucose and 4 % by-product from the starch substrate. Saccharification of cornstarch is reported as being widely performed at 55–60 °C and pH 5.0 (Mojović et al. 2006), although optimum saccharification is documented to take place at 60 °C and pH 4.5 (http://umpir.ump.edu.my/863/1/Siti_Nor_Shadila_Alias.pdf). In another study, however, Aggarwal et al. (2001) found that high temperatures retard saccharification and that the optimum conditions were 45 °C and pH of 5.0.

Recent technological improvements have eliminated one enzymatic step, the separate saccharification for ethanol production. Elimination of this step avoids high osmolarity stress in the initial stage of yeast fermentation and reduces the risk of contamination during the fermentation process (Nikolić et al. 2010; Grafelman and Meagher 1995). This process is also known as simultaneous saccharification and fermentation (SSF) (Figs. 11.3 and 11.7). In SSF, the saccharifying enzymes hydrolyze the liquefied starch into fermentable sugars, while concurrently yeast fermentation is used to ferment the sugars into ethanol. The ethanol industries mainly use commercially available glucoamylase, acid fungal α-amylase, and pullulanase as the saccharification enzymes. Glucoamylase is an exo-acting enzyme that hydrolyzes starch liquefact at the nonreducing end of the α-1,4-glycosidic bond. To accelerate glucoamylase activity in the liquefact slurry, pullulanase, a debranching enzyme, is used for its unique role in hydrolyzing the 1-6 linkages of the amylopectin branch to produce a linear, free dextrin chain.

DuPont has commercialized glucoamylases that are marketed as GA-L NEW and DISTILLASE® ASP. Both the products are effective in sequential saccharification and fermentation, as well as in the SSF process. GA-L NEW is intended for glucose production from liquefied starch using controlled fermentation with a selected strain of *Aspergillus niger*. GA-L NEW performs best at a pH of 4.0–5.0 and a temperature of 55–60 °C.

DISTILLASE® ASP is a blend of enzymes, predominantly containing glucoamylase supplemented with pullulanase and protease produced by controlled fermentation of genetically modified strains of *Bacillus licheniformis* and *Trichoderma reesei*, respectively. This enzyme blend was designed for ethanol production to perform under conditions of variable pH during saccharification. This enables the complete process (the liquefaction and saccharification and SSF) to be independent of the pH conditions, which saves chemicals, less uncertainty about product yield, and less acidity in thin stillage water. The process also ensures additional nutrients to yeasts having a subsidiary protease activity to produce FAN (free amino nitrogen) (Duan et al. 2011a). When used in the sequential saccharification/fermentation process, DISTILLASE® ASP is thermostable up to 70 °C and active in a broader pH range, 4.0–5.5. Due to these unique advantages, it also supports the goal of reducing microbial contamination. It is reported that pullulanase activity enhances the breakdown of long-chain branches of α-1,6-glycoside linkages to produce linear dextrins during SSF, which in turn enhances fermentable sugar production by the action of glucoamylase. The net result is a shortening of hydrolysis time by as much as 37 % (Gantelet and Duchiron 1999).

These industries are highly water and energy intensive. Increasing prices of crude oil and other fossil fuels have stoked worldwide interest in alternative fuel sources (Mann and Liu 1999; Karuppiah et al. 2008). Energy security is a critical priority for all countries because of the volatility, uncertainty, complexity, and ambiguity (VUCA) of the global fossil fuel market, due to high prices, declining production, and unstable geopolitical acts of war and terrorism. These issues underscore the vulnerability of currently dominant global energy needs to supply disruptions (Gopinathan and Sudhakaran 2009). Hence, fuel alcohol from starch needs constant process improvement in the biomass conversion process to make it economically viable.

The emerging very high gravity (VHG) fermentation technology is one such measure to increase fermentation rates and ethanol concentration and to minimize waste effluent (Bvochora et al. 2000). VHG technology is now widely used to increase the concentration of starchy materials in the feedstock (≥30% w/w DS) and to increase plant throughput (Devantier et al. 2005). Concurrently, VHG technology lowers energy cost per liter of alcohol, bacterial contamination risk, and capital costs. In addition, increased harvest of high-protein spent yeast residue is obtained with VHG fermentation (Bvochora et al. 2000). VHG requires selection of the right yeast species (*Saccharomyces cerevisiae*) that can produce high ethanol concentrations and not be retarded by high sugar and ethanol concentrations. VHG requires the maintenance of appropriate environmental and nutritional conditions for optimum performance. The drawback of VHG fermentation is that the increased ethanol recovered does not compensate for its higher cost relative to other processes (Gohel and Duan 2012a).

A 6 % yeast inoculum size was found to be critical for reducing the fermentation time during VHG fermentation for ethanol production (Breisha 2010). More rapid yeast cell growth was associated with shorter lag phase in the growth cycle (Breisha 2010). At low (<5 %) yeast inoculum sizes, longer lag phases and lower ethanol production rates during the initial stage of yeast fermentation were observed along with higher contamination (Akin et al. 2008; Sharma et al. 2007).

Urea is widely used by ethanol producers worldwide as an organic nitrogen source for yeast. The downside is that ethanol reacts with urea to produce ethyl carbamate which is carcinogenic to humans (Ough et al. 1988; Pretorius 2000; Kitamoto et al. 1991). To replace urea, DuPont has produced acid fungal proteases for whole starch-containing feedstock. This product is commercially available as FERMGEN™. This protease hydrolyzes grain proteins into amino acids, peptides, and free amino nitrogen (FAN) essential for yeast growth. Furthermore, it plays a key role in hydrolyzing the protein matrices in the grain kernel that bind the various fractions, releasing "hard-to-hydrolyze" starch. The protease also accelerates ethanol production and enables higher ethanol yield from grain-based substrates than processes that do not use protease (Duan et al. 2011a).

Protease also shortens the fermentation cycle by 8–12 h and has improved the fermentation efficiency of VHG fermentation without compromising ethanol recovery. VHG fermentation is enhanced when hydrolyzed wheat protein is used as a source of FAN (Thomas and Ingledew 1990). In barley mash fermentation, FAN enhances yeast growth and multiplication (Kim et al. 2008). It is reported that expression of aspartyl protease on the cell surface of industrial ethanol-producing yeast results in higher yeast cell counts with increased rate of growth. A recombinant strain of *S. cerevisiae* exhibiting a higher yield of ethanol and lower residual sugar compared with the parental strain has been reported by Guo et al. 2010.

Gohel and Duan (2012a) documented a three-step conventional liquefaction process followed by SSF under yeast fermentation conditions using 30 % and 35 % dry solid concentrations of Indian broken rice (starch # 68.23 %) and pearl millet (starch # 60.64 %) feedstocks. These studies were performed with and without addition of acid fungal protease platform (FERMGEN™). At the end of the SSF process, ethanol yield with 30 % and 35 % DS of Indian broken rice feedstock was observed to be ~15.14 % and 16.23 % (v/v at 20 °C) without FERMGEN™ versus 16.03 % ± 0.02 % and 16.41 % ± 0.08 % (v/v at 20 °C), respectively, with the addition of FERMGEN™. The corresponding yields with pearl millet feedstock were ~12.14 % and 12.67 % (v/v at 20 °C) at 55 h in 30 % and 35 % DS concentrations, respectively, without FERMGEN™ and 14.13 % and 14.56 % (v/v at 20 °C) with FERMGEN™. These results showed that the addition of protease plays a key role in producing more ethanol. Wu et al. (2006) used a three-step conventional process of ethanol production from US pearl millet with 65.30 % starch and 25 % dry solid concentration. Their process involved liquefaction at 95 °C for 45 min followed by 80 °C for 30 min, saccharification at 60 °C for 30 min, and finally yeast fermentation, with an ethanol yield of ~11 % v/v at 20 °C, a fermentation efficiency of 90 %, and residual starch concentration of 3.45 %. Zhan et al. (2006) used the conventional process for US sorghum containing 68.8 % starch and 25 % DS concentration and obtained 10.72 % v/v ethanol yield.

No-Cook Process in Ethanol Production

Following the success of VHG fermentation in ethanol manufacture, the ethanol-producing industries have aggressively searched for solutions to reduce steam consumption used in the liquefaction process. However, steam consumption for distillation cannot be circumvented. DuPont-Genencor Science has powered the development of STARGEN™ technology, a granular starch hydrolyzing enzyme (GSHE) to hydrolyze no-cook starch directly into fermentable sugars under yeast fermentation conditions without using steam (Fig. 11.3). It is an optimized blend of amylases that hydrolyze raw (granular) starch in a no-cook process. These amylases are derived from genetically modified strains of *Trichoderma reesei*. The no-cook process enables all these biological processes in a single step without requiring any steam to cook the starchy materials (Bellissimi and Ingledew 2005; Bothast and Schlicher 2005; Gohel and Duan 2012b). The single step is performed in a fermenter under yeast fermentation conditions, reducing capital investment and physical space used for separate tanks for the liquefaction and saccharification processes. The space can be utilized to install more fermentation tanks instead to produce more ethanol within the existing plant facility. This GSHE process has the additional advantages of higher efficiency of starch conversion into ethanol due to minimal sugar loss that is inevitably incurred in a high-temperature cooking process due to heat-catalyzed Maillard reaction between amino acids and reducing sugars. Thus, the GSHE process yields less biomass (1.95 kg per 100 kg starch) due to reduced yeast stress. In the conventional jet cooking process, the biomass of yeast produced is 3.88 kg per 100 kg starch (Shetty et al. 2007). This is because the Maillard reaction products cannot be utilized by yeast for fermentation into ethanol (Göğüş et al. 1998). The free soluble sugars that are spared from the Maillard reaction by eliminating the cooking process additionally play a role in ethanol production from yeast fermentation.

The conventional process involving higher liquefaction temperatures has about 81–90 % fermentation efficiency, compared to 97–98 % with the GSHE process using Indian broken rice and pearl millet feedstocks (Gohel and Duan 2012b). In the conventional process, Indian ethanol producers with a plant capacity of 110–130 MT of Indian broken rice (68 % starch, 28 % dry solids) feedstock per day consume 49.5 MT steam for the liquefaction process, followed by SSF with yeast fermentation, producing 10 % v/v at 20 °C, 410 L ethanol per MT of Indian broken rice with 86 % fermentation efficiency (http://www.pcbassam.org/EIAREPORT/EIA_Radiant/2%20Chapter%20The%20Project.pdf). In the GSHE process, the daily consumption of 49.5 MT steam in liquefaction is saved, which directly impacts the overall ethanol production cost. The STARGEN™ technology works effectively with all starchy feedstocks (Gohel and Duan 2012b).

Specialty Syrup Production

Most of the corn-milling processes use pure starch for the production of specialty syrups, in addition to wheat- and cassava-extracted starch. Some syrup producers also use dry-mill process when they use grain like broken rice and millet, as the starch within the grains is difficult to separate. The grade of specialty syrups produced depends on the method used to hydrolyze the starch and how long the hydrolysis is permitted.

Different grades vary by physical and functional properties, such as sweetness, compressibility, and viscosity. The extent of starch hydrolysis is characterized by DE determination. Maltodextrin and 32, 42, and 60 DE glucose syrups are widely used specialty syrups used in food and confectionery industries (Tharanathan 2005). These syrups are produced through enzymatic as well as acid hydrolysis methods (Underkofler et al. 1965; Aiyer 2005).

Maltodextrin Production

Maltodextrin $[(C_6H_{10}O_5)_nH_2O]$ is a polymer of saccharides, nutritive and not sweet, and consists of glucose units primarily linked by α-1,4-glucosidic bonds, with DE (dextrose equivalent) values between 8 % and 30 % (Storz and Steffens 2004; Dokic-Baucal et al. 2004; Gibiński 2008; Muntean et al. 2010). Maltodextrin is more soluble in water than native starches and is cheaper than other major edible hydrocolloids, and its solutions have a bland flavor and smooth texture in the mouth (Dokic-Baucal et al. 2004). It is used for multifaceted purposes in food products, including bulking, caking resistance, texture and body improvement, film formation, binding of flavor and fat, serving as an oxygen barrier, imparting a surface sheen, aiding dispersion and solubility, increasing soluble solids, crystallization inhibition, controlling the freezing point, fillings, and as product extenders. Maltodextrin can be used as an anticaking agent and suitable filler in the production of spray-dried foods (Chronakis 1998; Setser and Racette 1992; Alexander 1992).

Maltodextrin is produced by enzymatic as well as acid hydrolysis methods (Wang and Wang 2000). The enzymatic method has distinct advantages, circumventing the need to remove salts formed during acid neutralization, operating in a wider pH range and at lower temperatures (with obvious energy savings), with easier process control (Haki and Rakshit 2003). It is difficult to produce syrup or powder of less than 30 % DE through acid hydrolysis process due to the formation of crystalline-stable nondegradable starch. The differences in structure of maltodextrins derived from different starchy materials also determine their physicochemical properties, and the characterization of maltodextrins and source identification assume importance to match each application (Wang and Wang 2000; Moore et al. 2005).

The traditional industrial method of maltodextrin production involves conventional high-temperature liquefaction of whole rice grain flour, corn, or wheat starch using thermostable α-amylase (Fig. 11.8) to produce 8–10 % DE liquefact, followed by saccharification using maltogenic enzymes such as acid α-amylase, beta-amylase, pullulanase, and glucoamylase to achieve a DE level of 14–30 % maltodextrin with specific properties as desired in the current market. Maltodextrins of 20–35 % DE are widely produced in powder as well as syrup forms with more than 78 % Bx. (Brix).

Various commercially available thermostable α-amylases such as SPEZYME® FRED, CLEARFLOW® AA, and SPEZYME® ALPHA from DuPont are used by the starch industries in the liquefaction process, followed by β-amylases for manufacturing of a variety of maltodextrin syrups or powder. The beta-amylases used are OPTIMALT® BBA (produced from barley); acid fungal amylase, CLERASE® L (produced by a selected strain of *Aspergillus* oryzae var); pullulanase, OPTIMAX® L 1000 (produced from controlled fermentation of a genetically modified strain of *Bacillus licheniformis*); and glucoamylase, GA-L NEW (produced from controlled fermentation of a selected strain of *Aspergillus niger*).

Specialty Glucose Syrup

Currently, glucose syrup produced by acid hydrolysis of starch of 30 % and 42 % DE is widely used in many industries because of its special composition, although enzyme hydrolysis is becoming popular and common in the starch sugar industry. The industrial process of acid hydrolysis of starch at high pressures uses hydrochloric acid as a catalyst. After achieving

Table 11.1 Specialty glucose syrup production by acid and enzymatic hydrolysis

		Degree of polymerization (% DP)				
Processes	Syrup types	DP1	DP2	DP3	DP4	DP4+
Acid hydrolysis	DE30	9.50	9.00	8.50	8.00	65.00
	DE42	17.50	14.00	11.50	10.00	47.00
	DE55	30.00	18.00	13.50	9.50	29.00
Enzyme hydrolysis	DE34	9.404	9.029	15.17	11.41	54.99
	DE43	17.58	14.46	20.49	10.56	36.91
	DE54	31.49	13.42	16.87	6.618	31.61

30–42 % DE, the starch hydrolysate is neutralized to a pH of 4.6–4.8 using sodium carbonate which also precipitates out the lipids and proteins (Howling 1992). These impurities are removed through skimming or centrifugation. Residual solid impurities of fat and protein flakes are removed from the acid hydrolysate liquor by sieving it through a series of pressure cloth filters. The remaining impurities such as color precursors, protein or protein hydrolysate, peptides, amino acids, and undesired flavors are eliminated by refining with powdered or granular activated carbon techniques followed by ion-exchange resins (Cotillon 1992). After refining, the syrup liquor is adjusted to ~80 % dry solids and cooled. The final product called glucose syrup is packed in containers. Acid hydrolysis has the disadvantage that acid is corrosive and challenging to handle. Furthermore, it produces undesirable by-products such as 5-hydroxymethylfurfural (http://umpir.ump.edu.my/863/1/Siti_Nor_Shadila_Alias.pdf) and tints the product with a brownish color if dextrose reversion is carried out for extended acid hydrolysis reaction to produce 55 % DE syrup (Bozell 2001). Starch processing by acid hydrolysis has additional disadvantages including poor product quality, environmental pollution, complex process handling, high capital expenditure periodically due to corrosion of the process tank, high steam pressure, and lower product yields. However, industries continue to use acid hydrolysis to achieve the desired DP (degree of polymerization, number of dextrose molecules linked together) profile (Table 11.1), to meet the specific property profile needs of the confectionary industry engaged in chocolate or candy production.

DuPont Genencor Science has powered the development of enzymatic methods to produce 32 %, 42 %, and 60 % DE glucose syrup with the desired DE and DP profiles that were best produced by the acid hydrolysis method (Table 11.1). In the enzymatic processes, any of the DuPont's α-amylases, as discussed earlier, are used in liquefaction to produce 12–14 % DE liquefact, followed by two enzymatic saccharification reactions with acid α-amylases, GC 626 (derived from a strain of *Aspergillus kawachii* expressed in *Trichoderma reesei*) and glucoamylase marketed as GA-L NEW (produced by controlled fermentation using a selected strain of *Aspergillus niger*). Various dosages of this combination are used to achieve the desired DE and DP profiles formerly produced using acid hydrolysis. This technology has been found successful with any source of starch or starchy materials. Enzymatic technology is environmentally friendly, with much less acid water produced; it also does not cause corrosion of the process tanks, produces high yields, and is amenable for control of DE and DP profile targeted through inactivating the enzymes by reducing pH and/or increasing the temperature.

High-Glucose Syrup (>95 DE) Production

High-glucose syrup is widely used as a raw material in sorbitol production through chemical catalytic hydrogenation process (Ahmed et al. 2009); high-fructose syrup production through enzymatic glucose isomerase process (Lee et al. 1990); biochemicals such as lactic acid, lysine, and MSG fermentation (Duan 2009); and ethanol production through yeast fermentation process (Dombek and Ingram 1987). It is also used for dextrose monohydrate production through crystallization of >95 DE glucose syrup (Hull 2009). The process of crystallization allows only dextrose to crystallize leaving behind other sugars dissolved in mother liquor. The dextrose crystals are recovered and washed using a centrifuge and dried to produce a very pure product. Dextrose is less sweet than sucrose, which is useful in food processing industries where less sweetness is desired. A >95 DE is the ultimate product of starch hydrolysis using an ideal enzymatic process. A liquefact with 12–14 % DE is suitable for saccharification to produce such syrup (Hull 2009). Liquefaction is ideally a continuous process, whereas saccharification is most often conducted as a batch process. Saccharification is followed by a treatment with a blend of various concentrations of bacterial pullulanase and fungal glucoamylase marketed as products marketed under OPTIMAX® brand. These enzymes accelerate the reaction and can produce higher glucose yields (>95 %) at 38 % DS (dry solids). Saccharifying at higher solid levels substantially reduces evaporation costs at the plant level, in addition to enabling increased throughput without loss in yield to meet seasonal demands. These enzymes produce a better substrate for isomerization into fructose or for hydrogenation into sorbitol. The enzymes also reduce refining costs and permit saccharification at high concentrations of dissolved solids.

High-Fructose Syrup

High-fructose syrup is also called as glucose-fructose syrup. A variety of high-fructose syrups such as HFCS-42, HFCS-55, and HFCS-90 are produced for various applications (Parker et al. 2010). HFCS-42 has approximately 42 % fructose and 53 % glucose and is mostly used in beverages, processed foods, cereals, and baked goods; HFCS-55 has approximately 55 % fructose and 42 % glucose mostly used in soft drinks; and HFCS-90 contains approximately 90 % fructose and 10 % glucose, used in specialty applications (Marshall and Kooi 1957). Among all three, HFCS-42 and HFCS-55 are most widely used to replace sugar because of having more than 40 % of sweetening value relative to the caloric value. HFCS is so sweet that it is cost-effective for companies to use small quantities of HCFS in place of other more expensive sweeteners or flavorings. High-fructose-containing syrups are prepared by enzymatic isomerization of dextrose with glucose isomerase (Bhosale et al. 1996). The starch is first converted into dextrose by enzymatic liquefaction and saccharification. The dextrose syrup feed is processed through immobilized glucose isomerase (GI) columns in a continuous process to produce HFCS-42 (Gromada et al. 2008; Illanes et al. 1992). Syrup with 55 % fructose is blended using enriched fructose syrup with fructose of more than 90 % together with 42 % fructose syrup. More than 90 % concentrate is produced using simulated moving bed chromatography (Ching and Ruthven 1985).

Immobilization of glucose isomerase (IGI) offers several advantages for industrial and biotechnological applications, including repeated use, ease of separation of reaction products from the biocatalyst, improvement of enzyme stability, continuous operation in a packed-bed reactor, and ready alteration of the properties of the enzyme (Seyhan and Dilek 2008). GI obtained from different sources such as *Flavobacterium*, *Bacillus*, and some *Streptomyces* and *Arthrobacter* species is immobilized on different support materials such as DEAE cellulose (Chen and Anderson 1979; Huitron and Limon-Lason 1978), polyacrylamide gel (Demirel et al. 2006; Strandberg and Smiley 1971), and alginate beads (Rhimi et al. 2007).

GENSWEET™ IGI is an immobilized glucose isomerase [EC 5.3.1.5, D-xylose ketol isomerase] from DuPont, produced by the controlled fermentation of a selected strain of *Streptomyces rubiginosus*. This enzyme is cross-linked using polyethylenimine and glutaraldehyde, and granular particles are produced by extrusion/marumerization technology, followed by drying. This immobilized enzyme offers unique physical and functional properties primarily designed to offer predictable, consistent performance and tolerance to process variations, including variation in substrate quality. This enzyme requires Mg^{2+} (25–100 ppm) and metabisulfite (50–175 ppm) as an activator. Prior to loading this enzyme into the column, it requires a hydration process, achieved by suspending the GENSWEET™ IGI enzyme in isofeed (substrate) at pH 7.6–8.0 and 54–60 °C at a ratio of 1 kg dry enzyme per 1.5 gal of syrup and mildly agitating for 1–2 h. The hydrated enzyme is transferred to a column, preferably with a diaphragm pump to avoid excess abrasion of the particles. The column upflow is initiated with isofeed, gradually increasing the feed rate to about 0.9–1.0 bed volumes per hour over a 2 h period. The column upflow is continued at this rate for 2 h or until froth generated by excessive agitation or pumping is removed. The upflow is gradually decreased to zero over 30 min. Then the column downflow is initiated with isofeed by gradually increasing flow over 4 h from zero to desired operational flow. The benefits of using GENSWEET™ IGI in this process include well-controlled and consistent performance, more rapid hydration and less discoloration of the glucose produced, flexibility of plant operation, less reduction in upflow pressure, and reduced channeling.

Maltose Syrup Production

Maltose is a naturally occurring disaccharide, consisting of two glucosyl residues linked by an α-1,4-glucosidic linkage, and is the smallest in the family of oligosaccharides. It is the main component of maltosugar syrup (Sugimoto 1977). Maltose is the main component of high-maltose syrup. The syrup is classified based on the content of maltose. Maltose syrup, containing different levels of maltose, can be produced from liquefied starch using enzymatic processes. Maltose syrups are produced on a large scale in syrup, powder, and crystal form with several grades of purity. Various maltose syrups are drawing considerable interest for commercial applications, because it is less susceptible to crystallization and is relatively nonhygroscopic. Commercial applications for maltose syrups are possible in the brewing, baking, soft drink, canning, confectionery, and other food and beverage industries. Ultrapure maltose is used as an intravenous nutrient. Catalytic reduction of maltose results in maltitol, a low-calorie sweetener. Recently, high-maltose syrup has become a key raw material for industrial production of a new class of sugars, i.e., isomalto-oligosaccharides (IMO) (Duan et al. 2011b). These sugars are receiving increased attention as health (Bifidobacterium growth factors) and functional food ingredients.

Corn, potato, sweet potato, and cassava starches as well as whole rice flour are known raw materials for maltose manufacture. In enzymatic manufacturing of syrups, the first step of starch liquefaction is common to all, but it is important to get it right in order to achieve the right DE and DP profile for the next saccharification enzyme which is used to manufacture a variety of maltose syrups or specialty syrups (http://www.agfdt.de/loads/st07/gangabb.pdf). This is because of the variety of maltogenic enzymes used in saccharification based on the target sugar composition desired. To produce 40–50 % maltose syrup, using a single maltogenic enzyme, β-amylase or fungal α-amylase, the liquefact DE should be in the range of 12–14 %. Higher concentration maltose syrup (50–60 %) can be produced either by using β-amylase alone or with pullulanase with a liquefact DE in the range of 10–12 %. High concentration maltose syrup (>80 %) is produced using β-amylase, acid α-amylase, and pullulanase with a liquefact DE of 4–5 %. DuPont has marketed several liquefaction enzymes such as discussed earlier to achieve desired DE liquefacts that can be saccharified with the same maltogenic saccharification enzymes, such as β-amylase, OPTIMALT® BBA; acid fungal amylase, CLERASE® L; and pullulanase, OPTIMAX® L 1000, for producing a range of maltose syrups.

Functional Oligosaccharides

Oligosaccharides are an important group of polymeric carbohydrates with 2–10° of polymerization that are found either free or in combined forms in all living organisms. Structurally, oligosaccharides are composed of 2–10 monosaccharide residues linked by glycosidic bonds that are readily hydrolyzed by acids or enzymes to release the constituent monosaccharides (Nakakuki 1993). Functional oligosaccharides have recently received more attention in recent years because of their role in the microecology of intestinal flora and their potential application in health sector (Zivkovic and Barile 2011; Tuohy et al. 2005; Qiang et al. 2009). The major functional oligosaccharides are xylo-oligosaccharides, fructo-oligosaccharides, and isomalto-oligosaccharides (Lai et al. 2011; Oku and Nakamura 2003). Xylo-oligosaccharide (XO) is a kind of functional oligosaccharide that is considered a safe health additive. Xylo-oligosaccharide is composed of 2-7 xylopyranoses linked with β-1,4-glycosidic bonds. The xylopyranoses include xylobiose, trisaccharide, and other oligosaccharides (Zhou et al. 2009). Fructo-oligosaccharides are produced from inulin by endoinulinases, used as potent prebiotics and dietary fibers, and also possess other beneficial functionalities (Guiraud et al. 1987; Sangeetha et al. 2005; Singh and Singh 2010). Furthermore, the completely hydrolyzed product, i.e., fructose, produced with exoinulinase is emerging as a safe sweetener in the food industry. Recently, inulin has emerged as a promising substrate for the enzymatic synthesis of fructo-oligosaccharides and high-fructose syrup (Singh and Singh 2010).

Isomalto-Oligosaccharides

Isomalto-oligosaccharides also known as IMO contain 40 % α-1,6-glucosidic linkages. IMOs include isomaltose, panose, isomaltotriose, and higher branched sugars. Isomalto-oligosaccharides (IMOs) are receiving growing attention due to their biological functions/role as prebiotics that can enhance the growth of *Bifidobacteria* in the large intestine of humans and animals and reduce the cariogenic effect (causing dental caries) of sucrose (Kaneko et al. 1995). Isomalto-oligosaccharides have been produced by using the transglucosylation activity of enzymes obtained from microorganisms (Kuriki et al. 1993). The IMO-producing enzyme that catalyzes the transglucosylation of maltose to form isomalto-oligosaccharides is called transglucosidase. IMO is commercially one of the most important polysaccharide categories with an estimated market demand of about 200,000 t per year worldwide (van Dokkum et al. 1999). The conventional method of producing IMO from starch involves a three-step enzymatic process, liquefaction using thermostable α-amylase, followed by saccharification using pullulanase and beta-amylase to produce maltose, which in the third step is used as the substrate for the transglucosidase enzyme to produce IMO (Pan and Lee 2005). The transglucosidase catalyzes both hydrolytic and transfer reactions on incubation with α-D-gluco-oligosaccharides. Transfer occurs most frequently to HO-6 (hydroxyl group 6 of the glucose molecule), producing isomaltose from D-glucose and panose from maltose. The enzyme can also transfer to HO-2 or HO-3 of D-glucose to form kojibiose or nigerose or back to HO-4 to form maltose (McCleary et al. 1989). The action on maltose produces equimolar concentration of panose and glucose. As a result of transglucosidase reactions, the malto-oligosaccharides are converted to isomalto-oligosaccharides, a new class of polysaccharides containing high proportions of glucosyl residues linked by α-D-1,6 linkages from the nonreducing end. Being a non-fermentable sugar, IMO is widely used as a bulking agent in animal feed to increased body weight, in dental care due to anti-cariogenic activity, and in baking due to anti-spoiling (preventing staleness) properties. DuPont has commercialized the transglucosidase enzyme in the form of purified D-glucosyltransferase (transglucosidase, EC 2.4.1.24) free from glucoamylase activity, produced through controlled fermentation using a selected strain of *Aspergillus* (Li et al. 2005). In molasses, non-fermentable sugars including raffinose and stachyose are converted to sucrose, galactose, glucose, and fructose, which can subsequently be fermented into alcohol.

Maltotetraose Syrup

Maltotetraose, a linear tetramer of α-D-glucose, has many uses in the food and pharmaceutical industries because of its uniquely low sweetness (equivalent to 20 % sucrose), resistance to retrogradation, retention of desired levels of moisture in foods, and high viscosity compared to sucrose, thus improving the texture of processed foods. It reduces sweetness without affecting the inherent taste and flavor of foods. G4 syrup (high-maltotetraose syrup) exhibits a lower rate of Millard reaction as it has less glucose and maltose content. This syrup does not lower the freezing point of water as much as sucrose or high-fructose syrup. Hence, it can be used to alter the freezing points of frozen foods. G4 syrup imparts gloss and can be used in industries such as a paper sizer (Aiyer 2005). In addition to nutritional and taste properties, its antimicrobial property was also discovered. Feeding a maltotetraose-rich corn syrup inhibits the growth of intestinal putrefactive bacteria such as *C. perfringens* and Enterobacteriaceae (Kimura and Nakakuki 1990). Commercial G4-forming amylase produced by *Pseudomonas saccharophila* was expressed in *Bacillus licheniformis* that performs efficiently in the presence of pullulanase at 60–65 °C and pH 5.0–5.5 to produce >45 % DP4 G4 syrups (Duan et al. 2010a).

Summary and Conclusions

Industrial enzymes provide green and sustainable solutions for various starch industries in the midst of growing environmental anxiety. Commercialization of industrial enzymes calls for continuous technological innovation to identify and characterize new catalysts from natural sources as well as directed evolution with optimal performance for selected applications, further modification for enhanced performance, and increased expression in suitable model systems. The enzyme bio-industry sector has played a significant role in the current commercial status of biotechnology at global scale. The future will witness more novel applications of industrial enzymes in far more arenas than anticipated today.

The global market for industrial enzymes will continue to expand as new uses for enzymes are discovered in the chemical industry at large. Several factors will contribute to this growth: (1) protein engineering and direct evaluation with high-throughput screening, (2) improved knowledge of enzyme mechanisms, (3) reduction in the production costs of industrial enzymes, and (4) improved means for enzyme immobilization and bioprocess engineering.

References

Aehle W (2007) Enzymes in industry: production and applications, 3rd edn. Wiley-VCH Verlag GmbH and Co. KGaA, Weinheim

Aggarwal NK, Nigam P, Singh D et al (2001) Process optimization for the production of sugar for the bioethanol industry from sorghum, a nonconventional source of starch. World J Microbiol Biotechnol 17:411–415

Ahmed MJ, Khadom AA et al (2009) Optimization hydrogenation process of D-glucose to D-sorbitol over Raney nickel catalyst. Eur J Sci Res 30:294–304

Aiyer PV (2005) Amylases and their applications. Afr J Biotechnol 4:1525–1529

Akin H, Brandam C, Meyer XM et al (2008) A model for pH determination during alcoholic fermentation of a grape must by *Saccharomyces cerevisiae*. Chem Eng Process 47:1986–1993

Alexander RJ (1992) Maltodextrins. Production, properties and applications. In: Schenk EW, Hebeda RE (eds) Starch hydrolysis products. VCH, New York

Baron M, Chamayou A, Marchioro L et al (2005) Radicalar probes to measure the action of energy on granular materials. Adv Powder Technol 16:199–211

Bellissimi E, Ingledew WM (2005) Analysis of commercially available active dry yeast used for industrial fuel ethanol production. J Am Soc Brew Chem 63:107–112

Bhosale SH, Rao MB, Deshpande VV (1996) Molecular and industrial aspects of glucose isomerase. Microbiol Mol Biol Rev 60:280–300

Blanchard PH (1992) Technology of corn wet milling and associated processes. Elsevier, Amsterdam, chaps. 1 and 3

Bornscheuer UT, Pohl M (2001) Improved biocatalysts by directed evolution and rational protein design. Curr Opin Chem Biol 5:137–143

Bothast RJ, Schlicher MA (2005) Biotechnological processes for conversion of corn into ethanol. Appl Microbiol Biotechnol 67:19–25

Böttcher D, Bornscheuer UT (2010) Protein engineering of microbial enzymes. Curr Opin Microbiol 13:274–282

Bozell JJ (2001) Chemicals and materials from renewable resources, ACS symposium series. American Chemical Society, Washington, DC

Breisha GZ (2010) Production of 16% ethanol from 35% sucrose. Biomass Bioenerg 34:1243–1249

Buchholz K, Kasche V, Bornscheuer UT (2005) Biocatalysts and enzyme technology. Wiley-VCH, Weinheim

Bull AT, Goodfellow M, Slater JH (1992) Biodiversity as a source of innovation in biotechnology. Annu Rev Microbiol 46:219–252

Bull AT, Bunch AW, Robinson GK (1999) Biocatalyst for clean industrial products and processes. Curr Opin Microbiol 2:246–251

Bvochora JM, Read JS, Zvauya R (2000) Application of very high gravity technology to the cofermentation of sweet stem sorghum juice and sorghum grain. Ind Crop Prod 11:11–17

Cauvain SP, Young LS (2001) Baking problems solved. Woodhead Publishing Ltd., Cambridge/England, pp 25–26

Cawley RW, Mitchell TA (1968) Inhibition of wheat α-amylase by bran phytic acid. J Sci Food Agric 19:106–108

Chandel AK, Rudravaram R, Rao LV et al (2007) Industrial enzymes in bioindustrial sector development: an Indian perspective. J Comm Biotechnol 13:283–291

Chaplin MF, Bucke C (1990) The large-scale use of enzymes in solution. In: Chaplin MF, Bucke C (eds) Enzyme technology. Cambridge University Press, Cambridge, UK

Chen WP, Anderson AW (1979) Purification, immobilization and some properties of glucose isomerase from *Streptomyces flavogriseus*. Appl Environ Microbiol 38:1111–1119

Ching CB, Ruthven DM (1985) An experimental study of counter-current adsorption system. Chem Eng Sci 40:877–891

Christophersen C, Andersen E, Jakobsen TS et al (1997) Xylanase in wheat separation. Starch-Stärke 49:5–12

Chronakis LS (1998) On the molecular characteristics, compositional properties and structural-functional mechanisms of maltodextrins. Crit Rev Food Sci Nutr 38:599–637

Cotillon M (1992) Refining: carbon treatment, reactive precoats and ion exchange. In: Schenck F, Hebeda R (eds) Starch hydrolysis products. VCH, New York

Crabb WD, Shetty JK (1999) Commodity scale production of sugars from starches. Curr Opin Microbiol 2:252–256

Daiuto E, Cereda M, Sarmento S et al (2005) Effects of extraction methods on yam (*Dioscorea alata*) starch characteristics. Starch-Stärke 57:153–160

De Cordt S, Hendrickx M, Maesmans G et al (1994) The influence of polyalcohols and carbohydrates on the thermostability of α-amylase. Biotechnol Bioeng 43:107–114

Demirel G, Ozcetin G, Sahin F et al (2006) Semiinterpenetrating polymer networks (IPNs) for entrapment of glucose isomerase. React Funct Polym 66:389–394

DeSantis G, Jones JB (1999) Chemical modification of enzymes for enhanced functionality. Curr Opin Microbiol 10:324–330

Devantier R, Pedersen S, Olsson L (2005) Characterization of very high gravity ethanol fermentation of corn mash. Effect of glucoamylase dosage, pre-saccharification and yeast strain. Appl Microbiol Biotechnol 68:622–629

Dokic-Baucal L, Dokic P, Jakovljevic J (2004) Influence of different maltodextrins on properties of O/W emulsions. Food Hydrocoll 18:233–239

Dombek KM, Ingram LO (1987) Ethanol production during batch fermentation with *Saccharomyces cerevisiae*: changes in glycolytic enzymes and internal pH. Appl Environ Microbiol 53:1286–1291

Duan G (2009) Impact of the industrial enzyme progress on the production of chemicals. Chin J Biotechnol 25:1808–1818

Duan G, Qian Y, Rafael SF, Shetty J (2010a) Improved production of maltotetraose syrup using a *Pseudomonas saccharophila* moltoteraohydrolase variant and a debranching enzyme. WO/2010/132157

Duan G, Qian Y, Shetty J et al (2010b) Starch separation process. PCT WO2010/129648

Duan G, Dunn-Coleman N, Lantero O et al (2011a) Acid fungal protease in fermentation of insoluble starch substrates. US8075694 B2, December 13 2011

Duan G, Shetty J, Vadakoot J et al (2011b) Grain composition containing prebiotic iso-maltooligosaccharides and method of making and using same. US Patent 7,993,689 B2

Dzogbefia VP, Buamah R, Oldham JH (1999) The controlled fermentation of cocoa (*Theobroma cacao* L): enzymatic process and associated physicochemical changes in cocoa sweatings. Food Biotechnol 13:1–12

Englyst HN, Kingman SM, Cummings JH (1992) Classification and measurement of nutritionally important starch fractions. Eur J Clin Nutr 46:33–50

FAO (2002) Partnership formed to improve cassava, staple food for 600 million people. Food and Agriculture Organization of the United Nations, Rome, http://www.fao.org/english/newsroom/news/2002/10541-en.html

Fernando S, Adhikari S, Chandrapal C et al (2006) Biorefineries: current status, challenges, and future direction. Energy Fuel 20:1727–1737

Gantelet H, Duchiron F (1999) A new pullulanase from a hyperthermophilic archaeon for starch hydrolysis. Biotechnol Lett 21:71–75

Gibiński M (2008) Production of oat hydrolysates with a low degree of starch saccharification. Pol J Food Nutr Sci 58:295–300

Gibreel JR, Sandercock J, Lan LA et al (2009) Fermentation of barley by using *Saccharomyces cerevisiae*: examination of barley as a feedstocks for bioethanol production and value-added products. Appl Environ Microbiol 75:1363–1372

Göğüş F, Bozkurt H, Eren S (1998) Kinetics of Maillard reactions between the major sugars and amino acids of boiled grape juice. LWT- Food Sci Technol 31:196–200

Gohel V, Duan G (2012a) Conventional process for ethanol production from Indian broken rice and pearl millet. Bioprocess Biosyst Eng. doi:10.1007/s00449-012-0717-1

Gohel V, Duan G (2012b) No-cook process for ethanol production using Indian broken rice and pearl millet. Int J Microbiol. doi:10.1155/2012/680232

Gohel V, Singh A, Maisuria V, Phadnis A, Chhatpar HS (2006) Bioprospecting and antifungal potential of chitinolytic microorganisms. Afr J Biotechnol 5:54–72

Gopinathan MC, Sudhakaran R (2009) Biofuels: opportunities and challenges in India. In Vitro Cell Dev Biol-Plant 45:350–371

Gough S, Brady D, Nigam P et al (1997) Production of ethanol from molasses at 45°C using alginate-immobilized Kluyveromyces marxianus IMB3. Bioprocess Eng 16:389–392

Grafelman DD, Meagher MM (1995) Liquefaction of starch by a single-screw extruder and post-extrusion static-mixer reactor. J Food Eng 24:529–542

Gromada A, Fiedurek J, Szczodrak J (2008) Isoglucose production from raw starchy materials based on a two stage enzymatic system. Pol J Microbiol 57:141–148

Guiraud JP, Bourgi J, Stervinou M et al (1987) Isolation of a respiratory-deficient *Kluyveromyces fragilis* mutant for the production of ethanol from Jerusalem artichoke. Biotechnol Bioeng 29:850–858

Guo ZP, Zhang L, Ding ZY et al (2010) Improving the performance of industrial ethanol-producing yeast by expressing the aspartyl protease on the cell surface. Yeast 27:1017–1027

Gupta R, Beg QK, Lorentz P (2002) Bacterial alkaline proteases: molecular approaches and industrial application. Appl Microbiol Biotechnol 59:15–32

Haki GD, Rakshit SK (2003) Developments in industrially important thermostable enzymes: a review. Bioresour Technol 89:17–34

Herrera S (2004) Industrial biotechnology- a chance at redemption. Nat Biotechnol 22:671–675

Howling D (1992) Glucose syrup: production, properties and applications. In: Schenck F, Hebeda R (eds) Starch hydrolysis products. VCH, New York

Huitron C, Limon-Lason J (1978) Immobilization of glucose isomerase to ion change materials. Biotechnol Bioeng 20:1377–1391

Hull P (2009) Glucose syrups: technology and applications. Wiley, West Sussex

Illanes A, Zúñiga ME, Contreras S et al (1992) Reactor design for the enzymatic isomerization of glucose to fructose. Bioprocess Biosyst Eng 7:5199–5204

Iyer PV, Ananthanarayan L (2008) Enzyme stability and stabilization—aqueous and non-aqueous environment. Process Biochem 43:1019–1032

Jenkins PJ, Donald AM (1998) Gelatinisation of starch: a combined Saxs/waxs/dsc and Sans study. Carbohydr Res 308:133–147

Johnston DB, Singh V (2001) Use of protease to reduce steep time and SO2 requirements in a corn wet-milling process. Cereal Chem 78:405–411

Johnston DB, Singh V (2004) Enzymatic milling of corn: optimization of soaking, grinding, and enzyme incubation steps. Cereal Chem 81:626–632

Kaneko T, Yokoyama A, Suzuki M (1995) Digestibility characteristics of isomaltooligosaccharides in comparison with several saccharides using the rat jejunum loop method. Biosci Biotechnol Biochem 59:1190–1194

Karuppiah R, Peschel A, Grossmann IE et al (2008) Energy optimization of an ethanol plant. AICHE J 54:1499–1525

Kim S, Dale BE (2004) Global potential bioethanol production from wasted crops and crop residues. Biomass Bioenerg 26:361–375

Kim TH, Taylor F, Hicks KB (2008) Bioethanol production from barley hull using SAA (soaking in aqueous ammonia) pretreatment. Bioresour Technol 99:5695–5702

Kimura T, Nakakuki T (1990) Maltotetraose, a new saccharide of tertiary property. Starch-Stärke 42:151–157

Kitamoto K, Oda K, Gomi K et al (1991) Genetic engineering of a sake yeast producing no urea by successive disruption of arginase gene. Appl Environ Microbiol 57:306–306

Klibanov AM (1983) Stabilization of enzymes against thermal inactivation. Adv Appl Microbiol 29:1–28

Kłosowski G, Mikulski D, Czupryński B et al (2010) Characterisation of fermentation of high-gravity maize mashes with the application of pullulanase, proteolytic enzymes and enzymes degrading non-starch polysaccharides. J Biosci Bioeng 109: 466–471

Kuriki T, Yanase M, Takata H et al (1993) A new way of producing isomalto-oligosaccharide syrup by using the transglycosylation reaction of neopullulanase. Appl Environ Microbiol 59:953–959

Lai P, Chi-Juan S, Chiun-CR W (2011) Effects of oligosaccharides on phase transition temperatures and rheological characteristics of waxy rice starch dispersion. J Sci Food Agric. doi:10.1002/jsfa.4712

Lee C, Saha BC, Zeikus JG (1990) Characterization of thermoanaerobacter glucose isomerase in relation to saccharidase synthesis and development of single-step processes for sweetener production. Appl Environ Microbiol 56:2895–2901

Li F, Vadakoot J, Duan G, Shetty JK (2005) Grain composi-tions containing prebiotic isomalto-oligosaccharides and methods of making and using same. Genencor International Inc. EP1601699 A2. Accessed 7 Dec 2005

Linko P, Hakulin S, Linko YY (1983) Extrusion cooking of barley starch for the production of glucose syrup and ethanol. J Cereal Sci 1:275–284

Liu YH, Lu FP, Li Y et al (2008) Characterisation of mutagenised acid-resistant alpha-amylase expressed in *Bacillus subtilis* WB600. Appl Microbiol Biotechnol 78:85–94

Mann JG, Liu YA (1999) Industrial water reuse and wastewater minimization. McGraw-Hill, New York

Marrs B, Delagrave S, Murphy D (1999) Novel approaches for discovering industrial enzymes. Curr Opin Microbiol 2:241–245

Marshall RO, Kooi ER (1957) The enzymatic conversion of d-glucose to d-fructose. Science 125:648–649

Martin CH, Nielsen DR, Solomon KV et al (2009) Synthetic metabolism: engineering biology at the protein and pathway scales. Chem Biol 16:277–286

Maurer KH (2004) Detergent proteases. Curr Opin Biotechnol 15:330–334

McCleary BV, Gibson TS, Sheehan H et al (1989) Purification, properties, and industrial significance of transglucosidase from *Aspergillus niger*. Carbohydr Res 185:147–162

Modilal MRD, Shenoi RR, Sebastian B et al (2011) Isolation, purification and characterization proteases from vegetable waste. Int J Pharm Life Sci 1:1–11

Mojović L, Nikolić S, Rakin M et al (2006) Production of bioethanol from corn meal hydrolyzates. Fuel 85:1750–1755

Montesinos T, Navarro JM (2000) Production of alcohol from raw wheat flour by Amyloglucosidase and *Saccharomyces cerevisiae*. Enzyme Microb Technol 27:362–370

Moore GRP, Luciana RC, Amante ER (2005) Cassava and corn starch in maltodextrin production. Quim Nova 28:596–600

Muntean E, Muntean N, Paizs C et al (2010) Evaluation of effective parameters on enzymatic hydrolysis at low starch concentrations. Bull UASVM Agric:322–327. doi:10.1590/S0100-40422005000400008

Mussatto SI, Dragone G, Guimarães PMR et al (2010) Technological trends, global market, and challenges of bio-ethanol production. Biotechnol Adv 28:817–830

Nakakuki T (1993) Oligosaccharides: production, properties and applications, vol 3, Japanese technology reviews. Gordon and Breach, Switzerland

Neves MA, Kimura T, Shimizu N et al (2006) Production of alcohol by simultaneous saccharification and fermentation of low-grade wheat flour. Braz Arch Biol Technol 49:481–490

Nikolić S, Mojović L, Rakin M et al (2010) Ultrasound-assisted production of bioethanol by simultaneous saccharification and fermentation of corn meal. Food Chem 122:216–222

Oberoi R, Beg QK, Puri S et al (2001) Characterization and wash performance analysis of an SDS-stable alkaline protease from a *Bacillus* sp. World J Microbiol Biotechnol 17:493–497

Oku T, Nakamura S (2003) Comparison of digestibility and breath hydrogen gas excretion of fructo-oligosaccharide, galactosyl-sucrose, and isomalto-oligosaccharide in healthy human subjects. Eur J Clin Nutr 57:1150–1156

Olempska-Beer ZS, Merker RI, Ditto MD et al (2006) Food-processing enzymes from recombinant microorganisms – a review. Regul Toxicol Pharm 45:144–158

Ough CS, Crowell EA, Gutlove BR (1988) Carbamyl compound reactions with ethanol. Am J Enol Vitic 39:303–307

Pan YC, Lee WC (2005) Production of high-purity isomalto-oligosaccharides syrup by the enzymatic conversion of transglucosidase and fermentation of yeast cells. Biotechnol Bioeng 89:797–804

Parker K, Salas M, Nwosu VC (2010) High fructose corn syrup: production, uses and public health concerns. Biotechnol Mol Biol Rev 5:71–78

Pohit S, Biswas PK, Kumar R et al (2009) International experiences of ethanol as transport fuel: policy implications for India. Energy Policy 37:4540–4548

Polaina J, MacCabe AP (2007) Industrial enzymes, structure, functions and applications. Springer, Dordrecht

Pretorius IS (2000) Tailoring wine yeast for the new millennium: novel approaches to the ancient art of winemaking. Yeast 16:675–729

Qiang X, YongLie C, QianBing W (2009) Health benefit application of functional oligosaccharides. Carbohydr Polym 77:435–441

Rahman MM, Rakshit KS (2003) Improved extractability of sweet potato starch using commercial hydrolytic enzymes. American society agricultural biological engineers publication paper no. 036070. ASAAE annual meeting, Michigan

Ramírez EC, Johnston DB, McAloon AJ et al (2009) Enzymatic corn wet milling: engineering process and cost model. Biotechnol Biofuel 2:2. doi:10.1186/1754-6834-2-2

Rhimi M, Messaud EB, Borgi MA et al (2007) Co-expression of L-arabinose isomerase and D-glucose isomerase in *E. coli* and development of an efficient process producing simultaneously D-tagatose and D-fructose. Enzyme Microb Technol 40:1531–1537

Rong Y, Sillick M, Gregson CM (2009) Determination of dextrose equivalent value and number average molecular weight of maltodextrin by osmometry. J Food Sci 74:C33–C40

Rosendal P, Nielsen BH, Lange NK (1979) Stability of bacterial alpha-amylase in the starch liquefaction process. Starch-Stärke 31:368–372

Roy I, Gupta MN (2004) Hydrolysis of starch by a mixture of glucoamylase and pullulanase entrapped individually in calcium alginate beads. Enzyme Microb Technol 34:26–32

Sangeetha PT, Ramesh MN, Prapulla SG (2005) Recent trends in microbial production, analysis, and applications of fructooligosaccharides. Trends Food Sci Technol 16:442–457

Sayaslan A (2004) Wet-milling of wheat flour: industrial processes and small-scale test methods LWT. Food Sci Technol 37:499–515

Setser CS, Racette WL (1992) Macromolecule replacers in food products. Crit Rev Food Sci Nutr 32:275–297

Seyhan S, Dilek TA (2008) Catalytic efficiency of immobilized glucose isomerase in isomerization of glucose to fructose. Food Chem 111:658–662

Shanavas S, Padmaja G, Moorthy SN et al (2011) Process optimization for bioethanol production from cassava starch using novel eco-friendly enzymes. Biomass Bioenerg 35:901–909

Shapouri HJ, Duffield J, McAloon A et al (2004) The 2001 net energy balance of corn-ethanol. In: Tumbleson M (ed) Corn: feedstock of the future (Corn utilization and technology conference proceedings). Corn Utilization and Technology, Indianapolis

Sharma V, Rausch KD, Tumbleson ME et al (2007) Comparison between granular starch hydrolyzing enzyme and conventional enzymes for ethanol production from maize starch with different amylase: amylopectin ratios. Starch-Stärke 59:549–556

Shetty J, Chotani G, Duan G et al (2007) Cassava as an alternative feedstocks in the production of renewable transportation fuel. Int Sugar J 109:3–11

Shetty J, Strohm BA, Singh V (2010) Phytase in enzymatic wet milling process. WO2010027846 A1. Accessed 11 May 2010

Singh V, Johnston DB (2002) Pasting properties and surface characteristics of starch obtained from an enzymatic corn wet-milling process. Cereal Chem 79:523–527

Singh V, Johnston D (2004) An enzymatic process for corn wet milling. Adv Food Nutr Res 48:151–171

Singh RS, Singh RP (2010) Production of fructooligosaccharides from inulin by endoinulinases and their prebiotic potential. Food Technol Biotechnol 48:435–450

Sivaramakrishnan S, Gangadharan D, Nampoothiri KM et al (2006) Alpha-amylase from microbial sources-an overview on recent developments. Food Technol Biotechnol 44:173–184

Sriroth K, Chollakup R, Chotineeranat S et al (2000) Processing of cassava waste for improved biomass utilization. Bioresour Technol 71:63–69

Steinke JD, Johnson LA (1991) Steeping maize in the presence of multiple enzymes. I. Static batch wise steeping. Cereal Chem 68:7–12

Storz E, Steffens K (2004) Feasibility study for determination of the dextrose equivalent (DE) of starch hydrolysis products with near-infrared spectroscopy (NIRS). Starch-Stärke 56:58–62

Strandberg GW, Smiley KL (1971) Free and immobilized glucose isomerase from *Streptomyces* phaechromogenes. Appl Microbiol 21:588–593

Sugden D (1997) Wheat starch and gluten manufacturing. World Grain 3:20–23

Sugimoto K (1977) Denpunkagaku handbook. Asakura, Tokyo, pp 450–461

Suresh K, Sree NK, Rao LV (1999) Utilization of damaged sorghum and rice grains for ethanol production by simultaneous saccharification and fermentation. Bioresour Technol 68:301–304

Sutherland JD (2000) Evolutionary optimisation of enzymes. Curr Opin Chem Biol 4:263–269

Szulczyk KR, McCarl BA, Cornforth G (2010) Market penetration of ethanol. Renew Sustain Energy Rev 14:394–403

Tester RF, Karkalas J, Qi X (2004) Starch structure and digestibility enzyme-substrate relationship. World's Poult Sci J 60:186–195

Tharanathan RN (2005) Starch – value addition by modification. Crit Rev Food Sci Nutr 45:371–384

Thomas KC, Ingledew WM (1990) Fuel alcohol production: effects of free amino nitrogen on fermentation of very-high-gravity wheat mashes. Appl Microbiol Biotechnol 56:2046–2050

Tuohy KM, Rouzaud GCM, Brück WM, Gibson GR (2005) Modulation of the human gut microflora towards improved health using prebiotics – assessment of efficacy. Curr Pharm Des 11:75–90

Underkofler LA, Danault LJ, Hou EF (1965) Enzymes in starch industry. Starch-Stärke 17:179–184

Van der Maarel MJEC, Van der Veen B, Uitdehaag JCM, Leemhuis H, Dijkhuizen L (2002) Properties and applications of starch-converting enzymes of the α-amylase family. J Biotechnol 94:137–155

van Dokkum W, Wezendonk B, Srikumar TS et al (1999) Effect of nondigestible oligosaccharides on large bowel functions, blood lipid concentrations & glucose absorption in young healthy male subjects. Eur J Clin Nutr 53:1–7

Vessia A (2007) Cassava: the food of the poor for future food security. Le Scienze Web News. http://www.lswn.it/en/nutrition/news/cassava_the_food_of_the_poor_for_future_food. Accessed 25 Jan 2010

Wang Y-J, Wang L (2000) Structure and properties of commercial maltodextrins from corn, potato, and rice starches. Starch-Stärke 52:296–304

Warnecke F, Hess MA (2009) A perspective: metatranscriptomics as a tool for the discovery of novel biocatalysts. J Biotechnol 142:91–95

Weegels PL, Marseille JP, Hamer RJ (1992) Enzymes as a processing aid in the separation of wheat flour into starch and gluten. Starch-Stärke 44:44–48

Wu X, Wang D, Bean SR, Wilson JP (2006) Ethanol production from pearl millet using *Saccharomyces cerevisiae*. American society agricultural biological engineer meeting: 057077

Yan S, Wu X, MacRitchie F, Wang D (2009) Germination-improved ethanol fermentation performance of high-tannin sorghum in a laboratory dry-grind process. Cereal Chem 86:597–600

Yeh W-K, Yang H-C, McCarthy JR (eds) (2010) Enzyme technologies: metagenomics, evolution, biocatalysis and biosynthesis. Wiley, Hoboken

Zeleznak KJ, Hoseney RC (1986) The role of water in the retrogradation of wheat starch gels and bread crumb. Cereal Chem 63:407–411

Zhan X, Wang D, Bean SR, Mo X, Sun XS, Boyl D (2006) Ethanol production from supercritical-fluid-extrusion cooked sorghum. Ind Crops Prod 23: 304–310

Zhou E, Xiaoliang P, Xiuzhi T (2009) Application study of xylo-oligosaccharide in layer production. Mod Appl Sci 3:103–107

Zivkovic AM, Barile D (2011) Bovine milk as a source of functional oligosaccharides for improving human health. Adv Nutr 2:284–289

About the Editors

Dr. Pratyoosh Shukla is working as Head, Department of Microbiology, at Maharshi Dayanand University, Rohtak, India.

His research interests are in the fields of enzyme technology, protein bioinformatics, and microbial biotechnology. He has more than 12 years of research and teaching experience in well-reputed universities of India and abroad. He is the author of four book chapters and one patent, has edited *3 Biotech* (Springer journal), and published more than 30 peer-reviewed international papers in highly reputed international journals and more than 60 conference technical papers in biotechnology-related fields. He has also served as the technique committee member in some international and national conferences. He has successfully carried out four R&D projects as Principal investigator and/or Coinvestigator.

He received several awards, including Prof. S.B. Saksena, F.N.A., Award in life sciences (1999); Best Presentation Award (Senior Category, 2006) by the National Council for Science and Technology Communication (NCSTC), India; NRF-DUT Post-doctoral Fellowship Award in Enzyme Biotechnology (2008); and Danisco India Award in Probiotics & Enzyme Technology (2010), and was also selected as Scientist in the Southern Ocean Antarctica Expedition (2011), DST-Fast Track Young Scientist (2012), etc.

Prof. Brett I. Pletschke is currently a Professor and Head of Biochemistry in the Department of Biochemistry, Microbiology and Biotechnology at Rhodes University, Grahamstown, South Africa. Prof. Pletschke has served as the Vice President and President of SASBMB (South African Society for Biochemistry and Molecular Biology) and is currently the Immediate Past President of SASBMB. In this capacity, he has acted as a voting member for SASBMB at IUBMB on occasions. Prof. Pletschke's research interest is focused on the phenomenon of enzyme-enzyme synergy, using lignocellulose as a suitable model substrate. He was awarded a Rhodes University Alty Teaching Award in 2006 and was nominated for the Vice Chancellor's Distinguished Research Award in 2008. He has delivered several plenary talks at international conferences, published more than 50 papers in peer-reviewed international journals or books, and has supervised or graduated 29 M.Sc./Ph.D. students.

P. Shukla and B.I. Pletschke (eds.), *Advances in Enzyme Biotechnology*,
DOI 10.1007/978-81-322-1094-8, © Springer India 2013